CONTROL OF QUANTUM SYSTEMS

Control of Quantum Systems

Editor:

Nelson Bolívar, Ph.D

ArclerPress

www.arclerpress.com

Control of Quantum Systems
Editor: Nelson Bolívar, Ph.D

© 2017 Arcler Press LLC
708 3rd Avenue, 6th Floor
New York
NY 10017
United States of America
www.arclerpress.com

ISBN: 978-1-68094-385-6

Library of Congress Control Number: 2016951058

About the Editor

Nelson Bolívar, Ph.D

Nelson Bolivar has a PhD. in physics from the University of Lorraine in France finished in 2014. His expertise is in quantum systems and condensed matter. His interest includes spintronic devices and correspondences between general relativity and condensed matter. He is currently an associate professor at the Central University of Venezuela

Contents

List of Contributors

Steffen J. Glaser[1]
[1]Department Chemie, TU-M¨unchen (TUM), Lichtenbergstrasse 4, 85747 Garching, Germany

Ugo Boscain
CNRS, CMAP Ecole Polytechn que, Palaiseau, France and Team GECO, INRIA Saclay, Ile-de-France ´

Tommaso Calarco
Institute for Complex Quantum Systems and Center for Integrated Quantum Science and Technology, University of Ulm, Albert-Einstein-Allee 11, 89069 Ulm, Germany

Christiane P. Koch
Theoretische Physik, Universit¨at Kassel, Heinrich-Plett-Str. 40, 24132 Kassel, Germany

Walter K¨ockenberger
Sir Peter Mansfield Magnetic Resonance Centre, University of Nottingham, University Park, NG7 2RD, UK

Ronnie Kosloff
Institute of Chemistry, the Fritz Haber Research Center for Molecular Dynamics, The Hebrew University, 91904 Jerusalem, Israel

Ilya Kuprov
School of Chemistry, University of Southampton, University Road, Southampton, SO17 1BJ, UK

Burkhard Luy
Institut f¨ur Organische Chemie and Institut f¨ur Biologische Grenzfl¨achen, Karlsruher Institut f¨ur Technologie (KIT), Fritz-Haber-Weg 6, 76131 Karlsruhe, Germany

Sophie Schirmer
College of Science, Swansea University, Singleton Park, Swansea SA2 8PP, Wales, UK

Thomas Schulte-Herbr¨uggen
Department Chemie, TU-M¨unchen (TUM), Lichtenbergstrasse 4, 85747 Garching, Germany

Dominique Sugny
Laboratoire Interdisciplinaire Carnot de Bourgogne (ICB), UMR 6303 CNRS-Universit´e de Bourgogne-Franche Comt´e, 9 avenue A. Savary, BP 47 870, 21078 Dijon Cedex, France
Institute for Advanced Study, Technische Universit¨at M¨unchen, Lichtenbergstrasse 2a, 85748 Garching, Germany

Frank K. Wilhelm
Theoretische Physik, Saarland University, 66123 Saarbr¨ucken, Germany

WU ReBing
Department of Automation, Tsinghua University, Beijing 100084, China
Center for Quantum Information Science and Technology, TNList, Beijing 100084, China

ZHANG Jing
Department of Automation, Tsinghua University, Beijing 100084, China
Center for Quantum Information Science and Technology, TNList, Beijing 100084, China

LI ChunWen
Department of Automation, Tsinghua University, Beijing 100084, China
Center for Quantum Information Science and Technology, TNList, Beijing 100084, China

LONG GuiLu
Center for Quantum Information Science and Technology, TNList, Beijing 100084, China
Department of Physics, Tsinghua University, Beijing 100084, China

TARN TzyhJong
Center for Quantum Information Science and Technology, TNList, Beijing 100084, China
Electrical and Systems Engineering Department, Washington University in St. Louis, St. Louis, Missouri 63130, USA

Zbigniew Puchała
Institute of Theoretical and Applied InformaticsPolish Academy of Sciences

ZHANG Jun
Joint Institute of UM-SJTU, Shanghai Jiao Tong University and Key Laboratory of System Control and Information Processing, Ministry of Education, Shanghai 200240, China

Łukasz Pawela
Institute of Theoretical and Applied InformaticsPolish Academy of Sciences

Zbigniew Puchała
Institute of Theoretical and Applied InformaticsPolish Academy of Sciences
Institute of Physics, Jagiellonian University

Michael H Goerz
[1]Theoretische PhysikUniversität Kassel

K Birgitta Whaley
Department of ChemistryUniversity of California

Christiane P Koch
Theoretische PhysikUniversität Kassel

Guy Bensky
Department of Chemical Physics Weizmann Institute of Science

Robert Amsüss
Vienna Center for Quantum Science and Technology (VCQ) Atominstitut, TU-Wien

Johannes Majer
Vienna Center for Quantum Science and Technology (VCQ) Atominstitut, TU-Wien

David Petrosyan
Institute of Electronic Structure and Laser, FORTH

Jörg Schmiedmayer
Vienna Center for Quantum Science and Technology (VCQ) Atominstitut, TU-Wien

Gershon Kurizki
Department of Chemical Physics Weizmann Institute of Science

Łukasz Pawela
Institute of Theoretical and Appl ed InformaticsPolish Academy of Sciences

Zbigniew Puchała
Institute of Theoretical and Appl ed InformaticsPolish Academy of Sciences

LI HaiOu
Key Laboratory of Quantum Information, Chinese Academy of Sciences, University of Science and Technology of China, Hefei 230026, China

YAO Bing
Key Laboratory of Quantum Information, Chinese Academy of Sciences, University of Science and Technology of China, Hefei 230026, China

TU Tao
Key Laboratory of Quantum Information, Chinese Academy of Sciences, University of Science and Technology of China, Hefei 230026, China

GUO GuoPing
Key Laboratory of Quantum Information, Chinese Academy of Sciences, University of Science and Technology of China, Hefei 230026, China

WANG Hai
State Key Laboratory of Quantum Optics and Quantum Optics Devices, Institute of Opto-Electronics, Shanxi University, Taiyuan 030006, China

XIE ChangDe
State Key Laboratory of Quantum Optics and Quantum Optics Devices, Institute of Opto-Electronics, Shanxi University, Taiyuan 030006, China

PENG KunChi
State Key Laboratory of Quantum Optics and Quantum Optics Devices, Institute of Opto-Electronics, Shanxi University, Taiyuan 030006, China

Harumichi Nishimura
School of ScienceOsaka Prefecture University

Masanao Ozawa
Graduate School of Information ScienceNagoya University

E. Jonckheere
Center for Quantum Information Science and Technology, University of Southern California, Los Angeles, CA 90089, USA

F. C. Langbein
School of Computer Science and Informatics, Cardiff University, Cardiff CF24 3AA, UK

S. G. Schirmer
College of Science (Physics), Swansea University, Singleton Park, Swansea SA2 8PP, UK

CONG Shuang
Deptartment of Automation, University of Science and Technology of China, Hefei 230027, China

LIU JianXiu
Deptartment of Automation, University of Science and Technology of China, Hefei 230027, China

PAN Yu
Key Laboratory of Systems and Control, Institute of Systems Science, Academy of Mathematics and Systems Science, Chinese Academy of Sciences, Beijing 100190, China

XI ZaiRong
Key Laboratory of Systems and Control, Institute of Systems Science, Academy of Mathematics and Systems Science, Chinese Academy of Sciences, Beijing 100190, China

CUI Wei
Key Laboratory of Systems and Control, Institute of Systems Science, Academy of Mathematics and Systems Science, Chinese Academy of Sciences, Beijing 100190, China

Łukasz Pawela
Institute of Theoretical and Applied Informatics, Polish Academy of Sciences, Bałtycka 5, 44-100 Gliwice, Poland

Zbigniew Puchała
Institute of Theoretical and Applied Informatics, Polish Academy of Sciences, Bałtycka 5, 44-100 Gliwice, Poland
Institute of Physics, Jagiellonian University, ulica prof. Stanisława Łojasiewicza 11, 30-348 Kraków, Poland

Preface

The control of quantum physical systems refers to those whose behavior is dominated by the laws of quantum mechanics. The effective capabilities of manipulating matter and light at those scales is steadily growing and new experimental and technological possibilities are at reach, the interest in developing systematic theories for controlled quantum mechanical systems is rapidly increasing.

The advantage offered by quantum information processes such as in spintronic devices has given to quantum control a new perspective, the ability of employing quantum systems to store, manipulate and retrieve information. It requires an unprecedented degree of control, and further motivates the development of control schemes specifically designed to the quantum mechanical setting. Over the past fifteen years, significant developments have been made in utilizing quantum attributes of light and matter to assume unprecedented control over the dynamics of atomic and molecular systems. This growth reflects a confluence of factors including the better understanding of quantum mechanics as a tool for chemistry and physics, the development of new laser devices increasing our ability to manipulate light, and the recognition that coherent laser light can be used to imprint information on atoms and molecules for practical purposes. In this book we cover some aspects of the basics and controlling of quantum systems using a theoretical and applied approaches, from the very grounds of the theory to some hybrid approaches to quantum control. Hope this book can guide the reader into the deeps of this fascinating field.

Editor

Nelson Bolívar, Ph.D

INTRODUCTION

A quantum system is a portion of the whole Universe (environment or physical world) which is taken under consideration to make analysis or to study for quantum mechanics pertaining to the wave-particle duality in that system and everything outside this system (i.e. environment) is studied only to observe its effects on the system. A quantum system involves the wave function and its constituents, such as the momentum and wavelength of the wave for which wave function is being defined. Quantum feedback control is a class of methods to prepare and manipulate a quantum system in which that system's quantum state or trajectory is used to evolve the system towards some desired outcome. Just as in the classical case, feedback occurs when outputs from the system are used as inputs that control the dynamics (e.g. by controlling the Hamiltonian of the system). The feedback signal is typically filtered or processed in a classical way, which is often described as measurement based feedback. However, quantum feedback also allows the possibility of maintaining the quantum coherence of the output as the signal is processed (via unitary evolution), which has no classical analogue.

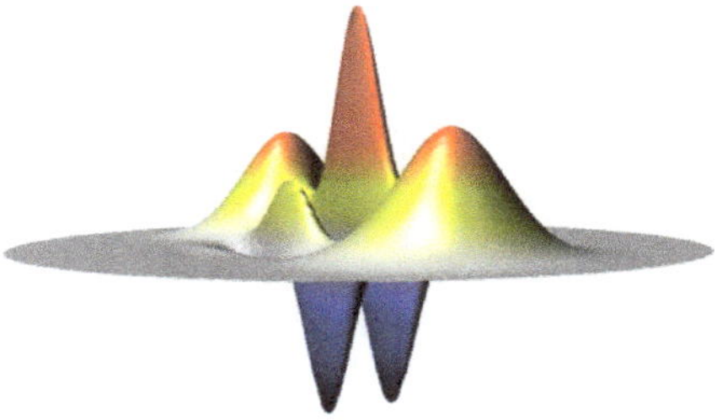

Open quantum system

In physics, an open quantum system is a quantum-mechanical system which interacts with an external quantum system, the environment. In reality, no quantum system is completely isolated from its surroundings, so every quantum system is open to some extent, causing dissipation in the quantum system. Techniques developed in the context of open quantum systems have proven powerful in fields such as quantum optics, quantum measurement theory, quantum statistical mechanics, quantum information science, quantum cosmology and semi-classical approximations. The environment we wish to model as part of our open quantum system is typically very large, making exact solutions impossible to calculate. However, by making key approximations it is possible to find out how the quantum system behaves in the presence of the environment. There are two approximations commonly made in the field of open quantum systems: the Born approximation and the Markov approximation. The Born approximation assumes that the system-bath coupling is relatively weak, and the bath is very large, so that the bath is negligibly affected by the system. The Markov approximation assumes that the bath has no memory of past events.

Classical Control Systems

Formally speaking, a control system consists of a dynamical system interacting with a controller, a device that influences the state of the dynamical system toward some desired end. The objective may be to regulate the flow of an industrial process, money, energy, information, and so on. In a "closed-loop," or feedback, control system, the controller uses outputs from the dynamical system to monitor and influence its interaction with that dynamical system. For a linear dynamical system, for example, such a situation could be described by the following equation:

$$d\mathbf{x} = A\mathbf{x}dt + Bd\mathbf{W} + C\mathbf{u} \ , \tag{1}$$

where x is a vector describing the state of the system, dW is a vector of Gaussian noise sources, and u is the vector of inputs determined by the controller. The matrix A gives the system's deterministic motion, and B and C describe, respectively, how the noise and input vectors are coupled into the system. A separate equation, namely describes the continuous measurement of system outputs by the controller. In each small time interval *dt*, the controller obtains the measurement result dy.

$$d\mathbf{y} = H\mathbf{x}dt + Rd\mathbf{V} \ , \tag{2}$$

That result is directly related to the true state of the system by some linear transformation H, but it also includes a Gaussian noise process V, which serves to represent imperfections in the measurement. Examples of control systems can be found in many applications. For instance, servomechanisms are control systems that use small control inputs to produce changes in large mechanical systems. In effect, the larger systems are "slaved" to the output of the servomechanisms (for example, liquid levels in reservoirs are controlled by float valves). Feedback circuits are used in ingenious ways in electronic amplification to manipulate input and output impedances and to improve the linearity, distortion, and frequency bandwidth of the output signal relative to the input signal. In an "open-loop" control system, the controller does not monitor the output of the dynamical system. A dynamical model for the system is assumed, and control is applied with the idea that the desired outcome will actually be achieved. Open-loop strategies are useful in situations in which the system dynamics are known precisely and vary only slowly. Processes with long measurement dead times are sometimes better suited to open-loop control methods than to feedback methods. Open-loop control strategies are applied in situations as diverse as the maximization of returns from financial investments, optimal determination of aircraft flight paths, and controlled dissociation of molecules with a large number of sequential, parallel, and nested control loops that are both open and closed.

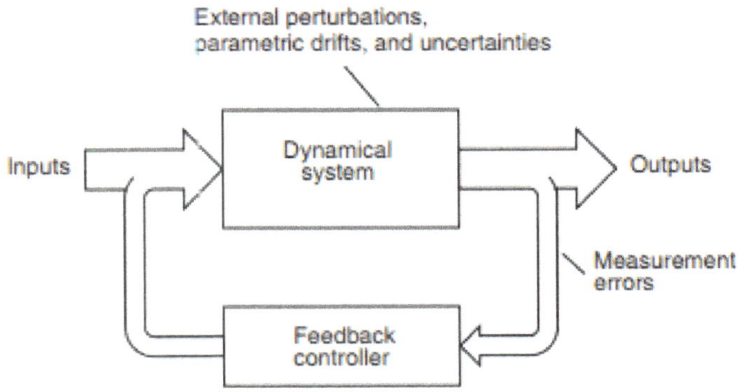

Figure 1 shows how to implement closed-loop control for a dynamical system. One must be able to measure some of the dynamical variables of the system under control (the outputs) and use them to influence some other variables (the inputs). In other words, given the output variables, the controller implements a particular control strategy to influence the state of the dynamical system by appropriately varying the inputs. Robust controllers take into account variations in system parameters and fluctuations from the external environment to produce control strategies with guaranteed stability bounds. Control systems can involve many different interacting physical systems.

For example, closed- and open-loop strategies can be combined as in the fast closed loop systems used to stabilize the slower, inherently unstable open-loop dynamics of modern fighter aircraft.

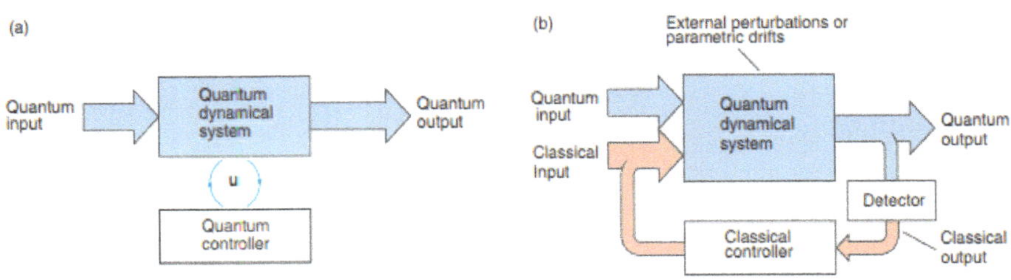

Figure 2. Directly and Indirectly Coupled Quantum Feedback (a) Both the dynamical system and the controller are quantum systems coupled through a unitary interaction. A quantum variable is coupled to the quantum controller, and a quantum input path from the controller goes directly back to the quantum system. (b) A quantum dynamical system can be viewed as having two sets of inputs, one relating to the variation in the classical parameters describing the Hamiltonian and the other representing fully quantum inputs. Similarly, the output channel can be divided into a quantum and a classical channel. The classical channel is, in fact, a piece of the quantum channel that has become classical after observation. The controller analyzes the classical record to form an estimate of the dynamical system's state and uses this information to implement the appropriate control.

Developing Control in Quantum Systems

The general picture of control systems outlined in the previous section appears to be extendable to quantum systems. Certainly, open-loop control problems are conceptually straightforward in the quantum context. One begins with the time evolution operator of the quantum system—the Schrödinger equation for the wave function, the Liouville equation for the density matrix, or more complicated dynamical evolution equations for the density matrix characterizing a system coupled to an environment. A theory for time-dependent variations in the evolution operator is then developed in such a way that the wave function or the density operator at some time is close to some target value. This target value does not have to be unique, nor in fact is the time evolution to that value unique. The approach just outlined applies equally well to classical probabilistic evolutions: Although quantum and classical systems are dynamically distinct, the principles for open-loop control are in fact very similar. Controlling chemical reactions by laser-produced electromagnetic fields that are time dependent is a well-known open-loop quantum control problem. In the frequency-resolved approach to control, the quantum interference between different evolutionary paths is being manipulated; in the time-resolved approach, the dynamics of wave packets produced by ultrafast laser pulses leads to control. For some specific control of the chemical reactions, one can optimize the temporal and spectral structure of those laser pulses. The fundamental differences between classical and quantum systems become real issues, however, in the field of closed-loop control. Quantum systems can have two distinct types of feedback control: directly and indirectly coupled quantum feedback (see Figure 2). As illustrated in Figure 2(a), in a system with directly coupled quantum feedback, a quantum variable of the system is coupled to the quantum controller, and a quantum input path from the controller goes directly back to the quantum system. When the quantum feedback is indirect, as shown in Figure 2(b), the quantum dynamical system under control is an observed system. It therefore generates a classical output, also known as the measurement record, which the controller may analyze to provide a best estimate of the original quantum state of the system. The controller then feeds back a classical signal to vary parameters in the quantum evolution operator in accord with the chosen control strategy. Hybrid couplings using both direct and indirect quantum feedback channels are easy to envisage: The channel from the system output to the controller input may be directly coupled whereas the channel from the controller output to the system input may be coupled indirectly through a classical path.

Inherent Noise Generation in Quantum Feedback Control

Unlike classical systems, quantum systems may be easily disturbed when information about them is extracted. Measurement disturbs a quantum system through the following intrinsic property of quantum mechanics: Obtaining accurate knowledge about one observable of a quantum system necessarily limits the information about an observable conjugate to the first. For example, particle position and momentum are conjugate observables, and the uncertainties inherent in the knowledge of both are codified by the famous Heisenberg uncertainty relation. If the chosen feedback control strategy involves measurement, one must take into account the effects of the measurement on the evolution of the quantum system.

Strategies for Quantum Feedback Control

The differences between classical and quantum measurements profoundly affect the design of feedback control algorithms. A classical controller extracts as much information from the system as possible. In quantum control, irreducible disturbances are inherent to any measurement, and therefore the measurement strategy becomes a significant part of the feedback algorithm. For example, just as the inputs to the system change with time, the measurements too may need to be varied with time so that the best control should be achieved.

Applications of Quantum Control Atomic optics is one field in which it should be possible to test quantum feedback control in the near future. It has already been demonstrated that a single atom can be trapped inside an ultralow-loss optical cavity (mirror reflectivity is R = 0.9999984 in experiments at Caltech) in the strong coupling quantum regime

Classical Simulation of Quantum Systems?

The field of quantum computing originated with a question posed by Richard Feynman. He asked whether or not it was feasible to simulate the behavior of quantum systems using a classical computer, suggesting that a quantum computer would be required instead. Saleh Rahimi-Keshari from the University of Queensland, Australia, and colleagues have now demonstrated that a quantum process that was believed to require an exponentially large number of steps to simulate on a classical computer could in fact be simulated in an efficient way if the system in which the process occurs has sufficiently large loss and noise.

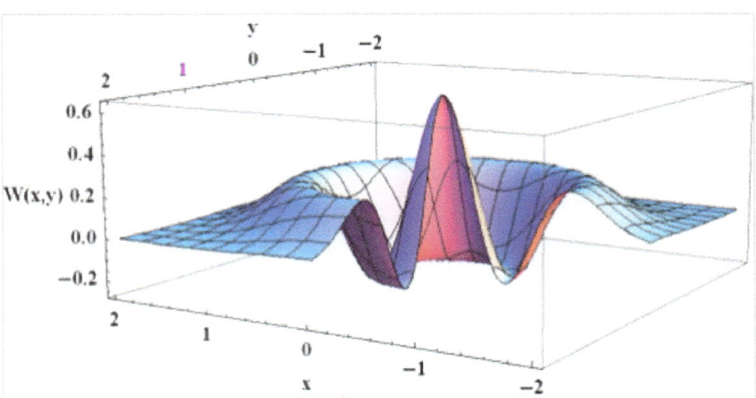

The quantum process considered by Rahimi-Keshari, is known as boson sampling, in which the probability distribution of photons (bosons) that undergo a linear optical process is measured or sampled. In experiments of this kind, NN single photons are

sent into a large network of beams splitters (half-silvered mirrors) and combined before exiting through MM possible output channels. The calculation of the probability distribution for finding the photons in each of the MM output channels is equivalent to calculating the permanent of a matrix. The permanent is the same as the more familiar determinant but with all of the minus signs replaced with plus signs. On any classical computer, the number of computational steps required to calculate the permanent of a matrix is believed to increase exponentially with increasing values of MM, which would make the problem impossible to solve for large values of MM.

1

CHAPTER

QUANTUM OPTIMAL CONTROL

Steffen J. Glaser, Ugo Boscain, Tommaso Calarco, Christiane P. Koch, Walter K¨ockenberger, Ronnie Kosloff, Ilya Kuprov, Burkhard Luy, Sophie Schirmer, Thomas Schulte-Herbr¨uggen, Dominique Sugny, and Frank K. Wilhelm

Keywords: Quantum Information

FOREWORD

The authors of this paper represent the QUAINT consortium, a European Coordination Action on Optimal Control of Quantum Systems, funded by the European Commission Framework Programme 7, Future Emerging Technologies FET-OPEN programme and the Virtual Facility for Quantum Control (VF-QC). This consortium has considerable expertise in optimal control theory and its applications to quantum systems, both in existing areas, such as spectroscopy and imaging, and in emerging quantum technologies, such as quantum information processing, quantum communication, quantum simulation and quantum sensing. The list of challenges for quantum control has been gathered by a broad poll of leading researchers across the communities of general and mathematical control theory, atomic, molecular-, and chemical physics, electron and nuclear magnetic resonance spectroscopy, as well as medical imaging, quantum information, communication and simulation. 144 experts in these fields have provided feedback and specific input on the state of the art, mid-term and long-term goals. Those have been summarized in this document, which can be viewed as a perspectives paper, providing a roadmap for the future development of quantum control. Because such an endeavour can hardly ever be complete (there are many additional areas of quantum control applications, such as spintronics, nano-optomechanical technologies etc.), this roadmap is designed as a living document that is available at the homepage of the VF-QC, [1], where additional aspects, as well as new developments and ideas, will be included.

Citation: Steffen J. Glaser, Ugo Boscain, Tommaso Calarco, Christiane P. Koch, Walter Köckenberger, Ronnie Kosloff, Ilya Kuprov, Burkhard Luy, Sophie Schirmer, Thomas Schulte-Herbrüggen, Dominique Sugny, Frank K. Wilhelm, quantum optimal control, doi:10.1140/epjd/e2015-60464-1.

INTRODUCTION

It is control that turns scientific knowledge into technology. The general goal of quantum control is to manipulate dynamical processes at the atomic or molecular scale, typically using external electromagnetic fields. The objective of quantum optimal control is to devise and implement shapes of pulses of external fields, or sequences of such pulses, that accomplish a given task in a quantum system in the best way possible. Quantum control builds on a variety of theoretical and technological advances, from the fields of mathematical control theory and numerical mathematics to better electronic devices, such as arbitrary wave form generators with sub-nanosecond time resolution or stronger magnetic fields. The challenge of manipulating nature at the quantum level has a huge potential for current and future applications. Quantum systems and processes cover a wide range from atomic and molecular physics, chemistry and materials (such as semiconductors, superconductors) to bio-systems and medicine. Useful applications range from magnetic resonance imaging and spectroscopy and the precise control of chemical reactions to emerging "second generation" quantum technologies. Quantum optimal control is a part of the effort to engineer quantum technologies from the bottom up, and many striking examples of surprising and non-intuitive – but extremely efficient and robust – quantum control techniques have been discovered in recent years. Examples of important current applications are precise measurement of magnetic fields with nanometer resolution using NV centers in diamond, state engineering in Bose-Einstein condensates and high-fidelity quantum gates in superconducting quantum processors. In a similar way to the first generation of quantum-based technologies that brought forward semiconductor transistors, lasers, magnetic resonance imaging and spectroscopy, the currently emerging second generation of quantum technologies based on superposition, entanglement and many-body quantum states are expected to generate new and disruptive technologies spintronic devices, quantum metrology, quantum computing technology, as well as novel instruments for elucidating chemical reaction dynamics and material properties. Quantum control is thus a strategic cross-sectional field of research, enabling and leveraging current and future quantum technology applications. While the instrumental details of manipulating the behavior of all those systems may differ (lasers, radio waves etc.), the control, identification and system design problems encountered share the same mathematical formalism that is distinct from classical control theory. Advancing quantum control therefore requires bringing together researchers from different application areas to forge a community, creating a common language and identifying common challenges. Further development of this field of research offers many beneficial effects for today's and tomorrow's society, related to health, secure communication, accurate navigation systems, efficient harvesting of solar power, the search for resources, efficient energy storage and transportation, quantum machines, precision sensing and monitoring of the environment. The paper is organized as follows: Section 2 is focussed on mathematical optimal control theory. It is followed by a description of the state of the art, as well as mid-and long-term perspectives for quantum control applications in atomic, molecular, and chemical physics (Sect. 3), magnetic resonance (Sect. 4) and quantum information and communication (Sect. 5). Prospects for applications and commercial exploitation are outlined in Section 6 and conclusions are given in Section 7.

GENERAL ASPECTS AND MATHEMATICS OF OPTIMAL CONTROL

The recent advances in quantum control, by now recognized to be essential for continued development of quantum technologies (see, e.g., [2,3]) and reviewed in Sections 3 to 5 below, are based on powerful tools from mathematical control theory [4–13]. As it often happens on the interface between mathematics and physics, mathematical concepts not only prove fruitful for the solution of physical problems, but in turn specific physical features require further mathematical development. In the context of quantum control, these are, for example, entanglement and the nature of quantum measurements. Eventually, a new research domain, mathematical quantum control theory [14–23] has emerged (see also [11,24], Sect. 19.3). In general, quantum control theory needs to answer two fundamental questions: that of controllability (i.e., what control targets are accessible, cf. Section 2.1) and that of control design (i.e. how can a target be reached). Approaches to control design can be open-loop or closed loop. In the latter case, the specific nature of quantum measurements needs to be taken into account. This is covered in Section 2.3. The remaining control design approaches are reviewed in Section 2.2. Open-loop techniques include approaches based on the Pontryagin maximum principle [4] with solutions obtained analytically (Sect. 2.2.1) or numerically (Sect. 2.2.2). Optimal control does not make any assumptions about the system and experimental constraints and robustness requirements can be fully taken into account, i.e. it is generally applicable. However in some cases, adiabatic control and its variants (cf. Sect. 2.2.3) can provide a more straight forward approach for solving robustness issues and implementing constraints.

Controllability and simulability

State of the art

Controllability analysis determines whether a quantum system can be brought from any given initial state to any desired target state, or, more generally, from a given set of initial states to any set of target states. Adapting results from classical linear control systems [7,8] to bilinear systems with a non-switchable drift term, a rigorous Lie-framework was developed for closed systems [12,25–29]. Based on this work, for quantum systems with finite dimension, controllability by now is well understood [14,18,20,30–33][1]. Different notions of controllability have been introduced for pure states, mixed states, and evolution operator dynamics [34]. The main controllability test is based on the rank of the dynamical Lie algebra, which is generated by the drift and the different control Hamiltonians. The difficulty of using the rank condition in large systems has motivated a geometric approach based on graph theory [35], yielding eventually a complete set of symmetry criteria for controllability [20].

In infinite-dimensional systems, the mathematics is much more intricate and the few existing results are con- fined to quantum systems with a discrete spectrum. An important result is a general obstruction property to exact controllability [33,36,37]. This was recently amended by positive results about exact [38–40] and approximate controllability [32,41–44], based on Galerkin techniques. For the specific case of a generalized Jaynes-Cummings model, i.e., several two-level systems coupled to a harmonic oscillator, symmetry methods were used to assess controllability [45,46]. Simultaneous controllability concerns the control of a continuum of finite dimensional quantum systems by only a few control fields. This is also known as ensemble controllability. Approximate and exact controllability results have been obtained recently in this direction [47–50]. This analysis is important for designing control fields which are robust to experimental imperfections [51–56]. Adiabatic techniques could be very useful to this purpose, see Section 2.2.3. In addition to experimental imperfections and fluctuations, decoherence may pose an obstacle to control. For open quantum systems, the control field usually cannot fully compensate dissipation, as rigorously shown for the case of Markovian dynamics [57]. These results for Markovian dynamics were recently generalized to a complete Lie semi-group picture [58,59]. In contrast, controllability of systems with non-Markovian dynamics presents by and large uncharted territory [60]. Reachability is expected to be larger for non-Markovian dynamics since nonMarkovianity implies information back-flow from the environment to the system [61]. Indeed, an explorative study showed non-Markovian map synthesis to be stronger than its restriction to the Markovian case (as anticipated in Ref. [62]). In contrast, and surprisingly so, for the simpler problem of state transfer under swichable Markovian noise this is not the case [63].

Finally, even if a system (A) is not fully controllable, its controlled dynamics may still suffice to generate a desired effective evolution as brought about by another quantum dynamical system (B). This is the paradigm of finite-dimensional quantum simulation. Adapting the tools from controllability analysis, one readily sees that system A can simulate system B if the system algebra of A encompasses that of system B. A recent generalisation of the results in reference [20] provides a complete set of symmetry criteria together with an algorithm for deciding the simulability on system-algebraic grounds [64]. Resorting to exactly solvable problems may help in this case [65].

Mid-term Prospects: Goals and Challenges

The major challenge is to better understand controllability in open quantum systems. For infinite-dimensional open systems with a discrete spectrum that undergo Markovian evolution, a rigorous understanding is nontrivial, but may be pursued by extending the standard LieGalerkin techniques to non-unitary evolution. With the final goal of identifying sufficient conditions for approximate controllability of open systems analogous to closed systems [32,41,43], a mid-term perspective is to perform a sound regularity analysis of a well-posed and defined mathematical control problem.

No rigorous controllability analysis so far has tackled open quantum systems which undergo non-Markovian dynamics. From a control point of view, non-Markovianity may be connected to beneficial aspects of the systemenvironment interaction, whereas the detrimental part is linked to those Markovian processes that cannot be remedied – an aspect important not only for controllability, but also for dissipative state engineering and quantum memories. Along similar lines, projective measurements may entail Zeno-type dynamics exploring directions that do not show up in the unobserved system [66].

Even for closed quantum systems, several highly relevant questions are still open. These include the exact or approximate controllability of the Schrödinger equation with mixed or continuous spectrum. This question is important because it covers dissociation and ionization processes, the control of which is a major goal in chemical physics, cf. Section 3.

Beyond controllability, a precise description of the reachable set (both in the closed and in the open case) and upper bounds on minimal time to reach target states are largely unknown. Also, a universal estimate for the time to control finite-dimensional quantum systems (with necessary drift) is still an open challenge.

Control Design

Geometric Optimal Control – State of the Art

Optimal control theory can be viewed as a generalization of the calculus of variations for problems with dynamical constraints. Its modern version was born with Pontryagin's Maximum Principle in the late 1950's [4], which generalises the classical Euler-Lagrange equations. Its development was boosted by the use of Kalman filters [7,67] in the Apollo programme. It is now a key tool in many applications, including quantum mechanics.

Solving an optimal control problem means finding a control law (e.g. a pulse sequence), such that the corresponding system trajectory satisfies given boundary conditions and equation of motion and minimizes a cost criterion, such as the energy or duration of the control. Pontryagin's Maximum Principle can be formulated in terms of Hamiltonian-like equations. In a typical workflow, one first obtains extremal trajectories by solving these equations. In the second step, one selects among the extremals those which minimize the cost. Although looking straightforward, the practical use of Pontryagin's Maximum Principle is far from trivial. Apart from integrating the Hamiltonian equations, the main difficulties are the presence of abnormal and singular extremals [68] and the problem of selection of optimal trajectories. For this reason methods from differential geometry and numerical analysis are often required. The latter are described in Section 2.2.2.

If the system is sufficiently simple, for example lowdimensional, the optimal control problem may be solved analytically. Typically, this requires understanding of the geometry of the control problem from which one can deduce the structure of the optimal solution, a proof of global optimality and physical limits, such as the minimal time to reach the target [5]. Mathematical tools that were developed recently [11–13,68–70] could tackle problems of increasing difficulty, including fundamental control problems for closed [14,71–75] and open quantum systems [76–81]. This method is able to treat quantum control problems ranging from two and three level quantum systems or two and three coupled spins to two-level dissipative quantum systems with dynamics governed by the Lindblad equation. Singular optimal arcs have been used to derive control fields in some example quantum systems [77,82–87]. This approach can also be advantageously combined with numerical optimization techniques in order to manipulate more complicated systems [88]. The Cartan decomposition method also turned out to be an efficient tool for the control of spin systems [89–91]. This decomposition leads to a reduction of the state space dimensionality and may enable its geometric description.

Numerical Optimal Control – State of the Art

If the set of control equations resulting from the maximum principle cannot be solved analytically, numerical optimal control theory provides a viable alternative. The available techniques include (i) gradient ascent algorithms [92] (which can be extended to second-order quasi-Newton and Newton methods [93–95]) and (ii) Krotov-type methods [96–98], also permitting this extension [99]. The main difference between these two approaches is that the control is updated (or replaced) for all times simultaneously in case (i) and sequentially in case (ii), which implies different convergence properties. The algorithms are comparatively easy to use and several program packages include optimal control modules, e.g. SIMPSON [100], SPINACH [101], DYNAMO [94], and QuTiP [102]. Modifications to account for experimental imperfections and limitations and to ensure robustness of the solution have been introduced [53,103–111], and numerical optimal control theory has been extended to open quantum systems [112–117]. It is also possible to restrict the control solution to a predefined analytical form. In this case, the control only depends on a limited set of parameters which are optimized [118–120]. Applications of numerical optimal control are discussed below in Sections 3 to 5 but by now have grown too numerous for a complete bibliography. A caveat for numerical algorithms that are derived from the maximum principle is that they require one to solve the system's equations of motion many times, at least twice per iteration step. This potentially hampers application to high-dimensional quantum systems. So far, three approaches have been pursued to cope with this issue. (i) optimization of the state space representation: for the high-temperature regime of ensemble NMR spectroscopy, the SPINACH package [101] uses state propagation with correlations limited to pertinent short-range ones (thus leading to efficient truncation of the underlying spin Lie algebra) for simulating unprecedentedly large spins systems such as entire biomolecules [121] with striking precision. (ii) gradient-free optimization: for many-body quantum systems, the chopped random basis (CRAB) method allows one to generate a parametrization of the control sequence and interfaces the tensor compression of the time-dependent density matrix renormalization group with parameter optimization [122,123]. CRAB is part of the QuTIP package [102]. The number of propagations may be significantly reduced compared to gradientbased optimization, provided a small basis is sufficient to represent the control. (iii) local control theory: it determines the control field from instantaneous dynamical properties of the system by requiring monotonic increase or decrease of a performance index [124,125]. Under wellestablished conditions [126], the system converges asymptotically towards the target. Local controls are reminiscent of the closed-loop Lyapunov method in stabilization, and correspondingly the approach has also been termed tracking. Due to the inherent nature of the quantum measurement process, this approach has been transformed into an open-loop control law in the quantum world. In this approach, the equation of motion needs to be solved only once.

Mathematical quantum optimal control theory shows how to design optimal control fields and describes under which conditions they exist [30,127]. It does, however, not investigate the complexity inherent to the search. First results on the search complexity [128] have been derived from an information theoretical analysis of quantum optimal control: indeed, connecting the classical information carried by the control field and the size of the effective Hilbert space explored by the dynamics, the (smoothed) complexity of the search is found to be polynomial and the minimal achievable error as well as the minimal duration of the optimal control field as a function of its bandwidth are bounded. The latter is a direct consequence of the finite classical channel capacity of the control field, similarly to the quantum speed limit that arises from the energy-time uncertainty relation. These results can be generalized to the presence of noise [128]. The search complexity is related to the description of the control landscape which specifies the control objective as a function of the control variables. Different results on the structure of the control landscape have been established recently [129–132]. One of the main claims has been that, provided there are no constraints on the control, the control landscape does not contain local traps in all controllable closed and some specific open quantum systems [17]. This proof has recently been contested and is still the object of a lively debate [131–133]. Once constraints are present, such as finite control field amplitude, this may induce traps in the control landscape [134,135]. Whether and how imposing robustness of the control influences the control landscape is an open question.

Control via Adiabatic Dynamics – State of the Art

While optimal control theory is a general approach that is particularly well suited for identifying fast controls, the target can also be tackled by adiabatic techniques [136–142]. The most widely known among these is Stimulated Raman Adiabatic Passage (STIRAP) [136]. Adiabatic techniques usually employ a sequence of very intense pulses over a comparatively long timescale which enforces adiabatic following of the system dynamics. The pulses can be frequency chirped according to the structure of the energy levels. Such processes are inherently robust to small variations of laser or system parameters and thus well-suited to open-loop control [50]. A main drawback of adiabatic control is the total time and energy requirements which cannot always be met in an experiment. To address this issue, shortcuts to adiabaticity were developed [143]. This approach can be viewed as an inverse engineering technique based on Lewis-Riesenfeld invariants. It has been applied to a variety of quantum systems [144–147] for which such an invariant exists. Even for an ensemble of quantum systems, analytical solutions based on this technique are not out of reach [146]. Similarly to enforcing robustness in ensemble optimization using optimal control [53,56], the basic idea consists in selecting among a family of (exact) solutions the ones that are effectively robust to some extent to variations of the system parameters.

Mid-term prospects: Goals and Challenges for Control Design

As with controllability analysis, the major challenge for control design is posed by open quantum systems because most systems of interest interact with their environment to a non-negligible extent. An important challenge is the extension of the standard adiabatic and shortcut techniques to open quantum systems. Beyond some preliminary results

[148,149], there are many open questions, for example, on how to generalize Lewis-Riesenfeld invariants to open system dynamics.

All open-loop control approaches discussed above assume sufficient knowledge of the quantum system. Correspondingly, for open quantum systems, system identification needs to include dissipative parameters. A first step to combine quantum control with implicit learning about the system parameters is the ADHOC technique [150]. Decoherence in open quantum systems can also be used as a resource in what has become known as dissipative state engineering [151–154]. Optimal control theory allows for tackling dissipative state engineering for quantum systems that are too complex for manual design of the driven-dissipative dynamics. This naturally includes incorporation of the noise as additional optimization parameter [63,155]. When added to coherent controls, timedependent Markovian noise (amplitude-damping) enables the control system to transform any initial state into any desired target state [63]. It can be easily integrated into toolboxes like DYNAMO [94], yet the implementation in realistic settings remains unexplored.

On the algorithmic level, the basics of numerical quantum optimal control are well established for both Markovian and non-Markovian open systems, as described above. The main challenge is the efficient numerical modeling of increasingly complex dynamics. A promising route is provided by stochastic methods that only require propagation of (several) pure states to unravel the true dynamics [115,156,157]. These methods are applicable to systems interacting with an environment or subjected to measurements that can be modelled by a stochastic process. An open question in this context is to find the most efficient way to control these systems such that the control is robust with respect to the stochastic parameter [158]. In view of numerical optimal control theory as an openloop technique, integration with applications is a crucial issue that has only very partially been addressed. In order to make the theory more useful for specific experiments, the gap between theory and experiments needs to be bridged. Despite recent progress, a systematic and efficient algorithm addressing this issue is still lacking. One option consists in combining open-loop optimal algorithms with closed-loop control techniques in order to reduce their respective drawbacks [150]. Using these two tools cooperatively is expected to provide flexible quantum control. A current challenge is to improve the computational speed and accuracy of the algorithms. A promising approach to this end is the combined use of numerical and geometric optimal control methods [88]. Analogously to open quantum systems, improving the computational speed will be an important prerequisite to attack control problems of increasing complexity, representing experimental settings in a realistic way. Eventually, the goal is to enable the problem-adapted design of controls in time-critical applications.

In view of the emerging better understanding of control complexity, a rigorous understanding of two crucial issues needs to be developed – interrelation between minimum control dimension and (effective) system dimension [120,128] and the question of traps in the quantum control landscape. Even if many quantum control problems have no traps, i.e. their control landscapes are well-suited to the local numerical optimization approaches, this is not always the case. One then needs to resort to global optimization techniques that are numerically expensive and generally do not perform very well. Whether and how global optimization methods, such as dynamic programming, may be (better) adapted to quantum control is also an open question. The efforts to solve more complicated control problems via geometric optimal control techniques must continue with a special emphasis on open quantum systems and on non-linear dynamics [159]. Another fundamental open question is whether it is possible to connect the Lewis-Riesenfeld invariants, used for the shortcuts to adiabaticity, with the formal framework of the Pontryagin Maximum Principle of optimal control theory. This would hold the promise of combining the best features of both methods.

Quantum Feedback Control Theory

State of the art

In contrast to the open-loop techniques discussed in Section 2.2, closed-loop strategies [19] are capable of coping with unpredictable disturbances. This approach is widely used in classical control theory, where information from the state of the system is fed back to the controller to correct the field in action. However, the perturbing effect of quantum measurements precludes direct application of classical concepts. It required development of new techniques tailored to quantum dynamics [15,160]. An indirect application is found in experimental closed-loop quantum control, where each cycle of the loop is realized experimentally, including preparation of a new sample in each cycle [161]. Various learning algorithms, such as genetic algorithms, have been developed to design the new control field after each iteration [17].

Control of a quantum system by feedback in a closed loop can be realized in two different ways [162] – using the measurement-based feedback approach [163] or the coherent feedback method [164]. As in classical control theory, the first option is based on a measurement process and real-time manipulation of the system depending on the measurement result. The measurement feedback approach, built on quantum filtering techniques [156], is at present mathematically well researched [15,165]. Quantum filtering consists in generating an estimate of the state of the system from the measurements performed on it

Measurement-based feedback has been shown to be efficient [166,167], making real-time control of quantum systems experimentally feasible [168,169]. In the second option, no measurement is used and the quantum system is directly connected to a quantum controller. Tests of experimental feasibility of coherent feedback control have recently led to impressive results [170–173].

Mid-Term Prospects: Goals and Challenges

It is the specific quantum nature of systems that makes them susceptible to measurement back-action. Thus, in the quantum domain, finding an experiment-class adapted balance between open-loop control techniques and closed loop feedback counterparts is one of the major challenges. Because measurement-based feedback is restricted to the processing time of the classical components, which have the slowest time scale, exploiting the limits of coherent feedback control is useful as a way of building further generations of quantum technologies. An advantage of coherent feedback over its measurement-based counterpart is the reduced noise produced by the control process since – there is no disturbance from a system measurement. Feedback control methods promise particular robustness and flexibility, but some important questions related to the quantum nature of the dynamics remains to be solved. In particular, there is no general theory showing that quantum controllers perform better than their classical counterparts, and if so, in which cases and under which experimental conditions. Many other questions are still open in quantum feedback control. They range from general theories about the role of weak measurements in the control of quantum systems to feedback control of non-Markovian dynamics and the influence of model uncertainties on the feedback control. Similarly, another interesting perspective is to explore whether programmable quantum processors [174,175] themselves could be used or programmed to steer quantum systems. Most of the developments of quantum feedback control have been made in the context of quantum optics [15,163,168,176]. Recent technological progress has enabled advances in superconducting circuits both theoretically [177] and experimentally [178,179]. A problem that is starting to attract attention is feedback control in quantum transport [180]. In summary, the objective is to understand how the current quantum control techniques developed for quantum optical systems can be applied to hybrid systems involving quantum dots, superconducting qubits and opto-mechanical resonators. The existing quantum feedback theory has to be adapted to these new types of dynamics.

Long-Term Vision

Quantum optimal control theory has reached a level of maturity that enables tackling the question of quantum control in a world that is never fully quantum, be it due to residual couplings to the environment or due to imperfections in the controls or system characterization. The ultimate goal is to develop a rigorous understanding of the fundamental limits, as well as the opportunities for quantum control, under real-life conditions, in terms of both controllability and control design. This includes in particular control of quantum systems undergoing nonMarkovian evolution, as found in condensed phase physics and chemistry. It will most likely require a combination of the techniques described above and a synergy of efforts, ranging from pure mathematics through algorithm design all the way to developing techniques for integrating theory and experiment. Meeting these challenges would provide the mathematical and algorithmic underpinnings for the application of quantum control in all of the three fields described below.

ATOMIC, MOLECULAR, AND CHEMICAL PHYSICS

State of the Art

Within the realm of atomic, molecular and chemical physics, quantum control was first discussed in the context of chemical reactions. It was termed 'coherent control' at the time and conceived as a way to control the outcome of a reaction using laser fields [181–183]. The basic idea was to employ interference of matter waves to enhance the desired outcome and suppress all others [184,185]. A way to create the desired matter wave interference is by tailoring laser fields [181–183]. A reaction is viewed as the following sequence of events – approach of the reactants; formation of a new chemical bond; intermediate dynamics of the generated molecular complex; stabilization into the target products, typically involving the breaking of another chemical bond. Each of these steps can in principle be controlled. The advent of femtosecond lasers and pulse shaping technology [186,187] in the 1990s allowed for experimentally testing the idea of controlling chemical reactions. In the lab, the laser pulse shape can most easily be determined in feedback loops combined with e.g. genetic algorithms [161]. Very soon after the theoretical proposals several pioneering experimental papers were published describing control of unimolecular dissociation or fragmentation, see [188–196] and references therein. Deciphering the control mechanisms underlying the optimized pulses is a complex task, and much still remains to be understood. A particularly contested question concerns the extent to which quantum coherence is involved [197–200], and at the moment, a striking example that relies on coherent quantum interference is missing. Control strategies include both weak- and strong-field scenarios. In the first case, a wavepacket is launched, and its ensuing (ro-)vibrational dynamics is exploited. In the strong-field regime, the laser pulse coherently controls the dynamics during the pulse while utilizing the effective modification of the energy levels of atoms [201–204] or, respectively, the potential energy landscape experienced by the molecules, via the dynamic Stark effect [205–207]. Weak field control of non-resonant excitation may employ optical interferences to control e.g. the population in the final state [208]. Weak field here is defined as low order perturbation theory being applicable. It was proven early in the development of quantum control, that for weak fields in an isolated quantum system, phase-only control is impossible for an objective which commutes with the drift Hamiltonian [209]. A qualitative explanation is that under such conditions there are no interfering pathways leading from the initial to the final stationary states. Experimental evidence has challenged this assertion, claiming demonstration of weak-field phase-only control for an excited state branching ratio [210]. The phenomena were attributed to the influence of the environment. A subsequent study showed that such controllability is solvent dependent [211]. A theoretical demonstration that phaseonly control is possible in weak field for an open quantum system soon followed [212]. It is still an open question if weak-field phase-only control is possible for targets that commute with the Hamiltonian in open quantum systems. For example, can population transfer from a ground to the excited surface be phase-controlled for a dye molecule in solution? While this has theoretically been shown to be impossible if the time evolution is Markovian [213], most solvents lead to non-Markovian system dynamics. In the context of coherent control of a chemical reaction, the bimolecular process of bond formation using femtosecond lasers remained more elusive [214–221] than bond breaking. Its coherent control was demonstrated only very recently [222]. Full control of a binary reaction – from its entrance channel of scattering reactants to the targeted products in a selected internal state – is still an open goal. Realizing this dream would create a new type of photochemistry with selective control of yields and branching ratios. Coherent control of bond formation in the gas phase turned out to be so difficult because it starts from an incoherent thermal ensemble. The laser pulse then needs to pick those scattering pairs which show some correlation in their translational or rotational motion. Averaging over rotations can be avoided by orienting or aligning the molecules using strong laser fields [223]. Adiabatic alignment occurs in-field and is achieved with nanosecond pulses. In contrast, femtosecond laser pulses create nonadiabatic alignment that persists after the pulse is over. This second option, where the alignment is produced in field-free conditions, is more interesting in view of the applications because the aligning laser pulse does not disturb the molecular system [224,225]. Spatial averaging in the gas phase implies an integral over the beam profile and blurs coherent effects as the atoms or molecules are exposed to different intensities. Spatial resolution is thus a prerequisite for control [226]. Subwavelength dynamic localization of the laser intensity can be achieved on the nanometre scale [227], lifting the restrictions of conventional optics. Optimal control theory can be used for spatial shaping of the laser fields used in nanoexcitation [228]. Another way to overcome thermal and spatial averaging is by cooling and trapping the atoms or molecules. Coherent control was suggested as a method to cool internal molecular degrees of freedom [112,229]. This has been realized experimentally for cooling vibrations of ultracold cesium dimers [230] and rotations of trapped aluminum hydride ions [231]. While restrictions imposed on the cooling efficiency by the molecular structure can be circumvented using optimal control [232], a persistent challenge to cooling internal molecular degrees of freedom are the timescale separation between vibrations and rotations as well as the enormous bandwidths that are required for strong bonds. In order to prepare molecules that are cold in their translational degrees of freedom, molecules are assembled from atoms which are much easier to cool. However, a major problem in creating molecules from atoms is the extreme change in time and length scale. Optimal control was studied to overcome this issue [233]. Initial experiments employed the simple scheme of a chirped pulse to compress two atoms to closer proximity [234]. This can be viewed as a first step in ultracold laser-induced, i.e., photoassociation. Association yielding ultracold molecules in a single internal quantum state has been demonstrated employing magnetic field ramps (magneto-association) followed by STIRAP-type protocols [235,236]. An experimental challenge still unfulfilled is complete coherent control of ultracold photoassociation. One of the important issues here, particularly in polyatomics, is the need for accurate ground and excited potential energy surfaces to design optimal pulses a priori, as well as to interpret the mechanism of pulses that are found by experimental optimization. For this reason there has been significant interest in recent years in inverting the information in optical spectroscopies in order to reconstruct excited state potential energy surfaces [237–239]. In addition to making or breaking chemical bonds, coherent control has demonstrated its versatility in studying energy transfer [240], light harvesting [241,242], and spectroscopy. The latter includes in particular non-linear and multi-dimensional spectroscopies [243–246] and microscopy [243,247,248]. In this context, coherent control allows to increase both the resolution and the specificity to a single molecule. Originally developed with phaselocked time-delayed laser pulses, non-linear and multidimensional spectroscopies utilize the interference created by these pulses. Laser pulse shaping not only lifts the constraint of having separate pulse beams in non-linear spectroscopy and microscopy [243–245], it also addresses naturally the requirement to decipher the interference which is at the basis of the non-linear spectroscopies. Eventually, this understanding has lead to the use of laser pulse shaping in optical imaging with compensation of the effect of scattering media such as biological tissue [249,250]; optical microscopy with enhanced chemical sensitivity and contrast [251–254]; chemical analysis and detection via optical discrimination [255–258]; cancer diagnosis [259]; and material processing [260]. Similarly, laser pulse shaping is expected to enhance chemical sensitivity in other detection methods such as mass spectroscopy [261,262]. Initially, the coherent control of molecules, be it in the context of chemical reactions or non-linear spectroscopies or energy transfer, considered the dynamics of the nuclear degrees of freedom, using femtosecond laser pulses as the main workhorse. More recently, the focus has shifted to controlling electron dynamics. This is being made possible by the development of advanced X-ray sources which probe the dynamics of electrons within atoms and molecules on attosecond time scales. Their potential for exploring the quantum nature of the nanoworld is unprecedented. For example, using XUV and X-ray light for multidimensional spectroscopy could probe valence excitations locally on different atomic sites in a molecule. This would be invaluable for understanding energy transfer in biological systems and quantum devices. The use of X-ray light sources is currently facing a number of challenges that can be tackled by quantum optimal control. First of all, the large energy of XUV and X-ray pulses results in a high probability of ionization, reflecting the problem of controllability when a continuum of states is involved. This has been addressed in a recent study, where optimal control theory was used to predict experimentally feasible pulses to drive XUV-Raman excitations through the ionization continuum [263]. Another control problem is the creation of the XUV light pulses themselves. In particular, high harmonic generation, where a very strong near-infrared femtosecond laser pulse accelerates an electron in such a way that it emits XUV light, is an ideal candidate for coherent control: theoretical predictions for optimum driving should be easy to adapt in experiment, given the existing pulse shaping capabilities. The challenge that high harmonic generation poses to optimal control theory is a frequency-domain target [103,264,265]. A conclusive answer whether shaping the femtosecond laser pulses can improve high harmonic generation has not yet been provided. Optimal control theory was first applied to chemical reactions using Krotov's method [266,267] and gradient ascent [268,269]. The theory was quickly extended to Liouville space [112,270] to treat condensed phase situations and cooling. Optimal control techniques have also been applied with success to molecular alignment and orientation in gas phase [223]. The design of optimal solutions has allowed to reach the best possible degree of alignment and orientation [271] under experimental constraints such as temperature [272] or collisions [273]. In addition to improving existing control strategies, optimal control has also been used to explore new regimes of alignment dynamics such as planar alignment [274]. The major experimental constraint in experiments with shaped femtosecond laser pulses is the fixed bandwidth. This can be accounted for in optimal control theory by including frequency filtering [105,111,275,276] or by imposing spectral constraints [109,277]. The effect of additional constraints is to pick a different solution out of the many solutions that typically exist in quantum optimal control [134,135,277]. Then a natural follow-up task to any optimization is to determine the control mechanism encoded in the optimized pulse. To date, a systematic approach tackling the control analysis is still lacking.

Mid-Term Prospects: Goals and Challenges

Thirty years after the conception of reaction control, it is fair to ask whether the idea of coherently controlling a chemical reaction can work at all. In this respect it is important to realize that the basic ingredients for achieving this goal have all been developed. The challenge that remains to be overcome is their assembly and application to a specific reaction. The dream of coherently controlling a chemical reaction all the way from its entrance channel to the reaction products thus seems to be within reach. It includes the controlled formation (or photoassociation) of a new chemical bond, the controlled dynamics of the intermediate complex, most likely involving a conical intersection, the controlled cleavage of another chemical bond as well as the stabilization of the reaction product. Realistically, demonstration of control over a complete reaction can be expected within the next few years for a sufficiently simple reaction complex, involving only a few atoms. Experimental techniques are currently developed to trap and cool a single molecule. These are particularly advanced for single molecular ions [278,279]. The molecules can be simple diatomics or large biological chromophors such as rhodopsin. Such trapped species are ideal candidates for control, inducing for example isomerization in rhodopsin. Instead of trapping and cooling molecules directly, they can also be assembled from cold atoms one by one. The number of different atoms which have been cooled and trapped has been increasing steadily. Atoms other than the first row alkali metals have the potential for complex chemistry with multipole bonds. This raises the challenge of assembling molecules from these constituents. The experimental obstacle is the development of pulse shaping techniques in the picosecond and nanosecond range. Another mid-term goal for quantum control is the control of electron dynamics. The ability to control electrons implies Angstrom-scale ultrafast imaging methods which can be realized in the form of laser-induced electron diffraction and high-harmonic spectroscopy. Specific mid-term goals that appear to be within reach using these tools include the control of subfemtosecond charge migration; the controlled generation of spin-polarized electrons from laser ionization; recognition of the absolute configuration of chiral molecules with shaped laser pulses; and, using high-harmonic spectroscopy of molecules, ultrafast imaging of structure and dynamics on sub-atomic length scales.

Long-Term Vision

Synthesis

The ultimate chemical synthetic challenge is to assemble large chiral molecules from elementary building blocks. Currently such syntheses are carried out in solution where the chemical products are stabilized by entropy generation caused by heat transfer to the environment. The vision is for synthesis by photoassociation via polarization shaped light where the final product is stablized by laser cooling and trapped by light.

Analysis

The vision is for a light field tailored to a specific molecule or functional group generating a specific physical out come such as light emission or ionization. Such a capability will enhance the threshold of detection of specific molecules, e.g. in hazard detection or in medical applications. Combined with spatial super-resolution, the resulting analytic methods are also expected to find applications in molecular-scale microscopy.

MAGNETIC RESONANCE

The optimal control of spin dynamics is at the heart of well established magnetic resonance technologies and of emerging new fields of quantum technologies. Nuclear magnetic resonance spectroscopy (NMR) [280–282], electron spin resonance spectroscopy (ESR) [283] and magnetic resonance imaging (MRI) [284] are based on the control of nuclear or electron spins with the help of time-dependent electromagnetic fields. Magnetic resonance is one of the most impressive success stories of quantum control and technology. The mathematical description of the state and dynamics of nuclear spins or electron spins is essentially identical to the canonical description of abstract quantum bits (qubits). In terms of theoretical and experimental control of spins or qubits, NMR had a long head start compared to other quantum technologies because the NMR community has actively explored and developed spin system control methods for more than 60 years. This was driven by many specific and practically important applications in physics, chemistry, biochemistry, biology and medicine. The interdisciplinary impact of quantum-control enabled magnetic resonance is impressively reflected by Nobel prizes in physics (Felix Bloch, Edward Purcell, 1952), chemistry (Richard Ernst, 1991; Kurt W¨uthrich; 2002), and medicine (Paul Lauterbur, Peter Mansfield, 2003) [285–287]. Today, NMR is arguably the most important tool in chemistry to determine molecular structure and dynamics of molecules. ESR is an essential technique in radical reaction chemistry, catalysis, electrochemistry and photosynthesis research. MRI is one of the most informative and frequently used modalities in medical diagnostics. The huge range of practical applications has generated a multi-billion dollar branch of the instrument manufacturing industry (Bruker, Siemens, Phillips, General Electric, JEOL, etc.). This in turn has resulted in the continuous development of more and more sophisticated instruments with superb flexibility in terms of the available control schemes: for example, arbitrary waveform generators and linear amplifies have been standard NMR and MRI equipment for more than three decades (and have more recently also become available with sub-nanosecond time resolution for ESR applications [108,288]). With their help, even very complex pulse shapes can routinely be implemented with high fidelity [289]. The excellent agreement between theory and experiments (as a result of the highly accurate theoretical description of the physics of coupled spins and the availability of very reliable hardware to implement virtually arbitrary pulse sequences and shapes) also has made NMR an attractive testing ground for the experimental demonstration of new control approaches for finite-dimensional quantum systems in general. Concepts of quantum control and sophisticated quantum-control design tools developed in the field of NMR have found many applications in other fields, such as in quantum information processing [290–292], optics (photon echos) [293], in neutron scattering and in the control of nano devices based on quantum dots, artificial atoms, etc. Two important emerging fields of magnetic resonance are hyperpolarization methods for bulk NMR and the control and measurement of individual spins or spin systems, e.g. of NV centers in diamond, with many potential applications in sensing and quantum information processing [294–296]. Hyperpolarization techniques (also known as spin cooling), [297–299] can generate highly polarized non-thermal spin states. Hence, the relatively low sensitivity of NMR (due to the small magnetic moments of nuclear spins and the resulting weak thermal polarization) can be overcome by using a variety of approaches. In particular, there are two different methods that have recently become increasingly popular in practical applications. The first one is based on transfer of the much higher thermal polarization from unpaired electrons onto the nuclear spin ensemble in a process called dynamic nuclear polarization (DNP) [300–302]. The second method involves the use of parahydrogen and a transfer of its highly populated singlet spin state onto nuclear spins in receptor molecules [303]. Both methods have been known for many years, but only recently significant progress has been made in terms of a full quantum description of the underpinning spin physics and the optimization of the required experimental hardware. The detection and control of individual nuclear spins close to nitrogen-vacancy (NV) centers in diamond [304,305] is an excellent example of a new area of optimal control of individual spin systems. In these experiments, single nuclear-spin detection is achieved by an efficient readout based on couplings of the nuclear spins to electron spin states and their detection using optical techniques. In the field of spintronics, spins of individual atoms (such as 31P donors) or quantum dots can be electrically detected and controlled by radio- or microwave pulses with high fidelity [306,307].

State of the Art

The design of pulse sequences that provide maximum sensitivity, selectivity, as well as maximum resilience to instrumental imperfections, is central in the ongoing effort to improve magnetic resonance technologies. Both analytical and numerical methods have been used to design multiple-pulse sequences (including composite or shaped pulses). Since the 1950s, the field of NMR has been a highly active "breeder" for methods to control finite-dimensional quantum systems. The wide spectrum of applications has motivated the development of more and more sophisticated tools for quantum control in a quest to push the experimental performance to their limits. Pulsed magnetic resonance techniques such as the two-pulse Hahn echo and the three-pulse stimulated echo [308] were important milestones on the path to multiple-pulse sequences, which often consist of tens or hundred thousands of individual pulses with defined amplitudes and phases. This was enabled by the development of theoretical tools such as average Hamiltonian theory (AHT) [309] and the symmetry analysis of composite pulses and multiple-pulse cycles and supercycles [282,310–312].

More recently, sophisticated theoretical approaches of geometric optimal control theory [4,12,68,69,71,82,89] have been applied to quantum control problems motivated by magnetic resonance applications. Although these analytical approaches are typically limited to low-dimensional quantum systems, they are able to provide the best possible solutions to a given problem by proving global optimality. The resulting physical performance limits are important theoretical results in their own right and also provide benchmarks for numerical and experimental optimization techniques. Geometric optimal control has been very successfully applied to highly non-trivial examples of uncoupled and coupled spins both in the absence and presence of relaxation (dissipation and decoherence). For example, based on geometrical optimal control analysis, the minimal time for quantum gates in systems of two and three coupled spins have been determined and experimentally implemented [71,89,313]. And in this context, the socalled quantum-gate design metric was discovered, which plays an important role for the design of so-called geodesic unitary gates [71,314]. For example, time-optimal control schemes for the simulation of effective trilinear couplings in systems with only next-neighbor couplings were derived based on geometric optimal control and experimentally demonstrated using NMR. This is a particularly impressive example, because in the limit of small time increments Δt_{sim} under the simulated trilinear Hamiltonian, the actual duration of the controls also approaches zero, whereas previous approaches based on AHT have a minimal duration of 1/(2J), i.e. for $\Delta t_{sim} \rightarrow 0$, the gain of the optimal-control schemes approaches infinity [71,314]. Examples of geometric optimal control applications involving singular arcs [68] in NMR are relaxation-optimized polarization transfer experiments [78,79,82,315], minimal-time controls for the saturation of spins [77], for maximizing contrast in MRI [88] and for maximizing the achievable signal-to-noise ratio per unit time [316,317]. Since the 1980s, numerical algorithms such as conjugate gradients, downhill simplex (Nelder Mead), genetic algorithms and simulated annealing have been applied to find efficient and robust quantum-control schemes for magnetic resonance applications [318–323]. Powerful optimal-control algorithms based on the Pontryagin Maximum Principle (PMP) were applied already since the mid 1980s to the design of shaped pulses for the manipulation of nuclear spin ensembles [324–326]. Potent variants of these PMP-based algorithms were also developed in the context of NMR applications for coupled spins in NMR spectroscopy and quantum information processing [92,93,327]. They are able to optimize pulse sequences for an ensemble of spin systems with realistic parameter ranges for detunings, scalings of the control field etc., i.e. they can take into account variations or uncertainties of experimental parameters. Numerical ensemble-control methods have resulted in extremely powerful control schemes with unprecedented resilience to pulse imperfections, while at the same time taking experimentally bounded control amplitudes and pulse energy limits into account.

In magnetic resonance, an efficient and user-friendly software landscape is starting to emerge – there are a few software packages that support sophisticated quantum mechanical magnetic resonance simulations as well as optimal control. In particular, SPINACH [101] supports all forms of magnetic resonance spectroscopy under the same roof, implements sophisticated spin relaxation theories as well as most mainstream optimal control algorithms that are presently used in magnetic resonance spectroscopy: GRAPE [92], GRAPE-BFGS [93,94], Krotov [96] and Krotov-BFGS [99]. Another advantage of SPINACH is the availability of polynomially scaling spin dynamics simulation algorithms that make previously intractable NMR and EPR simulation (and therefore control) problems accessible [121,328]. In the solidstate NMR community, the SIMPSON software package (SIMulation Program for SOlid state Nuclear magnetic resonance) has been the most extensively used generalpurpose software. It also includes an optimal control toolbox to facilitate robust experiment design [100]. MATPULSE [329,330], DYNAMO [94] and QuTiP [102] are versatile Matlab and Python based simulation and optimization programs. Other software packages for the simulation and/or optimization of spin dynamics include SIMONE [331], OCTOPUS(SI) [332,333], Gamma [334] and SPINEVOLUTION [335]. Both analytical and numerical methods have proven very effective for two- and three-level systems [72,74–77,80,159,336–348], two uncoupled spins [349,350], and two coupled spins [83,85,89,351–355]. Significant progress has been made in understanding of how to optimally control coupled spin systems with more than two spins [65,71,86,87,313–315,356–374]. Recent advances include robust broadband and band-selective pulses in NMR and ESR. Depending on their role in a given experiment, universal rotation (UR) pulses (e.g. for refocussing) or point-to-point (PP) pulses (e.g. for excitation or inversion of spins) are required. Systematic studies of the offset bandwidth (range of detunings) and robustness with respect to scaling of the control amplitude (width of B1 inhomogeneity distribution) of optimized PP [52,53] and UR pulses have been performed [55] for experiments where the maximum control amplitudes (e.g. in many applications of NMR or ESR spectroscopy) or the pulse energy (e.g. in MRI) are experimentally limiting factors.

For a desired fidelity, the bandwidth covered by a (composite or shaped) pulse can be made much larger than the maximum control amplitude (maximum Rabi frequency) at the price of increased pulse duration. Surprisingly, it was found empirically that the pulse duration typically scales only linearly with the desired offset range of operation. Hence, unprecedented fidelities can be achieved that are e.g. required for quantitative NMR applications or for quantum error correction schemes. In addition, it was demonstrated that coherence transfer elements can be designed that are not only resilient to offset and control amplitude variations but also to variations in coupling constants [355,375]. In general, UR pulses are significantly longer than PP pulses for the same error resilience, which was exploited in a new strategy to build complete sequences based on standardized UR pulses when necessary and standardized PP pulses whenever they are sufficient [376]. The recent application of optimal control methods to the problem of heteronuclear decoupling yields not only significantly improved performance [377,378] but also unprecedented flexibility in the design of tailored decoupling sequences [379–381].

Individual pulses were also optimized for Carr-Purcell-Meiboom-Gill echo train sequences [382,383]. Beyond individually optimized pulses, the simultaneous optimization of pulses provides signifi- cant performance gains by exploiting cooperative effects either in a single scan [384] or in multiple scans [385] with first applications in Ramsey-type experiments and in Hahn echo sequences. In liquid-state NMR, optimized broadband pulses have found applications in a number of experiments. Early examples are the CLIP-HSQC [386] and P.E.HSQC [387] for the measurement of one-bond and geminal couplings as well as more recent experiments involving homonuclear decoupling [388] and the rapid acquisition of heteronuclear correlation experiments [389]. All these experiments are of special interest as they are recorded in large numbers every day in chemical laboratories and improvements in robustness, accuracy, or acquisition time are of high interest. For the specific conditions of solid-state NMR and experiments in oriented systems, specific pulse sequences have been optimized [390–396]. In ESR spectroscopy, the first broadband optimal-control pulses have recently been developed and experimentally implemented [108,288]. In this context, the efficiency of novel approaches to take into account transient effects due to transfer line effects and limited resonator bandwidth was both numerically and experimentally demonstrated [108]. Motivated by magnetic resonance imaging applications, optimal control pulses for MRI have been developed [88,324–326,330,397–400].

Applications include improved spatially selective excitation schemes [106,401,402], pulses with minimal radio-frequency (RF) power and pulses that counteract RF inhomogeneity in parallel transmission at ultra-high field [330,398]. For chemical exchange saturation transfer (CEST) imaging, chemical exchange effects were taken into account in pulse sequence optimizaion [403]. For DNP, optimal polarization transfer schemes from electron to nuclear spins have been investigated for a small set of relatively simple model systems [302,404–407]. Optimal control was also used in conjunction with the use of parahydrogen to generate high nuclear spin polarization.

In particular, it was shown that the initial longitudinal two-spin order arising from the hydrogenation reaction can be equally distributed between several spins and converted into detectable magnetization [408]. Geometric and numerical tools from optimal control theory have not only provided pulse sequences of unprecedented quality and capabilities, but also new analytical insight and a deeper understanding of the mode of action of optimal pulses. Numerically optimized pulses can often be interpreted as robust implementations of analytically derived optimal trajectories [77]. These can be understood based on geometrical concepts [341]. In addition, useful tools have been developed to analyze complex pulses [409] and the resulting dynamics in coupled spin systems [405,410,411].

Mid-Term Prospects: Goals and Challenges

An important goal is to make optimal control algorithms easier to use, generally applicable and to further increase their speed. Depending on the applications, very different convergence rates are encountered and a systematic characterization of optimal control landscapes is still missing. With more efficient numerical optimizations methods, optimal control theory will make it possible to design problem-, sample- and patient-specific pulse sequences on the fly. The fast reoptimization of pulse sequences or of sequence elements, e.g. in response to the presence of magnetic susceptibility jumps, will significantly improve their performance. In addition to MRI applications, this could be important in production or process monitoring by NMR, where the pulse sequence should be able to adapt to the sample in the same way as shim currents currently do – examples are magnetic susceptibility and tuning variations in imaging, metabolomics and oil well logging [382]. Optimal control methods are expected to reduce the time that is required to determine structural and dynamical information on biomolecules (e.g. proteins). In this area, coping with large coupled spin networks, especially in the presence of relaxation, is computationally hard and further improved numerical/analytical approaches are highly desirable. Characterizing the experimental imperfections, fine-tuning the spectrometer and setting up the experiment is time-consuming and efficient closed loop feedback-based automatic procedures have to be developed, implemented and integrated with the adaptive design of pulse sequences to automate this process. In the field of medical imaging it is expected that optimal control methods will lead to more sensitive and more efficient pulse sequences, such that a patient has to spend less time in a scanner for an examination. This may be achievable by using multi-band excitation techniques and optimized image acquisition based on multiple transmit and receive channels. Optimal control methods are also expected to help in the extension of the clinical applicability of ultra-high-field scanners and to provide in-vivo spectroscopy with improved diagnostic value. Apart from better localized excitation, progress in the field of magnetization preparation (e.g. reducing the B0 sensitivity of fat saturation) might be achievable. Improved saturation pulses are expected to be useful for many different tasks, from chemical shift imaging (CSI) and chemical exchange saturation transfer (CEST) imaging [403] to single voxel spectroscopy. Also methods for improved quantification accuracy and biomarker imaging are highly desirable.

Small molecule NMR spectroscopy forms the basis of chemical analysis on a molecular level in solution. Applications are already manifold, but are expected to increase significantly during the next decade. Identification and quantification of compounds in complex mixtures is needed. Especially for quantification tasks, methods derived from optimal control will deliver more reliable and robust results. Optimal control sequences are likely to improve the detection level of side products, making chemical products safer and better defined. Quality control of food, pharmaceuticals and other products as well as metabolomics-type applications and personalized medicine will benefit from improved NMR. In the field of hyperpolarization, further hardware advances are expected, in particular in generating and modulating high-frequency microwave fields and it is envisaged that optimal control methods will play an important role in the development of more sophisticated experimental schemes to transfer the electron polarization to surrounding nuclear spins. Ideas on how to control electron-nuclear spin systems in a optimal way have already been published and discussed. This includes the manipulation of the nuclear spins using the anisotropy of the hyperfine interaction [412,413] and the exploitation of repeated generation of dipolar spin order to enhance polarization transport by spin diffusion [414]. Since more sophisticated catalysts are being developed for parahydrogenation reactions [298], it is also anticipated that pulse shapes derived from optimal control principles will be more frequently used to maximize the achievable polarization and to mediate polarization transfer to specific atoms.

Long-Term Vision

Important theoretical and practical aspects of optimal control in magnetic resonance are the physical limits of quantum dynamics. On the one hand, the questions of quantum state reachability in dissipative systems remain largely unresolved. On the other, many practical usage scenarios are time-constrained, and a more general understanding of the best possible performance in a give amount of time is highly desirable. A very desirable outcome of the continued progress in optimal control technology could be that, for a given level of performance, the use of optimal control sequences could significantly reduce the instrument costs as well as costs of sample preparation and purification. Sophisticated shim coils, frequency locks, complicated combinations of susceptibility matched materials in NMR probes and other expensive arrangements had originally been introduced to maximize spectral resolution and selectivity. If both could be achieved by tailored pulse sequences under less than ideal conditions, the complexity could be transferred from the instrument design to the mathematical optimization procedure. The concomitant reduction in hardware cost could result in more affordable instruments, e.g. for MRI examinations. Also, the integration of control design with image reconstruction and spectral calculation in NMR and with instrument design could result in better performance. The combination of open loop and feedback strategies may result in fast and fully automated tuneup procedures, which would further reduce the required experimental time. A long-term vision for magnetic resonance techniques is the detection of the nuclear spins of individual molecules, such as a protein. The ability to image the shape of a single molecule similar to the way we can image humans today would revolutionize structural biology and pharmacology. The recent developments in sensing using NV centers in diamonds (see Sect. 5) might eventually reach this goal, as well as generate completely new application areas, e.g. in medical diagnostics.

QUANTUM INFORMATION AND COMMUNICATION

State of the Art

Quantum technologies (see, e.g., [2]) exploit quantum coherence and entanglement as essential elements of quantum physics. Applications include high-precision measurements and sensing, which would reach unprecedented sensitivity, the simulation of physical and biological systems, which would be impossible to study otherwise, and quantum information processing, which would allow to solve computationally hard problems. Successful implementation of quantum technologies faces the challenge to preserve the relevant nonclassical features at the level of device operation. More specifically, each task of the device operation needs to be carried out with sufficient accuracy, despite imperfections and potentially detrimental effects of the surroundings. Quantum optimal control not only provides toolboxes that allow for identifying the performance limits for a given device implementation, it also provides the protocols for realizing device operation within those limits. Prominent tasks include the preparation of useful quantum states as well as implementation of quantum operations. The power of the quantum optimal control approach for implementing these tasks has very recently been demonstrated in a number of impressive experiments. For example, nonclassical motional states of a Bose-Einstein condensate were prepared with optimized control sequences for wavepacket interferometry [415], and the loading of an ultracold atomic gas into an optical lattice was improved [416]. With respect to quantum operations, quantum optimal control allowed for error resistant single-qubit gates with trapped ions [417] and for single qubit gates without the need for invoking the rotating wave approximation in nitrogen vacancy (NV) centers in diamond [418]. For the latter platform, optimal control is also at the basis of a spectroscopy protocol allowing to image nanoscale magnetic fields [295]. In quantum processor candidates based on superconducting circuits, leakage to non-computational states in the most common type of qubit, the transmon [419–421], was avoided and frequency crowding was accomodated [422,423] thanks to optimal control results. Closed-loop optimal control [150,424] enabled fine-tuning of gates that were determined manually, allowing them to reach consistent record fidelities within this platform. Control of donor qubits in Si has been achieved and characterized [425]. Control strategies for spins in semiconductors currently trade off conceptually simple single-spin schemes [426,427] with more robust and accessible two- and three spin techniques [428–430]. The short natural time scales suggest adiabatic schemes to be attractive [431–433] In view of scaling up control, the design and implementation of unitary maps have recently been demonstrated in a 16-dimensional Hilbert space, spanned by the electron and nuclear spins of individual cesium atoms [434].

The use of control methods in a broader sense has allowed for further significant experimental achievements, such as to improve of the coherence of a qubit, realized by the electron spin in an NV center, using dynamical decoupling [435]. A famous further example of high-end control techniques is the Paris experiment of stabilizing a quantum state with predefined photon number via real-time closed-loop feedback [168], which required to include the noise back-action of controls onto the system by way of stochastic differential calculus. These experimental achievements were preceded by a large number of theoretical predictions on how optimal control may improve or enable quantum state preparation, operation and readout. State preparation protocols include transport of atoms [436] and ions [437,438] as well as transport in a spin chain [86,356,359,363,369,439,440], photon storage [441], preparation of squeezed states [442], cluster-states [443], non-classical states in a cavity [444] or in spin chains [364,365], as well as preparation of a quantum register [122] and many-body entangled states [123,445,446] – to name just a few. Likewise, optimal control helped to implement high- fidelity quantum gates such as two-qubit gates with neutral atoms in dipole traps [447,448], on atom chips [449], or with Rydberg atoms [450,451], two-qubit gates between ions [452], between an ion and an atom [453], error-correcting qubit gates of electron and nuclear spins within single NV centers [454], or entangling gates between distant NV centers [296]. For superconducting qubits [455], two-qubit gates were optimized, starting from Cooper pair boxes [456–459] to modern transmonbased schemes [117,460]. In these optimizations, special attention was paid to robustness against noise [456,461] which can even be used as a tool for control [60]. Also, readout has been addressed [462,463]. In order to adapt to the strong filtering of control lines in superconducting qubits, transfer functions had to be taken into account [107,108,110,464] and experimental fluctuations and noise were accomodated [117,364,465]. Fidelity limits on two-qubit gates due to decoherence were studied for Markovian [56,114,117] as well as non-Markovian [116] time evolutions. It is noteworthy in this context that non-Markovian time evolutions also play a role in quantum simulation, e.g., of collision models [466].

In view of quantum computation, it has been suggested how to retain universality in spite of limited local control [65,364] by using environmental degrees of freedom [60]. The Jones polynomial, i.e., a central invariant in knot theory, can be evaluated, using an NMR spin ensemble at ambient temperature, in an algorithm equivalent to deterministic quantum computing with a single pure qubit [467–469]. In order to obtain these results, the quantum optimal control methodology had to be adapted to the requirements of Quantum Technologies. Optimization algorithms had to be derived for specific quantum gates [470–472], dissipative evolution as seen in the reduced system dynamics [56,113,114,461,473], or exploiting invariants in system-bath models [474], optimization up to local equivalence classes [475], which can also be used for arbitrary perfect entanglers [476,477] or optimizing for many-body entanglement [478]. Moreover, control techniques were adapted to non-linear dynamics as found in a BEC [479–481] and to general dynamics, functionals and couplings to be controlled [98]. Manybody systems can be optimized numerically with the chopped random basis (CRAB) method [122,123] by interfacing the time-dependent density matrix renormalization group (t-DMRG) with parameter optimization via e.g. the simplex algorithm. Other techniques specifically cover robustness against experimental fluctuations or noise [56,117,364,444] or filters in experimental implementation of controls [107,108,110,464].

Mid-Term Prospects: Goals and Challenges

The field of quantum technologies has matured to the point that quantum enhancement is explored beyond quantum computation only. Devices such as quantum simulators or quantum sensors are currently under active development. Control methods will be crucial to operate these devices reliably and accurately. This involves the device preparation, or reset, the execution of the desired time evolution, and the readout of the result. These tasks set the agenda for the next few years. More specifically, central mid-term milestones include the robust implementation of gates in a multi-qubit architecture, finding solutions to readout and fast reset limitations, automatization of key tasks of surface code error correction and optimal as well as robust generation of multi-particle entangled states for a variety of quantum technology platforms. All of these will require decoherence control. A main challenge for optimal quantum control is to reach convergence of numerical optimal control and experimentation. To date, either optimal control is used for computer-aided discovery of analytical schemes that can be remodeled in an experiment or the quantum processor itself is employed to calibrate gates. In order to better combine numerical optimal control and experimentation, the modeling of qubits as well as errors and other nonidealities of the system, in particular for open systems, needs to be improved and the robustness of pulses enhanced. Also, optimized pulses should initially be reduced to few parameters before more complicated and effective solutions can be pursued. On the other hand, pulse shaping platforms need improvement.

Coming to specific quantum technology platforms, applications of quantum control in superconducting qubits should follow the current European thrust towards analog and digital quantum simulation and lead to the preparation of entangled ground states, fast and accurate quantum gates and tools for quantum machines. For some instances, compatibility with quantum error correction is desired. For trapped ions, it seems realistic to combine quantum gates with ion transport in segmented traps, using optimal control techniques, which can also be applied to the systematic optimization of pulse sequences for efficient generation of complex operations. In the field of cold atoms, challenges for optimal control are twofold.

On the side of quantum simulations, the goal is to enable high- fidelity preparation and manipulation of many-body quantum states of increasing complexity, with and without local control. On the side of quantum communication, the goal is to enable efficient inter conversion between flying qubits and quantum memories via coherent atom-photon coupling, with and without cavities. In the case of color centers in diamond, with major applications in quantum sensing, the goal is to enhance the sensitivity of the defect spins employed as quantum probes via improved protection from environmental noise e.g. through dynamical decoupling techniques.

In general, control techniques are expected to contribute to decoupling and dissipative state engineering [151–153], for instance in view of enhancing the lifetime of quantum memories.

In order to improve the lifetime of a quantum register, control can also be used to implement error-correcting gates and circuits [454]. Moreover, while quantum compilation, i.e., the translation of a unitary gate into the machine language of pulses and evolutions, can readily be done via optimal control up to some 10 qubits [482], a scalable assembler of elementary gates (up to 10 qubits) into many qubits is an open problem that may benefit from tensor-contraction techniques. Both numerical optimal control and closed-loop control are expected to be useful for tackling these goals.

Numerical optimal control has the advantage of versatility, whereas closed-loop control can easily be tuned to specific tasks such as determining parameter uncertainties. A hybridization of both approaches is conceivable as well. The main difficulties that need to be overcome to reach the above mentioned mid-term milestones are a sufficiently accurate modelling of complex quantum dynamics to build control on top, integration of tomography and system identification with optimal control, efficient ways to take into account experimental constraints and uncertainties, and bridging the gap between the quantum control community and the communities of the respective quantum technology platforms.

Long-Term Vision

Several current quantum technology platforms show a strong scaling potential. Thus in the long term, control schemes need to be made scalable. This represents a severe challenge, but meeting this challenge will make quantum control a basic building block of every quantum technology and ensure, at the same time, their proper functioning in a world that is only partially quantum.

Take the examples of superconducting qubits, NC centers or spins in Si, where fabrication is a key task that could and should be improved by control techniques. The controlled adjustment of fabrication parameters should be simple, and the qubits should to a certain extent be robust to the influences of the rest of the architecture they are placed in. Independent of a specific platform, error correction at large, for instance by toric codes [483,484], is one of the strategic long-term goals that is expected to benefit from control techniques given recent advances by randomized benchmarking [424]. To this end, system-identification protocols matched with optimal control modules will be of importance. A pioneering step in this direction was made by the ADHOC technique [150] that combines open-loop control as a first step with closedloop feedback learning (with implicit parameter identifi- cation). Moreover, taking quantum control algorithms to match with tensor-contraction techniques in order to address quantum-many body systems (where first steps have been made by CRAB [122,123]) is expected to pave the way to more accurate handling of experimental quantum simulation setups.

In short, quantum control will be the means to get the most performance out of an imperfect system and combine challenging physics at the few-qubit level with engineering at the multi-qubit level. This should aim for example at enabling quantum simulations that are impossible on classical computers. In addition, the realization of the following long-term goals, using optimal control techniques, seems challenging yet conceivable: demonstrating the practical usefulness of engineered quantum states, for example in quantum metrology; implementing reliable strategies for the control of mesoscopic systems; exploring the dynamics of quantum many-body systems beyond equilibrium; and understanding the microscopic origin of thermodynamic laws. In other words, the long-term goal of quantum optimal control for quantum technologies is to develop a software layer enhancing the performance of quantum hardware for tasks in computing, simulation, communication, metrology and sensing beyond what is achievable by classical means, enabling the achievement of quantum supremacy [485].

PROSPECTS FOR APPLICATIONS AND COMMERCIAL EXPLOITATION

Quantum-control enabled technologies have the potential for truly revolutionary innovation. More sophisticated control techniques would make current technologies more powerful and also help to create novel technologies, e.g. highly sensitive magnetic detectors, microscopic temperature measurement devices, molecular imaging tools etc. Better control of quantum systems has the potential to significantly reduce instrument costs, replacing large NMR spectrometers with small and portable devices with many new fields of applications.

Quantum optimal control applications broadly fall into two classes: applications to novel quantum technologies and applications to chemistry-related areas such as spin resonance. Novel optimal control strategies have already been implemented in commercial NMR spectrometers and implementations of optimal control sequences in MRI are being pursued. This trend is expected to continue. Other chemistry-related applications where coherent control plays out powerfully are imaging, optical microscopy and various variants of spectroscopy as well as chemical analysis where shaped laser pulses improve the resolution and enhance the specificity to a particular molecule. Areas of application range from remote chemical detection all the way to cancer diagnosis.

On the other end of the spectrum, there is presently an emerging industrial effort in quantum computing lead by IBM, Google, and Microsoft. The first two companies have invested into the development of superconducting qubits and both of them use optimal control techniques [424,459]. This is no coincidence – optimal control can have impact in systems that have reached some technical maturity in research laboratories, which is the point at which industry gets interested. Further industrial perspectives will be linked to further development of the quantum technologies industry. Early convergence could happen in quantum sensing, which takes sensing ideas similar to those appearing in spin resonance and combines them with ideas from quantum technologies.

CONCLUSIONS

Quantum control is a key facilitator for spectroscopy and imaging as well as AMO physics and emerging quantum technologies for computation, simulation, metrology, sensing and communication. For all these applications, it is crucial to reach the required precision given experimental limits on control amplitudes, power, timing, accuracy of instruments as well as the ever-present interaction with the environment. Optimal control theory provides a framework to identify which quantum tasks can be accomplished with what precision in the presence of decoherence and experimental imperfections and limitations. Quantum control systems theory will require the integration of control aspects at many different levels. Future quantum technologies will rely on integrated architectures of hybrid quantum systems [486] with e.g. nuclear spins for long-term storage, quantum-nanomechanical devices for sensing and photons for the communication of quantum states. This will require also the integration of quantum mechanics in engineering education and vice versa. It will be necessary to establish strong links of quantum control experts to quantum engineering and to the manufacturing of quantum devices. Due to its interdisciplinary nature with applications in many fields, future advances in the optimal control of quantum systems will require the combined effort of people with expertise in a wide range of research fields. Only the close link of basic research, development and applications will open scientifically and economically rewarding perspectives and will foster the innovation potential of emerging quantum technologies in an optimal way. The Virtual Facility for Quantum Control (VF-QC) under the umbrella of the EU Coordinated Action for Quantum Technologies in Europe (QUTE-EUROPE) marks an important step in this direction. The primary goal of the VF-QC is to provide a common structure for the growing quantum control community in Europe, for the promotion of quantum control and the provision of expertise to other scientific communities, to policy makers and the general public. Establishing common terminology, common standards and common visions are crucial prerequisites to maximizing the beneficial impact of quantum control methods on current and future technology, economics and society.

This strategic report has grown out of the Pan-European Coordination and support action on Optimal Control of Quantum Systems (QUAINT), funded by the European Commission under grant number 297861. We would like to thank the board of referees handling QUAINT and all colleagues who have participated in the community consultation for their comments and suggestions. In particular we acknowledge contributions from Tobias Brixner, Andreas Buchleitner, Daniel Burgarth, Simone Montangero, Enrico Prati, David Tannor, and Zdenˇek Toˇsner. We are indebted to Per

J. Liebermann for assisting the writing process as well as Frank Langbein, Peter K. Schuhmacher and Andrea Dumont for managing the community consultation.

REFERENCES

1. The QUAINT website, www.quantumcontrol.eu (2015)
2. J.P. Dowling, G. Milburn, Phil. Trans. R. Soc. Lond. A 361, 1655 (2003)
3. M. Devoret, Quantum Machines: Measurement and Control of Engineered Quantum Systems, Ecole de ′ Physique des Houches 2011 (Oxford University Press, Oxford, 2014)
4. L.S. Pontryagin, V.G. Bol'tanskii, R.S. Gamkrelidze, E.F. Mischenko, The Mathematical Theory of Optimal Processes (Pergamon Press, New York, 1964)
5. E.B. Lee, L. Markus, Foundations of Optimal Control Theory (Wiley, New York, 1967)
6. A.E. Bryson, Y.C. Ho, Applied Optimal Control: Optimization, Estimation, and Control (Taylor and Francis, 1975)
7. R. Kalman, P.L. Falb, M.A. Arbib, Topics in Mathematical System Theory (McGraw-Hill, New York, 1969)
8. E. Sontag, Mathematical Control Theory (Springer, New York, 1998)
9. D.E. Kirk, Optimal Control Theory: An introduction (Dover Publications, New York, 2004)
10. J.M. Coron, Control and Nonlinearity (Am. Math. Soc., Providence RI, 2007)
11. A.A. Agrachev, Y.L. Sachkov, Control theory from the geometric viewpoint (Springer-Verlag, Berlin, 2004), Vol. 87
12. V. Jurdjevic, Geometric Control Theory (Cambridge University Press, Cambridge, 1997)
13. A. Bressan, B. Piccoli, Introduction to the mathematical theory of control, AIMS Series on Applied Mathematics (American Institute of Mathematical Sciences, Springfield, MO, 2007), Vol. 2
14. D. D'Alessandro, Introduction to Quantum Control and Dynamics, Applied Mathematics & Nonlinear Science (Chapman and Hall, Boca Raton, 2008)
15. H.M. Wiseman, G.J. Milburn, Quantum Measurement and Control (Cambridge University Press, Cambridge, 2010)
16. S. Cong, Control of Quantum Systems: Theory and Methods (Wiley & Sons, Singapore, 2014)
17. C. Brif, R. Chakrabarti, H. Rabitz, New J. Phys. 12, 075008 (2010)
18. G. Dirr, U. Helmke, GAMM-Mitt. 31, 59 (2008)
19. D. Dong, I.A. Petersen, IET Control Theory A. 4, 2651 (2010)
20. R. Zeier, T. Schulte-Herbr¯uggen, J. Math. Phys. 52, 113510 (2011)
21. C. Altafini, F. Ticozzi, IEEE Trans. Automat. Control 57, 1898 (2012)
22. A. Borzi, Nanoscale Syst. Math. Model. Theor. Appl. 1, 93 (2012)

23. B. Bonnard, D. Sugny, Optimal Control with Applications in Space and Quantum Dynamics, AIMS on Applied Mathematics (American Institute of Mathematical Sciences, Springfield, 2012), Vol. 5
24. V. Jurdjevic, H. Sussmann, J. Differ. Eq. 12, 313 (1972)
25. H. Sussmann, V. Jurdjevic, J. Differ. Eq. 12, 95 (1972)
26. R.W. Brockett, SIAM J. Control 10, 265 (1972)
27. R.W. Brockett, SIAM J. Appl. Math. 25, 213 (1973)
28. D. Elliott, Bilinear Control Systems: Matrices in Action (Springer, London, 2009)
29. F. Albertini, D. D'Alessandro, IEEE Trans. Automat. Control 48, 1399 (2003)
30. I. Kurniawan, G. Dirr, U. Helmke, IEEE Trans. Automat. Contr. 57, 1984 (2012)
31. U. Boscain, M. Caponigro, T. Chambrion, M. Sigalotti, Commun. Math. Phys. 311, 423 (2012)
32. G. Turinici, in Mathematical Models and Methods for Ab Initio Quantum Chemistry (Springer Berlin Heidelberg, 2000), Lecture Notes in Chemistry, Vol. 74, pp. 75–92
33. S.G. Schirmer, H. Fu, A.I. Solomon, Phys. Rev. A 63, 063410 (2001)
34. C. Altafini, J. Math. Phys. 43, 2051 (2002)
35. M.H. Goerz, G. Gualdi, D.M. Reich, C.P. Koch, F. Motzoi, K.B. Whaley, J. Vala, M.M. M¨uller, S. Montangero, T. Calarco, Phys. Rev. A 91, 062307 (2015)
36. F. Platzer, F. Mintert, A. Buchleitner, Phys. Rev. Lett. 105, 020501 (2010)
37. S. Sklarz, D. Tannor, Phys. Rev. A 66, 053619 (2002)
38. U. Hohenester, P.K. Rekdal, A. Borz`ı, J. Schmiedmayer, Phys. Rev. A 75, 023602 (2007)
39. G. J¨ager, D. Reich, M.H. Goerz, C.P. Koch, U. Hohenester, Phys. Rev. A 90, 033628 (2014)
40. A.G. Fowler, J.M. Martinis, Phys. Rev. A 89, 032316 (2014)
41. R. Barends et al., Nature 508, 500 (2014)
42. J. Preskill, e-prints arXiv:1203.5813, 25th Solvay Conference on Physics, 19-22 October 2011. (2012)
43. G. Kurizki, P. Bertet, Y. Kubo, K. Mølmer, D. Petrosyan, P. Rabl, J. Schmiedmayer, Proc. Natl. Acad. Sci. USA 112, 3866 (2015)
44. M.D. Grace, J. Dominy, R. Kosut, C. Brif, H. Rabitz, New J. Phys. 12, 015001 (2010)
45. M.M. M¨uller, D.M. Reich, M. Murphy, H. Yuan, J. Vala, K.B. Whaley, T. Calarco, C.P. Koch, Phys. Rev. A 84, 042315 (2011)
46. P. Watts, J. Vala, M.M. M¨uller, T. Calarco, K.B. Whaley, D.M. Reich, M.H. Goerz, C.P. Koch, Phys. Rev. A 91, 062306 (2015)
47. K. Singer, U. Poschinger, M. Murphy, P. Ivanov, F. Ziesel, T. Calarco, F. Schmidt-Kaler, Rev. Mod. Phys. 82, 2609 (2010)

48. H.A. F¨urst, M.H. Goerz, U.G. Poschinger, M. Murphy, S. Montangero, T. Calarco, F. Schmidt-Kaler, K. Singer, C.P. Koch, New J. Phys. 16, 075007 (2014)

49. T. Caneva, M. Murphy, T. Calarco, R. Fazio, S. Montangero, V. Giovannetti, G.E. Santoro, Phys. Rev. Lett. 103, 240501 (2009)

50. M. Murphy, S. Montangero, V. Giovannetti, T. Calarco, Phys. Rev. A 82, 022318 (2010)

2
CHAPTER

CONTROL PROBLEMS IN QUANTUM SYSTEMS

WU ReBing, Zhang Jing, LI ChunWen, Long Guilu &
Tarn Tzyhjong

Keywords: Quantum control, control theory, quantum physics

INTRODUCTION

Quantum control refers to the design of control strategies in systems that obey the principles of quantum mechanics, e.g. microscopic systems with few atoms or photons that were addressed early in the lecture "Plenty of Room at the Bottom" given by Richard Feynman in 1959 [1]. This dream has been pursued since 1960s right after the invention of lasers, which illuminated the hope to coherently control chemical reaction processes, but it was not until the end of 20th century when a burst of successes occurred in controlling ultrafast quantum dynamics. Nowadays, quantum control protocols are being introduced to more fields such as quantum information processing [2] and nanomaterials and nanodevices [3]. The inherent power afforded quantum control results from the unique nonclassical features of quantum mechanics. Essentially, orthogonal eigenstates of a quantum observable (corresponding classically identifiable states) can form superpositions and thereby span a much larger space of quantum states available for coherent manipulation; quantum tunneling allows crossing of energy barriers without as much work as in classical mechanics, thus potentially improve sensitivity and controllability. However, there is no free lunch on the other side; these advantages can only be exploited under extreme spatial (atomic or subatomic) and temporal (femtosecond and attosecond) physical scales. To realize the power of quantum control, vulnerable coherence and entanglement in quantum states need to be protected and measured before being destroyed by intermediate interacting environments. The earliest quantum control studies from the 1960s to the 1980s focused on (macroscopic) quantum ensembles in plasmas and laser devices, nuclear accelerators, and nuclear

Citation: ReBing Wu, Jing Zhang, ChunWen Li, GuiLu Long, TzyhJong Tarn Control problems in quantum systems doi:10.1007/s11434-012-5193-0.

power plants [4] (mostly in former Soviet Union), in which the systems were modeled as quantum harmonic oscillators, which have little difference with those from classical systems [5, 6]. In the 1980s, Tarn's group completed a series of studies on general (linear or nonlinear) quantum systems in regard to modeling [7], controllability [8], invertibility [9], and quantum nondemolition filter problems [10]. In Europe, Belavkin's investigations [11–13] on the optimal estimation of quantum signals in quantum channels shows that quantum feedback control and optimal control is in principle feasible using quantum filters and nondemolition measurements. Photonic control of quantum chemical processes might be the most productive area to date [14–18]. After discovering the IVR (intramolecular vibrational resonance) phenomenon that hinders the control of molecular dynamics, investigators realized that vibrational (electronic) molecular dynamics can only be controlled at femtosecond (attosecond) scales. Only at such intervals can the dominant degrees of molecular freedom be manipulated against IVR via coherent constructive and destructive superpositions. Early single-parameter control schemes include the Brumer-Shapiro phase control scheme [19, 20] and Tannor-Rice time-delay scheme [21]. Later on, optimal control theory (OCT) was applied to molecular spectroscopy [22, 23] as well as more complex cluster systems [24]. In actual system implementations, the introduction of model-independent learning algorithms [25] led to successful breakthroughs in the laboratory [18, 26]. By 2010, over 150 experiments have been successfully carried out in the laboratory, and a much greater number of numerical simulations have been reported [27]. In the AMO 2010 report composed by National Research Council of United States [28], it was pointed out that the future development of quantum control will challenge extremely critical physical circumstances including (1) systems in microscopic scales (e.g. nanoscale); (2) extremely low temperatures; (3) ultrafast dynamics; (4) high-energy limits; (5) information theoretic limits. Fields touching on any of these exciting physics will open up numerous opportunities for interdisciplinary research.

The field of quantum control itself is evolving rapidly. Figure 1 shows the statistics of research articles that were published over the past twenty years (data obtained from Web of Science, Thomson Reuters Co.). It can be seen that past quantum control studies were mostly in the areas of physics and chemistry, but in recent years research from engineering fields has grown very quickly. Moreover, several monographs have been published on quantum control with authors coming from chemistry, quantum optics and control theory [14,15,29,30] and the interested reader can learn more about the state-of-art in quantum control from various review articles [2, 27, 31].

More recently, interdisciplinary studies on quantum control have also attract much attention in China [32–37]. In this review, we aim to give a general review over the fundamental problems and theory from the viewpoint of system control. In Section 1, we will describe how to model quantum control system under semiclassical controls. Then, in Section 2, we discuss the fundamental analysis problems of quantum control including stability and controllability. In Section 3, control design methods will be introduced, particularly on optimal control design methods. Finally, perspectives will be presented.

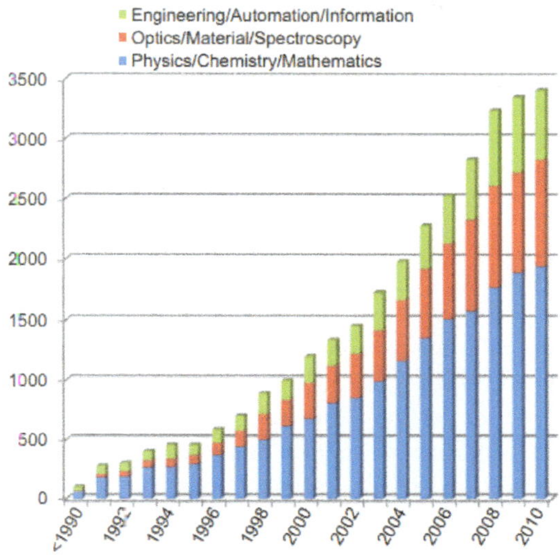

Figure 1: Statistics of the number of SCI publications on quantum control in the science and engineering fields (Source: Web cf Science).

MODELING OF QUANTUM CONTROL SYSTEMS

Physical objects in nature are usually classified as either matter or fields; thus the goal of quantum control can likewise be classified as the control of matter and the control of fields. The control of matter employs well-controlled light sources to effectively manipulate the dynamics of matter (e.g. molecules, spins, electrons or macroscopic subjects such as quantum dots and superconducting devices). The control of field refers to the preparation of optical states (e.g. singlephotons or squeezed states) for high-precision measurements or information processing, or the generation of new light sources (of high frequency or high intensity) for more advanced control or measurement of matter [38], such as via higher-order harmonic generation (HHG) [39]. To control matter or fields, the object needs to interact with the control, in which the interaction mechanism of matter or field plays the central role. In practice, coherent (laser) fields are often employed to manipulated quantum dynamics. If the photon number contained in the field is large (~105/cm3 in each mode), the field can be approximated with high precision by a classical electromagnetic field, called the semiclassical approximation. Moreover, if the size of matter being radiated is far smaller than the wavelength of light, the interaction with the field depends only on the position r0 of the center of mass. Under an appropriately chosen gauge, the interaction Hamiltonian can be written as

$$H_{\text{int}} = \boldsymbol{E}(\boldsymbol{r}_0, t) \cdot \boldsymbol{D}, \quad (1)$$

where D is the dipole operator of the quantum system. If the size of matter is comparable with the wavelength, a coupled Maxwell equation representing the evolution of the field will be required. Although no direct interaction exists between electromagnetic fields, field control can be indirectly done by another external field via some linear or nonlinear medium. Under appropriate physical circumstances, a semi-classical control system over a time-varying quantum wave function $\Psi(t)$ can be expressed in the following general form:

$$i\hbar\frac{\partial\Psi(t)}{\partial t} = \left\{H_0 + \sum_{k=1}^{r} u_k(t)H_k\right\}\Psi(t), \tag{2}$$

where H_0 is the internal Hamiltonian and H_k (k = 1, ...r) are control Hamiltonians activated by associated time-dependent control parameters $u_k(t)$ that can be tuned by their frequency, phases, or profile. Here the system can be physically either matter or field. Such a system is referred to as a bilinear control system because the equation is linear in both the state and the control functions. From the view of system control theory, it is also convenient to define a target to evaluate the control performance. For example, a natural choice is the expectation value of some system observable

$$y(t) = \langle\Psi(t)|O|\Psi(t)\rangle,$$

which can be measured when the system is a quantum ensemble or the system is measured for a sufficient number of times. In quantum information science, the evolution operator itself is often a target of control, which leads to the following form of control system:

$$i\hbar\frac{\partial U(t)}{\partial t} = \left\{H_0 + \sum_{k=1}^{r} u_k(t)H_k\right\}U(t), \tag{4}$$

where the unitary evolution operator U(t) always starts from the identity operator. The corresponding output is often chosen as the fidelity with some desired quantum gate, i.e. $y = |\mathrm{Tr}(W^\dagger U(T))|$, where W is the target unitary transformation. In practice, the system is always under the influence of external environments and should be modeled as an open quantum system, but we will not go further into this aspect. Interested readers can refer to the literature [30]. In recent years, networked quantum systems drew much attention, which is crucial in building quantum structures and thus enabling construction of large-scale electronic circuits at the quantum level. More importantly, as the core of control theory, the implementation of feedback control relies on the possibility of integrating quantum components into feedback loops. In classical system theory [40], it is natural to treat each component of a composite system as an objective with inputs and outputs:

$$\dot{x}(t) = f[x(t), u(t)], \quad y(t) = g[x(t), u(t)],$$

where the input u(t) causally affects the evolution of the state x(t) through the dynamical equation, which subsequently in- fluences the variation of the output y(t) as a function of the state and control variables. However, this general structure needs to be carefully treated in quantum systems when the input and/or output are quantum variables, as they acts on different Hilbert spaces. The seminal paper [41] presented the Langevin (as well as master) equations for quantum systems for which the input and output are optical fields:

$$\dot{a} = -i\omega a - \frac{\gamma}{2} + \sqrt{\gamma} b_{\text{in}}, \quad b_{\text{out}} = b_{\text{in}} + \sqrt{\gamma} a.$$

Here the input and output act on the same Hilbert space, representing the incoming and outcoming free-space fields.

Based on the theory, systematic studies have been performed on linear quantum networks involving concatenation, series product, and parallel product, which are interesting topics that are quite different from classical control systems. In the space of coherent states, a beautiful parallel with classical harmonic oscillator networks can be observed [42], although more interesting are systems having nonlinear Hamiltonians at the single-photon level. For more details, readers may look into another paper written by Zhang and James on the topic of coherent feedback control.

SYSTEM ANALYSIS: STABILITY AND CONTROLLABILITY

A major motivation of system control theory is to achieve high control performance under limited resources and uncontrollable disturbances. However, a more fundamental requirement is to maintain the system at a desired working status; this calls for stability. The concern for stability in quantum system dates back to the very beginning of quantum physics when the answer for "why matter stays as it is?" was pursued [43]. For example, an atom consisting of a nucleus surrounded by electrons cannot exist stably in classical electrodynamics theory because electrons will eventually fall to the surface of the nucleus after losing energy via electromagnetic radiation. It can only be explained in quantum theory by the lower-bound eigenstructure of electronic states. During the past century, quantum stability was mainly studied by mathematical physicists. To engineer quantum systems, it is also of practical importance to keep quantum systems stable under open-loop or close-loop designs. A typical circumstance is when a system is desired to be stabilized on (or as close as possible to) the unit sphere of pure states against decoherence effects. To our knowledge, a quantum stability study that should be physically meaningful and convenient for engineering control design seems to be lacking. For open-loop control systems, a frequently discussed problem is the stability of systems under periodic [44] or chaotic [45] external fields, where the system is stable if the kinematic energy is bounded. For feedback control purposes, the concept of input-output stability is extended to quantum control systems [46]. There are still many open problems in parallel with various stability concepts in classical theory, e.g. input-output stability, input-to-state stability, stochastic stability and structural stability. As another fundamental concept in control theory, controllability refers to the ability of steering quantum systems between arbitrary states. The controllability issue was first attacked by Tarn et al. in the early 1980s with respect to quantum systems with finite-dimensional controllability Lie algebras; later this was extended to time-dependent systems [47] and systems with infinite-dimensional controllability Lie algebras [48]. A special class of such systems was also extensively studied in the control of molecular systems, where quantum systems were assumed to possess at first approximation a finite number of levels. Based on Lie algebraic analysis, the controllability of such systems received careful study in regard to the dipole structure [49]. Recently, an interesting field-atom system was studied [50] via methods established in differential geometry, showing that a two-level system interacted with a single optical mode is controllable over any finite-dimensional subspace without the need for the rotating-wave approximation, which was thought to be crucial in understanding the dynamics of strongly interacting quantum systems. To summarize, the study of controllability for finite-level quantum systems has been relatively complete, although that of infinite-level systems, especially systems with unbounded and continuous spectra, is extremely difficult. Moreover, it appears that controllability is only of interest to control theoreticians and mathematicians, whereas no direct relevance can be seen to actual control design. However, this is not the case [51], as controllability determines the complexity of finding optimal controls (i.e. search efforts). The more a system is controllable, the easier is the search to reach a global optimal solution. Thus, delving further into the fundamental study of controllability is deservedly worthwhile.

CONTROL DESIGN METHODOLOGIES

The design methods of quantum control can be roughly classified as open-loop and feedback methods. The open-loop design (see Figure 2 for schematics) determines the control before applying the control onto the system [27], while feedback design (see Figure 3 for schematics) adaptively adjusts the system dynamics during the evolution according to information acquired from the system [31]. The body of literature on open-loop control is much larger than that of feedback control, in contrast to classical control systems where feedback control is dominant. This is because the measurement required for feedback is difficult; one can even do it, though the trade-off between inevitable back-action and the inquired information will have to be considered. It is not even known whether feedback will eventually improve the control performances. Only recently has feedback attracted any attention. Here we will review the main design methods in openloop control design, and the closed-loop control methods will be surveyed by Zhang and James in this special topic. More specifically, we will focus on optimal control methods, which cover most of the relevant literature. There are also design methods based on physical intuition (e.g. [52,53]) and Lie algebraic decompositions (e.g. [54]), but we will not discuss these in any detail. Every control target can be described by one or multiple costs as functional(s) of the control fields, say J[u(·)]. For example, the frequently-used expectation value of some observable is an implicit functional of control, because the state depends on the control through the Schr¨odinger equation.

Figure 2: Schematics for quantum open-loop control system.

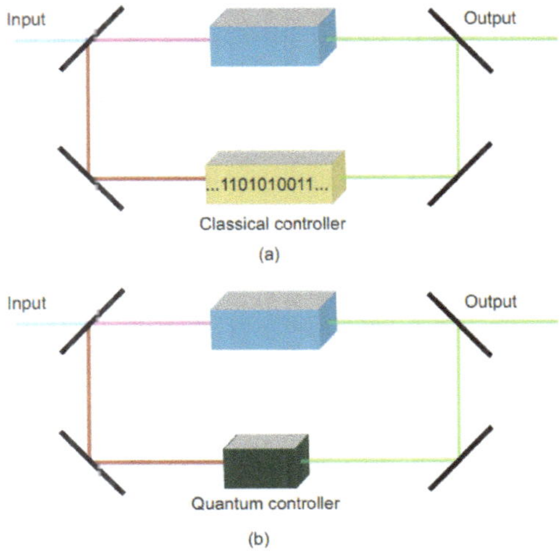

Figure 3: Statistics from quantum feedback control system. (a) Quantum measurements record information about a system as classical signals that are then processed to update control inputs. (b) A so-called coherent feedback loop where the system output is directly processed by a quantum controller instead of being measured.

In practice, one can also choose cost functionals as the state transition probability [53], gate fi- delity [55], decoherence suppression [56], fluence minimization [57], and the time expended in control [58]. The more prevalent optimization algorithms used in quantum control problems can be classified as iterative-type [55, 59] and gradient-type [60–62] algorithms. All these algorithms start from an initial guess of the control field that can be randomly chosen or come from the natural form of laboratory implementations. The iterative-type algorithms generally converge quickly although their convergence is heavily dependent on the initial guess; the gradient-type algorithms are more stable as long as the step-size is sufficiently small, but the drawback is that their convergence is very slow, especially near optimal solutions. In addition, a common requirement is that system Hamiltonians have to be precisely known, which is almost impossible in laboratory situations. Even when the Hamiltonian is available, the numerical integration of the time-dependent Schrodinger equation will usually be very heavy. To overcome such difficulties in laboratory, evolutionary-type algorithms [25, 63] have been proposed in recognition of the fact that the most difficult part (solving Schrodinger equation) can actually be done by nature herself, and what the algorithm needs to do is only to measure the output and to adjust the control according to the designed algorithm. In this way, one can also tune his control variables without knowing exactly what happens to the system dynamics. Theoretically, such algorithms are less efficient than Hamiltonian-dependent algorithms, but in practice it can evolve very rapidly without having to solve the Schrodinger equation. Both types of algorithms can suffer from the possible existence of so-called false traps, i.e. the search could converge to a control solution that is locally but not globally optimal. Such problems are generally more severe if optimization resources are limited, which appears to be the case for quantum control in the laboratory. However, it is surprising [64] that most of the reported simulation results observed quick convergence to global optimal solutions, and experiment yields were most remarkably improved, implying that the traps are not supposed to be ubiquitous in quantum control. This observation was supported via mathematical analysis of the so called quantum control landscape [64]. It was proved that if the system is controllable, then upon some mild assumptions there exist no traps on the landscape in maximizing observable expectation values [65], gate fidelity [66], and even the costs in open systems [67]. This opens up a new field in which investigators can obtain a deeper understanding of quantum control and optimization, and with such understanding it is possible to develop more efficient quantum control algorithms.

PERSPECTIVES

Quantum control is a new interdisciplinary field requiring collaborative research performed by researchers from different disciplines; such collaborations have already come into being. From chemistry to NMR, superconducting devices, and quantum dots, quantum control has shown itself powerful in many experimental setups. Although, in the past, quantum physics has been seen to be too hard for engineers to understand, and engineering methodologies have been only accepted reluctantly by physicists, the gaps in both has been greatly reduced during the past decades as experimental technologies have quickly develop. Motivated by explorations of the ultimate physical limits and the great impact on future technologies, the fusion of theory and practice in regard to quantum control will open up tremendous research opportunities.

REFERENCES

1. Feynman R. There's plenty of room at the bottom. Annual Meeting of the American Physical Society, California Institure of Technology, Pasadena, CA, 1959
2. Vandersypen L M, Chuang I L. NMR techniques for quantum control and computation. Rev Mod Phys, 2004, 76: 1037–1069
3. Grigorenko I, Rabitz H. Optimal control of the local electromagnetic response of nanostructured materials: Optimal detectors and quantum disguises. Appl Phys Lett, 2009, 94: 253107
4. Butkovskiy A, Samoilenko Y I. Control of quantum-mechanical processes and systems. In: Mathematics and its Applications. Boston: Kluwer Academic Publishers, 1990
5. Butkovskii A, Pustyl'nikova E. Controlling the coherent states of a quantum oscillator. Automation Remote Contr, 1982, 43: 1393–1398
6. Butkovskii A, Pustyl'nikova E. Control of coherent states of quantum systems with a quadratic hamiltonian. Automation and Remote Control, 1984, 45: 1000–1008
7. Tarn T J, Huang G M, Clark J W. Modelling of quantum mechanical control systems. Math Model, 1980, 1: 109–121

8. Huang G M, Tarn T J, Clark J W. On the controllability of quantum-mechanical systems. J Math Phys, 1983, 24: 2608–2618
9. Ong C K, Huang G M, Tarn T J, et al. Invertibility of quantum-mechanical control systems. Math Sys Theory, 1984, 17: 335–350
10. Clark J W, Ong C K, Tarn T J, et al. Quantum nondemolition filters. Math Sys Theory, 1985, 18: 33–55
11. Belavkin V P, Grishanin B A. Optimum measurement of quantum variables. Prob Info Transmission, 1972, 8: 259–265
12. Belavkin V P, Grishanin B A. Optimum estimation in quantum channels by the generalized heisenberg inequality method. Prob Info Transmission, 1972, 9: 209–215
13. Belavkin V P. Optimal quantum filtration of markovian signals. Prob Info Transmission, 1978, 7: 1–14
14. Rice S A, Zhao M. Optical Control of Molecular Dynamics. New York: John Wiley, 2000
15. Shapiro M, Brumer P. Principles of the Quantum Control of Molecular Processes. Hoboken: Wiley-Interscience, 2003
16. Bandrauk A D, Delfour M C, Le Bris C. Quantum control: Mathematical and numerical challenges. In: CRM Proceedings & Lecture notes, American Mathematical Society, Providence, 2003
17. Warren W S, Rabitz H, Dahleh M. Coherent control of quantum dynamics: The dream is alive. Science, 1993, 259: 1581–1588
18. Rabitz H, de Vivie-Riedle R, Motzkus M, et al. Whither the future of controlling quantum pheonomena? Science, 2000, 288: 824–828
19. Brumer P, Shapiro M. Control of unimolecular reactions using coherent light. Chem Phys Lett, 1986, 126: 541–546
20. Shapiro M, Brumer P. Laser control of product quantum state populations in unimolecular reactions. J Chem Phys, 1986, 84: 4103–4104
21. Tannor D J, Rice S A. Control of selectivity of chemical reaction via control of wave packet evolution. J Chem Phys, 1985, 83: 5013–5018
22. Tannor D J, Kosloff R, Rice S A. Coherent pulse sequence induced control of selectivity of reactions: Exact quantum mechanical calculations. J Chem Phys, 1986, 85: 5805–5820
23. Zewail A H. Femtochemistry: Atomic-scale dynamics of the chemical bond. J Phys Chem A, 2000, 104: 5660–5694
24. Bonacic-Koutechy V, Mitric R. Theoretical exploration of ultrafast dynamics in atomic clusters: Analysis and control. Chem Rev, 2005, 105: 11–65
25. Judson R S, Rabitz H. Teaching lasers to control molecules. Phys Rev Lett, 1992, 68: 1500
26. Dantus M, Lozovoy V. Experimental coherent laser control of physicochemical processes. Chem Rev, 2004, 104: 1813–1859

27. Brif C, Chakrabarti R, Rabitz H. Control of quantum phenomena: Past, present, and future. New J Phys, 2010, 12: 075008
28. Bucksbaum P, Eisenstein R. Controlling the Quantum World: The Science of Atoms, Molecules, and Photons. Washington, D.C.: The National Academies Press, 2007
29. D'Alessandro D. Introduction to Quantum Control and Dynamics. Boca Raton: Chapman & Hall/CRC, 2008
30. Wiseman H. Quantum Measurement and Control. Cambridge: Cambridge University Press, 2010
31. Dong D, Petersen I. Quantum control theory and applications: A survey. IET Contr Theo Appl, 2010, 4: 2651–2671
32. Cao W C, Liu X S, Bai H B, et al. Bang-bang control suppression of amplitude damping in a three-level atom. Sci China Ser G-Phys Mech Astron, 2008, 51: 29–37
33. Qi B. On the quantum master equation under feedback control. Sci China Ser F-Inf Sci, 2009, 52: 2133–2139
34. Cheng D Z. Advances in automation and control research in China. Sci China Ser F-Inf Sci, 2009, 52: 1954–1963
35. Yuan Q, Li J H. Controlled quantum state transfer via parity measurement. Sci China Ser G-Phys Mech Astron, 2009, 52: 1203–1207
36. Yan S L. Control of chaos in an external-cavity multi-quantum-well laser subjected to dual-wedges and optical dual-feedback. Chin Sci Bull, 2009, 54: 1158–1163
37. Wei H R, Di Y M, Yan W. Synthesis of some three-qubit gates and their implementation in a three spins system coupled with ising interaction. Sci China Phys Mech Astron, 2010, 53: 664–671
38. André A, Bajscy M, Childress L, et al. Quantum control of light using coherent atomic memory. In: Atomic Physics 19: XIX International Conference on Atomic Physics 2004. AIP Conf Proc, 2005, 770: 291–300
39. Bartels R A, Murnane M M, Kapteyn H C, et al. Learning from learning algorithms: Application to attosecond dynamics of high-harmonic generation. Phys Rev A, 2004, 70: 04340440.
40. Isidori A. Nonlinear Control Systems. New York: Springer-Verlag, 1995
41. Gardiner C W, Collett M J. Input and output in damped quantum systems: Quantum stochastic differential equations and the master equation. Phys Rev A, 1985, 31: 3761–3774
42. Gough J E, James M R, Nurdin H I. Quantum master equation and filter for systems driven by fields in a single photon state. In: Joint 50th IEEE Conference on Decision and Control (CDC) and European Control Conference (ECC), 2011
43. Lieb E. The stability of matter. Rev Mod Phys, 1976, 48: 553–569
44. Combescure M. The quantum stability problem for some class of time-dependent hamiltonians. Ann Phys, 1988, 185: 86–110

45. Casati G, Chirikov B, Guarneri I, et al. Dynamic stability of quantum chaotic motion in a hydrogen-atom. Phys Rev Lett, 1986, 56: 2437–2440

46. D'Helon C, James M R. Stability, gain, and robustness in quantum feedback networks. Phys Rev A, 2006, 73: 053803

47. Lan C, Tarn T J, Chi Q S, et al. Analytic controllability of time-dependent quantum control systems. J Math Phys, 2005, 46: 052102

48. Wu R B, Tarn T J, Li C W. Smooth controllability of infinite-dimensional quantum-mechanical systems. Phys Rev A, 2006, 73: 012719

49. Schirmer S G, Fu H, Solomon A I. Complete controllability of quantum systems. Phys Rev A, 2001, 63: 063410

50. Bloch A M, Brockett R W, Rangan C. Finite controllability of infinite-dimensional quantum systems. IEEE Trans Auto Contr, 2010, 55: 1797–1805

51. Wu R B, Hsieh M, Rabitz H. Role of controllability in optimizing quantum dynamics. Phys Rev A, 2011, 83: 062306

52. Chelkowski S, Bandrauk A D. Raman chirped adiabatic passage: A new method for selective excitation of high vibrational states. J Raman Spectr, 1997, 28: 459–466

53. Bergmann K, Theuer H, Shore B W. Coherent population transfer among quantum states of atoms and molecules. Rev Mod Phys, 1998, 70: 1003–1025

54. Schirmer S, Greentree A, Ramakrishna V, et al. Constructive control of quantum systems using factorization of unitary operators. J Phys A, 2002, 35: 8315

55. Palao J, Kosloff R. Quantum computing by an optimal control algorithm for unitary transformations. Phys Rev Lett, 2002, 89: 188301

56. Zhang J, Li C W, Wu R B, et al. Maximal suppression of decoherence in markovian quantum systems. J Phys A, 2005, 38: 6587

57. D'Alessandro D. Optimal contro of two level quantum systems. IEEE Trans Auto Contr, 2001, 46: 866–876

58. Khaneja N, Brockett R, Glaser S J. Time optimal control in spin systems. Phys Rev A, 2001, 63: 032308

59. Zhu W S, Rabitz H. A rapid monotonically convergent iteration algorithm for quantum optimal control over the expectation value of a positive definite operator. J Chem Phys, 1998, 109: 385–391

60. Khaneja N, Reiss T, Kehlet C, et al. Optimal control of coupled spin dynamics: Design of nmr pulse sequences by gradient ascent algorithms. J Magn Reson, 2005, 172: 296–305

61. Rothman A, Ho T S, Rabitz H. Quantum observable homotopy tracking control. J Chem Phys, 2005, 123: 134104

62. Rothman A, Ho T S, Rabitz H. Observable-preserving control of quantum dynamics over a family of related systems. Phys Rev A, 2005, 72: 023416

63. Shir O M, Beltrani V, Bäck T, et al. On the diversity of multiple optimal controls for quantum systems. J Phys B, 2008, 41: 074021

64. Rabitz H, Hsieh M, Rosenthal C. Quantum optimally controlled transition landscapes. Science, 2004, 303: 1998–2001

65. Wu R, Rabitz H, Hsieh M. Characterization of the critical submanifolds in quantum ensemble control landscapes. J Phys A, 2008, 41: 015006

66. Rabitz H, Hsieh M, Rosenthal C. The landscape for optimal control of quantum-mechanical unitary transformations. Phys Rev A, 2005, 72: 52337

67. Wu R, Pechen A, Rabitz H, et al. Control landscapes for observable preparation with open quantum systems. JMath Phys, 2008, 49: 022108

3
CHAPTER

LOCAL CONTROLLABILITY OF QUANTUM SYSTEMS

Zbigniew Puchała

Keywords: Quantum control, Local control, Hypergraph

INTRODUCTION

The controllability of a given quantum system is a fundamental issue of the quantum information science. It concerns whether it is possible to drive a quantum system into a previously fixed state. There have been proposed different notations of controllability of quantum systems, such as state controllability, equivalent state controllability or operator controllability [1,3,8]. In the case of finite dimensional quantum systems the criteria for controllability can be expressed in terms of Lie-algebraic concepts [2,8,9]. These concepts provide a mathematical tool, in the case of closed quantum systems, i.e. systems without external influences. However, the Lie-algebraic criteria may be difficult to check, especially when the dimension of the system in question is large. For this reason there has been proposed methods for verifing controllability based on a graph theory [6,7,15,16] and in many cases this setting provides an easier way for controllability inspection.

It is an important question whether the system is controllable when the control is performed only on a subsystem. This kind of approach is called a local-controllability and can be considered only in the case when the subsystems of a given system interact. For examples may serve coupled spin chains or spin networks [6–8]. In [7] there has been derived a graph infection criterion that ensures the controllability by relaxation and in [6] it has been shown, that it also can be used in the case of algebraic control. In this paper we generalize the graph infection criterion to the quantum systems that cannot be described with the use of a graph theory, like spin S = 1/2 extended XY model [13,14] or p-spin interaction model [4,10]. In order to provide new criteria for controllability,

Citation: Zbigniew Puchała, Local Controllability of Quantum Systems, DOI: 10.1007/s11128-012-0391-x.

we introduce the notation of hypergraphs and reformulate the infection property in this setting. The introduced hypergraphs infection criterion has a topological nature and therefore it is not connected to any particular experimental realization of quantum information processing. This paper is organized as follows. In Sect. 2 we provide a general description of a quantum mechanical control system. We introduce notation of controllability and provide a necessary and sufficient criteria of controllability of quantum systems. In Sect. 3 we introduce a notion of local controllability, provide a definition of hypergraph, define hypergraph infection property and finally give a new criterion for controllability. In Sect. 4 we provide the summary of the presented work and give some concluding remarks.

QUANTUM MECHANICAL CONTROL SYSTEMS

The dynamics of closed quantum systems can be described by a Schrödinger equation

$$\frac{d}{dt}|\psi(t)\rangle = -iH(t)|\psi(t)\rangle, \tag{1}$$

where $|\psi(t)$ is an element of the complex sphere S^{N-1} representing a pure state, H(t) is a matrix function, which is Hermitian for every t and called a Hamiltonian of the system. In this paper we assume, that the Hamiltonian is in the form $H_\delta + \sum_{i=1}^{M} h_i(t)H_i$, thus a quantum system can be described as

$$\frac{d}{dt}|\psi(t)\rangle = -i\left(H_\delta + \sum_{i=1}^{M} h_i(t)H_i\right)|\psi(t)\rangle, \tag{2}$$

where matrices H_δ, H_1, H_2,..., H_M are Hermitian. Term H_δ is called a drift term since it drives an evolution in no control is applied. For given controls, the Eq. (2) is (time-variant) linear, and thus has unique solution. In this case the system is bilinear [9], with specialization that drift and control matrices are skew-Hermitian. The solution of (2) can be given as

$$|\psi(t)\rangle = U(t)|\psi(0)\rangle, \tag{3}$$

where $|\psi(0)$ is an initial condition and U(t) is the solution of an operator equation

$$\frac{d}{dt}U(t) = -i\left(H_\delta + \sum_{i=1}^{M} h_i(t)H_i\right)U(t), \tag{4}$$

with an initial condition U(0) = 𝟙 (N × N identity matrix). The solution U(t) is an element of Lie group of unitary matrices U(N), if we assume that $\mathrm{tr}H_\delta = \mathrm{tr}H_i = 0$, then the solution is in a group of special unitary matrices SU(N). There are various notion of controllability for the system given in (2), [1,3,8]. Here we will consider two of them.

Definition 1 (Operator Controllable System) We call system (2) Operator Controllable if it is possible to drive an operator X in (4) to any value in U(N) (or SU(N)).

Definition 2 (State Controllable System) We call system (2) State Controllable if it is possible to drive the state $|\psi$ from the complex sphere S^{N-1} to any other state on the sphere.

One can also define Equivalent State Controllable System where it is possible to drive any initial state to any element on the complex sphere modulo a phase factor, but since from a physics point of view states that differ only by a phase factor are indistinguishable, thus the equivalent state controllability is equivalent to state controllability. The main theorem concerning controllability conditions on bilinear quantum systems follows from more general fact concerning controllability on Lie groups and was proved in the 1970's of the last century [12]. If we specify the theorem to the bilinear quantum systems it can be stated as follows.

Theorem 1 *Let us denote by $\mathcal{L}$ the Lie algebra generated by the matrices $iH_\delta, iH_1, \ldots, iH_M$.* i.e. $\mathcal{L} = \{iH_\delta, iH_1, \ldots, iH_M\}_{\mathcal{L}}$. *We have the following*

- *The system is operator controllable if and only if the Lie algebra $\mathcal{L}$ is an algebra $u(N)$ or $su(N)$.*
- *The system is state controllable if and only if $\mathcal{L}$ is $u(N)$, $su(N)$ or in the case of even N the algebra $\mathcal{L}$ is isomorphic to $sp\left(\frac{N}{2}\right)^2$.*

LOCAL CONTROLLABILITY

Let $V = C \cup \bar{C}$ be a given composite system, with Hamiltonian in the following form

$$H = H_\delta + \sum_{k=1}^{M} h_k(t) H_k, \tag{5}$$

where H_δ is a drift—in most situations coupling Hamiltonian on whole system, and H_k are local Hamiltonians acting on subsystem C, thus are in the form $H_k = H_C^k \otimes 1_{\bar{C}}$. The action of Hamiltonians H_k are governed by time depended parameters $h_k(t)$. By the Theorem 1, V is operator controllable if and only if i H_δ and i H_k are generators of the Lie algebra of skew-Hermitian operators on the composite system V, i.e. $\{iH_\delta, iH_1, iH_2, \ldots, iH_M\}_{\mathcal{L}} = u(V)$. In this paper we assume, that the control Hamiltonians generates the full unitary algebra on a subsystem C, i.e. $\{iH_1^C, iH_2^C, \ldots, iH_M^C\}_{\mathcal{L}} = u(C)$. In article [6] was given a sufficient criterion that guarantees that a many-body quantum system with drift described by a network can be controlled by properly manipulating the (local) Hamiltonian of one of its subsystems. The criterion is based on a topological properties of the graph defined by the coupling terms in a drift Hamiltonian H_δ. The applications of above result in the case of Heisenberg spin chains are presented in a paper [11].

System Specification

Assume, that we have a composite system $X = X_1 \cup X_2 \cup \cdots \cup X_n$. The associated Hilbert space we denote by $H^X = H^X_1 \otimes \cdots \otimes H^X_n$. Let i be an identifier of a system X_i and by X we denote a set of subsystems identifiers $X = \{1, 2,..., n\}$. The dimension of the Hilbert space H^X is equal to $N = dim(\mathcal{H}^{\mathcal{X}}) = \prod_{i=1}^{M} dim(\mathcal{H}^{\mathcal{X}_i})$. We say, that a Hermitian operator H acting on HX acts only on a subsystem i if H is in the form

$$H = \mathbb{1} \otimes \cdots \otimes \mathbb{1} \otimes H_i \otimes \mathbb{1} \otimes \cdots \otimes \mathbb{1}, \tag{6}$$

where, the operator H_i acts on a H^X_i. More generally we say, that a Hermitian operator H acting on H^X acts only on a subsystem $P \subset X$ if

$$[H, H_i] = 0, \tag{7}$$

for all operators Hi acting on subsystems $i \notin P$. As usual we denote by u(N) the Lie algebra of skew-Hermitian N × N matrices. If a composite system is specified, we will use a notation $u(X_i)$ for Lie algebra of N × N skew-Hermitian matrices acting on H^X_i. If i is an identifier of a subsystem X_i, we will write u(i) to denote this algebra. Similarly if $P \subset X$, by u(P) we denote a Lie algebra of N × N skew-Hermitian acting on a subsystem P.

Hypergraphs: Definitions and Properties

In this section we give a definition of a hypergraph and provide an infection property on a hypergraph. The infection property for hypergraphs has been adopted from papers [7,6], where it was defined for graphs and used to provide controllability conditions for quantum networks.

Definition 3 (Hypergraph) Let $X = \{x_1, x_2,..., x_n\}$ be a finite set of nodes, and let $E = (E_1, E_2,..., E_k)$ be a family of subsets of X, we will call them edges. If we have, that $E_i \neq \emptyset$ for $i = 1, 2, \ldots, k$ and $\bigcup_{i=1}^{k} E_i = X$, then we call $(X, \mathcal{E})$ a hypergraph (see [5]). In this paper we assume, that $X = \{1, 2,..., n\}$.

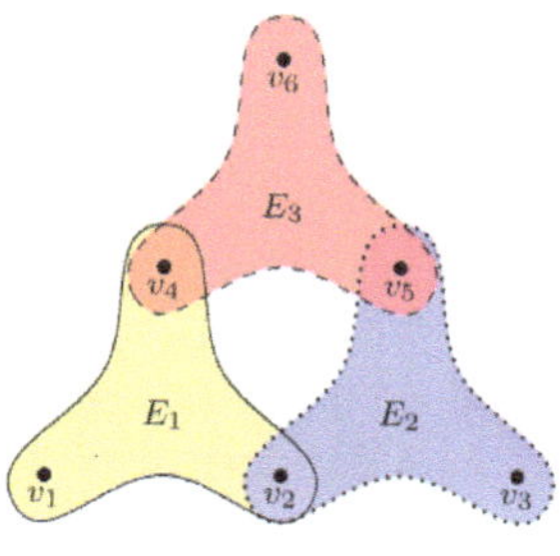

Figure 1: Graphical representation of hypergraph with nodes $X = \{1, 2,..., 6\}$ and edges $\mathcal{E} = \{E_1, E_2, E_3\} = \{\{1, 2, 4\},\{2, 3, 5\},\{4, 5, 6\}\}$

The infection process on a hypergraph can be stated as follows: if some set of nodes A is infected, then the infection spreads onto a healthy neighbours B if there exists edge E joining some elements of infected group $A_1 \subset A$ with B, $E = A_1 \cup B$. Moreover E is the only edge that joins elements from A_1 with healthy nodes.

Definition 4 (Infection spread) Assume that $A \subset X$ is infected, we say that infection can spread onto $B \subset X$ from A, where $A \cap B = \emptyset$ if

$$\exists_{E \in \mathcal{E}} E \cap A \neq \emptyset \text{ and } E \setminus A = B, \tag{8}$$

moreover if $x \in (E \cap A)$ and $x \in F$ for some $F \in E$, then $F = E$ or $F \subset A$. If there exists an initial set of nodes C that can infect whole hypergraph, we call such set infecting.

Definition 5 (Hypergraph infection property) For a hypergraph $(X, \mathcal{E})$ we call a subset $C \subset X$ infecting if there exist a sequence $C = P_1 \subsetneq P_2 \subsetneq \cdots \subsetneq P_m = X$ such that an infection can spread from Pi onto $P_{i+1} \setminus P_i$ for $i = 1, 2, \ldots, m-1$.

Example 1 To illustrate the above definitions we provide an example. Let us define a hypergraph $(X, \mathcal{E})$ with nodes $X = \{1, 2, \ldots, 6\}$ and edges $\mathcal{E} = \{E_1, E_2, E_3\} = \{\{1, 2, 4\}, \{2, 3, 5\}, \{4, 5, 6\}\}$. The graphical representation of the hypergraph is presented in Fig. 1.

If one assumes, that a set $A = \{1, 2, 4\}$ is infected, then it is easy to see, that the infection can spread e.g. onto $\{3, 5\}$ by an edge $E_2 = \{2, 3, 5\}$. Similarly if a set $\{1, 2, 3, 4, 5\}$ is infected, then infection can spread onto $\{6\}$ by an edge $E_3 = \{4, 5, 6\}$. The above gives us, that the hypergraph $(X, \mathcal{E})$ has an infection property, with an infecting set $C = \{1, 2, 4\}$ and a sequence $C = P_1 = \{1, 2, 4\} \subsetneq P_2 = \{1, 2, 3, 4, 5\} \subsetneq P_3 = \{1, 2, 3, 4, 5, 6\} = X$.

Propagation Property

To provide a controllability conditions we must assume that the drift Hamiltonian meets some criteria. To do so we introduce a notion of propagating property, which relates the underlying hypergraph with appropriate Lie algebras.

Definition 6 (Propagating property) Assume, that we have a composite system $X = X_1 \cup X_2 \cup \cdots \cup X_n$. Let $(X, \mathcal{E})$ be a hypergraph. The nodes of the hypergraph are the subsystems identifiers $X = (1, 2, \ldots, n)$. The edges are related to the Hermitian operator H, which acts on a H^X. The relation is given by

$$H = \sum_{E \in \mathcal{E}} H_E, \tag{9}$$

where HE are Hermitian operators acting on subsystem E. We say, that H has a propagating property if for all $E \in \mathcal{E}$,, and for all $\emptyset \neq E' \subset E$ we have

$$\{[iH_E, u(E')], u(E')\}_{\mathfrak{L}} = u(E). \tag{10}$$

Example 2 Assume that we have a hypergraph $(X, \mathcal{E})$ described in Example 1. We also assume, that a composite system is composed with six qubits and for an edge E = {e_1, e_2, e_3} of the hypergraph the Hermitian operator H_E is given by

$$iH_E = S_2^{e1}S_2^{e2}S_2^{e3} + S_3^{e1}S_3^{e2}S_3^{e3} + S_1^{e1}S_2^{e2}S_3^{e3} + S_2^{e1}S_1^{e2}S_0^{e3} + S_3^{e1}S_2^{e2}S_0^{e3}, \quad (11)$$

where S_j^m is jth Pauli matrix on subsystem m, in this case

$$S_j^{(m)} = \underbrace{\mathbb{1}\otimes\cdots\otimes\mathbb{1}}_{m-1}\otimes\sigma_j\otimes\underbrace{\mathbb{1}\otimes\cdots\otimes\mathbb{1}}_{6-m}, \quad (12)$$

with notation $\sigma_0 = 1$. The Hamiltonian $H = \sum_{E\in\mathcal{E}} H_E$ has a propagation property. It is quite cumbersome task to check this analytically, but it can be done using computer algebra systems with symbolic computation and a procedure to generate a basis of a dynamical Lie algebra,. Since the propagation property must be checked only on a small subsystem, this computation is fast and efficient.

Main Theorem

Now we can state the main theorem, which express a new controllability criterion in notion of hypergraph infection and propagating property. The proof of the theorem follows the similar line of argument that the proof of the theorem in [6].

Theorem 2 Let us assume, that the drift Hamiltonian H_δ has a propagating property, we also assume, that a subset C ⊂ X infects the hypergraph $(X, \mathcal{E})$. Then, the system is controllable if we perform, the control only on a subsystem C.

Proof First we will show by induction, that $u(P_j) \subset \{iH_\delta, u(C)\}_\mathfrak{L}$, where subsets P_j are defined in the hypergraph infection property (Def. 5). The first induction step is obvious, since $u(P_1) = u(C) \subset \{iH_\delta, u(C)\}_\mathfrak{L}$. Next, we assume, that for some j < m, we have

$$u(P_j) \subset \{iH_\delta, u(C)\}_\mathfrak{L}. \quad (13)$$

Since, we have, that infection spreads from P_j onto $P_{j+1}\backslash P_j$, there exist an edge $E \in \mathcal{E}$, such that $E \cap P_j \neq \emptyset$ and $E\backslash P_j = P_{j+1}\backslash P_j$. We define $E^{(j)} = E \cap P_j$ and write

$$\begin{aligned}[iH_\delta, u(E^{(j)})] &= \sum_{F\in\mathcal{E}}[iH_F, u(E^{(j)})] \\ &= [iH_E, u(E^{(j)})] + \sum_{F\in\mathcal{E}, F\neq E}[iH_F, u(E^{(j)})].\end{aligned} \quad (14)$$

Now using, the infection property of the graph ($E \in \mathcal{E}$), we have, that the last sum above can be restricted to $F \in \mathcal{E}, F \subset P_j$. Thus we have

$$[iH_E, u(E^{(j)})] = [iH_\delta, u(E^{(j)})] - \sum_{F\in\mathcal{E}, F\subset P_j} [iH_F, u(E^{(j)})]. \tag{15}$$

The last sum is an element of u(P_j) and the first term of the right hand side is an element of $\{iH_\delta, u(C)\}_\mathfrak{L}$, so we obtain, that $[iH_E, u(E^{(j)})] \in \{iH_\delta, u(C)\}_\mathfrak{L}$. Now using propagation property of the drift Hamiltonian, we have

$$\{[iH_E, u(E^{(j)})], u(E^{(j)})\}_\mathfrak{L} = u(E). \tag{16}$$

Since $\{u(P_j), u(E)\}_\mathfrak{L} = u(P_{j+1})$, thus $u(P_{j+1}) \in \{iH_\delta, u(C)\}_\mathfrak{L}$. At the last step of induction procedure j = k, we obtain

$$u(P_k) = u(X) \subset \{iH_\delta, u(C)\}_\mathfrak{L}, \tag{17}$$

and since u(X) is the maximal algebra which can be obtained, we have

$$u(P_k) = u(X) = \{iH_\delta, u(C)\}_\mathfrak{L}. \tag{18}$$

Now by Theorem 1, we obtain that the system is operator controllable.

To use the above theorem to check whether the system is controllable one first must test the propagation property for a drift Hamiltonian and ensure that the underlying hypergraph has an infection property. The second task is purely topological and can be done rather easily, while the first one—the propagation property must be checked only on a small subsystem, which reduces the complexity. Using Theorem 2 we obtain that the dynamical system described in Examples 1, 2 is controllable by performing the control only on a subsystem $\{1, 2, 4\}$. We remind here the assumption, which was made at the beginning of Sect. 3, that local control Hamiltonians generates the full unitary algebra on a specified subsystem.

CONCLUDING REMARKS

In this paper we provided a new controllability criterion of multipartite quantum systems based on notion of hypergraph. We generalized the graph infection criterion to the quantum systems that cannot be described with the use of a graph theory. To do so we have introduced a notion of local controllability, provided a definition of hypergraph, defined hypergraph infection property and finally gave a new criterion for controllability. The introduced criterion has a topological nature and therefore it is not connected to any particular experimental realization of quantum information processing.

ACKNOWLEDGMENTS

We acknowledge the financial support by the Polish National Science Centre under the grant number N N514 513340.

REFERENCES

1. Albertini, F., D'Alessandro, D.: Notions of controllability for quantum mechanical systems. In: Decision and control. Proceedings of the 40th IEEE conference on, vol. 2, pp. 1589–1594. IEEE (2001)
2. Albertini F., D'Alessandro D.: The Lie algebra structure and controllability of spin systems. Linear Algebra Appl. 350(1–3), 213–235 (2002)
3. Albertini F., D'Alessandro D.: Notions of controllability for bilinear multilevel quantum systems. Autom. Control, IEEE Trans. 48(8), 1399–1403 (2003)
4. Baxter R., Wu F.: Exact solution of an ising model with three-spin interactions on a triangular lattice. Phys. Rev. Lett. 31(21), 1294–1297 (1973)
5. Berge C.: Graphs and Hypergraphs, vol. 6. Elsevier, North Holland, NY (1976)
6. Burgarth D., Bose S., Bruder C., Giovannetti V.: Local controllability of quantum networks. Phys. Rev. A 79(6), 60–305 (2009)
7. Burgarth D., Giovannetti V.: Full control by locally induced relaxation. Phys. Rev. Lett. 99(10), 100–501 (2007)
8. D'Alessandro D.: Introduction to Quantum Control and Dynamics. Chapman & Hall, London (2008)
9. Elliott D.: Bilinear Control Systems: Matrices in Action. Springer, Berlin (2009)
10. Goldschmidt Y. et al.: Solvable model of the quantum spin glass in a transverse field. Phys. Rev. B Condens. Matter 41(7), 4858 (1990)
11. Heule R., Bruder C., Burgarth D., Stojanovió V.M.: Local quantum control of Heisenberg spin chains. Phys. Rev. A 82(5), 052–333 (2010). doi:10.1103/PhysRevA.82.052333
12. Jurdjevic V., Sussmann H.: Control systems on lie groups. J. Diff. Equ. 12(2), 313–329 (1972)
13. Lou P.: Thermal and spin transports in spin-1/2 extended xy chain. Phys. Status Solidi (B) 241(6), 1343–1349 (2004)
14. Titvinidze I., Japaridze G.: Phase diagram of the spin extended model. Eur. Phys. J. B-Condens. Matter Complex Syst. 32(3), 383–393 (2003)
15. Turinici G., Rabitz H.: Quantum wavefunction controllability. Chem. Phys. 267(1–3), 1–9 (2001)
16. Turinici G., Rabitz H.: Wavefunction controllability for finite-dimensional bilinear quantum systems. J. Phys. A Math. Gen. 36, 2565 (2003)

4
CHAPTER

GEOMETRIC METHOD IN QUANTUM CONTROL

Zhang Jun

Keywords: Quantum control, Cartan decomposition, steering, universal quantum circuit

INTRODUCTION

Control on quantum mechanical systems has received extensive interests in the past decade due to its fundamental importance in many physical applications of quantum phenomena, e.g. , NMR, nanotechnology, molecular systems, quantum optics, and quantum computation [1]. The early research on this topic can be traced back to 1980s. In [2, 3], Tarn et al. investigated modeling and controllability of quantum mechanical control systems. Rabitz et al. discussed optimal control in quantum systems with emphasis on the quantum state transitions [4, 5]. After Shor published his seminal paper on factoring algorithm in mid 1990s [6], quantum system control has attracted particular research efforts as an indispensable part of quantum information processing. The generation of quantum operation is a prerequisite condition for many real physical applications of quantum phenomena. For example, often required in quantum algorithm is to apply Quantum Fourier Transform, which amounts to control the physical system to generate a prescribed quantum operation from initial condition. Other examples include using robust control techniques to protect quantum coherence against environmental noises, and applying quantum feedback control to improve the control performance. Just as in the case of traditional control, the very first question that quantum system control has to address is how to design open-loop control law. In this paper, we survey the control synthesis technique on generating a desired quantum operation from a given quantum mechanical system. This corresponds to the universality problem in quantum information processing. We consider only discrete variable quantum systems. For continuous variable systems, readers are referred to [7]. Mathematically, a discrete

Citation: Jun Zhang, Geometric method in quantum control, DOI: 10.1007/s11434-012-5186-z.

quantum system can be described by quantum states, which are unit vectors in a complex vector space:

$$\{\psi \in \mathbb{C}^N : \|\psi\| = 1\}. \tag{1}$$

When N = 2, this corresponds to a 2-dimensional state called a quantum bit, or qubit. A qubit is the simplest quantum system, and it can be used to describe, for example, an electron moving in a magnetic field. A single-qubit state has two basis states denoted as |0and |1, and any single-qubit state can be expressed as a linear superposition of these two basis states. For an n-qubit state, it contains 2^n basis states, and similar to the single-qubit case, an arbitrary n-qubit state is a linear superposition of its basis states. To manipulate quantum system, we can apply quantum operations to the quantum state. For an n-qubit system, a quantum gate can be represented by a $2^n \times 2^n$ unitary matrix. Here are several single-qubit gates:

$$\sigma_x = \begin{bmatrix} 0 & 1 \\ 1 & 0 \end{bmatrix}, \quad \sigma_y = \begin{bmatrix} 0 & i \\ -i & 0 \end{bmatrix}, \quad \sigma_z = \begin{bmatrix} 1 & 0 \\ 0 & -1 \end{bmatrix}.$$

These are indeed the Pauli matrices in quantum physics. For two-qubit quantum operations, we can distinguish two different types. The first type is local gates, which consists of two single-qubit gates K and L that are applied to an individual qubit. The matrix representation of this local gate is $K \otimes L$, i.e. , the tensor product of K and L. Tensor product is also called Kronecker product in control theory. The other type is nonlocal gates, which are the gates that cannot be written as the tensor product of two single-qubit gates. Two important nonlocal gates are CNOT and SWAP gates [8]:

$$\text{CNOT} = \begin{pmatrix} 1 & 0 & 0 & 0 \\ 0 & 1 & 0 & 0 \\ 0 & 0 & 0 & 1 \\ 0 & 0 & 1 & 0 \end{pmatrix}, \quad \text{SWAP} = \begin{pmatrix} 1 & 0 & 0 & 0 \\ 0 & 0 & 1 & 0 \\ 0 & 1 & 0 & 0 \\ 0 & 0 & 0 & 1 \end{pmatrix}.$$

The main problem we consider here is to generate any arbitrary quantum operation from a quantum physical system. This can be treated as a geometric control problem on a Lie group. For the quantum system, the state space is the unitary Lie group that is composed of all the $2^n \times 2^n$ unitary matrices. The dynamics of quantum system is determined by the Schrodinger equation:

$$i\dot{U} = H(v)U, \quad U(t_0) = I. \tag{2}$$

From the perspective of control theory, this is a right invariant vector field on unitary Lie group. Note that the control field v is in the Hamiltonian. Now the generation of quantum gates from quantum system becomes a steering problem, that is, we aim at finding control field v such that the system trajectory generated by Schrodinger equation can achieve the desired target at certain final time. The control of an n-qubit system is a difficult problem. The reason is threefold. First, the dimension of the problem increases exponentially because the dimension of an n-qubit system is 2^n. Secondly, the dynamics is determined by the bilinear Schr¨odinger equation. Thirdly, the dynamics is defined on the unitary Lie group, which has complicated group structure. To simplify this problem, researchers have developed the approach of universal gate set [9]. It is well-known that in classical circuit design NAND is a universal gate because it can be used to implement any arbitrary logic gate. In quantum information processing, the idea is similar. We want to use a set of elementary quantum gates such that any large and complicated quantum gate can be decomposed as a combination of these elementary gates. The standard result in the literature is that any quantum gate on arbitrarily many qubits can be decomposed as a composition of single- and two-qubit gates. Therefore, to implement any arbitrary quantum gate, we need to implement only single- and two-qubit gates. The single-qubit gates are defined on the Lie group S U(2). The control of S U(2) is the same as that of S O(3), the rotation matrix group, because S U(2) is a double cover of S O(3). This is a well-studied problem in many other disciplines such as robotics [10], and its extension to one-qubit gates has been studied in details in, e.g. , [11]. In the rest of this paper, we will thus assume that the single-qubit gates can be readily generated, and we will focus on the implementation of two-qubit gates, which is a control problem on the Lie group S U(4).

BASIC CONTROL STRATEGY

We first give a brief introduction about the mathematical background of the control problems on S U(4), the set of all the two-qubit quantum operations. The Lie algebra su(4) has a direct sum decomposition $su(4) = \mathfrak{k} \oplus \mathfrak{p}$, where

$$\begin{aligned}\mathfrak{k} &= \text{span}\ \frac{i}{2}\{\sigma_x^1, \sigma_y^1, \sigma_z^1, \sigma_x^2, \sigma_y^2, \sigma_z^2\},\\ \mathfrak{p} &= \text{span}\ \frac{i}{2}\{\sigma_x^1\sigma_x^2, \sigma_x^1\sigma_y^2, \sigma_x^1\sigma_z^2, \sigma_y^1\sigma_x^2, \sigma_y^1\sigma_y^2, \sigma_y^1\sigma_z^2,\\ &\qquad \sigma_z^1\sigma_x^2, \sigma_z^1\sigma_y^2, \sigma_z^1\sigma_z^2\}.\end{aligned} \tag{3}$$

Here σ_x, σ_y, and σ_z are the Pauli matrices, and $\sigma_\alpha^1\sigma_\beta^2 = \sigma_\alpha \otimes \sigma_\beta$. From the Lie bracket computation, it is easy to obtain that

$$[\mathfrak{k}, \mathfrak{k}] \subset \mathfrak{k}, \quad [\mathfrak{p}, \mathfrak{k}] \subset \mathfrak{p}, \quad [\mathfrak{p}, \mathfrak{p}] \subset \mathfrak{k}. \tag{4}$$

Therefore $su(4) = \mathfrak{k} \oplus \mathfrak{p}$ is a Cartan decomposition, and any U ∈ S U(4) can be decomposed as

$$U = k_1 A k_2 = k_1 \exp\{\frac{i}{2}(c_1\sigma_x^1\sigma_x^2 + c_2\sigma_y^1\sigma_y^2 + c_3\sigma_z^1\sigma_z^2)\}k_2, \tag{5}$$

Where $k_1, k_2 \in SU(2) \otimes SU(2)$, and $c_1, c_2, c_3 \in \mathbb{R}$.

Our basic control strategy to generate any arbitrary two qubit operation is to separate the implementation of one and two-qubit gates. To explain this idea, we need to introduce the notion of local equivalence. Two unitary transformations U, $U_1 \in$ S U(4) are called local equivalent if they differ only by local operations: U = $k_1 U_1 k_2$, where k_1, $k_2 \in$ S U(2) ⊗ S U(2) are local gates. We denote this equivalence relation as U ~ U_1, and the equivalence class of U as [U]. It can be easily proved that this local equivalence relation defines an equivalence class on the whole space of all the two-qubit quantum operations. Now remember we assume all the 1-qubit gate can be implemented. Therefore, to implement a two-qubit gate U, we only need to implement a gate U_1 that is locally equivalent to U. Once we obtain U_1, we can just add on those two local gates k_1 and k_2 to achieve U exactly. Therefore, our original control problem on the unitary Lie group becomes a control problem on the local equivalence classes.

Before we can solve the control on the equivalence classes, we need to answer the following two questions: (1) Given two gates U and U_1, how to determine whether or not they are locally equivalent? (2) What is the structure of local equivalence classes? The first question was answered by Makhlin in [13], where a procedure was given to calculate two local invariants G_1 and G_2. The assertion is that if two-qubit operations U and U_1 are locally equivalent to each other if and only if they have identically the same local invariants. We include this procedure here for completeness. For a given two-qubit gate U, define UB = Q†UQ, where

$$Q = \frac{1}{\sqrt{2}} \begin{pmatrix} 1 & 0 & 0 & i \\ 0 & i & 1 & 0 \\ 0 & i & -1 & 0 \\ 1 & 0 & 0 & -i \end{pmatrix}. \tag{6}$$

Let $m = U_B^T U_B$. Then the local invariants are given by

$$G_1 = \frac{\text{trace}^2\,(m(U))}{16 \det U}, \tag{7}$$

$$G_2 = \frac{\text{trace}^2\,(m(U)) - \text{trace}\,(m^2(U))}{4 \det U}. \tag{8}$$

In the following table, we list the local invariants for several commonly-used two-qubit gates.

	G_1	G_2
Local gates	1	3
CNOT	0	1
C(Z)	0	1
SWAP	−1	−3

It is evident that CNOT and Controlled-Z (referred as C(Z)) are locally equivalent because they possess identical local invariants $G_1 = 0$ and $G_2 = 1$. To answer the second question, we insert Cartan decomposition (5) into Makhlin's procedure and obtain that [14]

$$\begin{aligned} G_1 &= \cos^2 c_1 \cos^2 c_2 \cos^2 c_3 - \sin^2 c_1 \sin^2 c_2 \sin^2 c_3 \\ &\quad + \frac{i}{4} \sin 2c_1 \sin 2c_2 \sin 2c_3, \\ G_2 &= 4\cos^2 c_1 \cos^2 c_2 \cos^2 c_3 - 4\sin^2 c_1 \sin^2 c_2 \sin^2 c_3 \\ &\quad - \cos 2c_1 \cos 2c_2 \cos 2c_3. \end{aligned} \tag{9}$$

Eq. (9) unravels the relation between the local invariants G_1 and G_2 and the coordinates $[c_1, c_2, c_3]$ of a two-qubit gate. From this relation, given a triplet $[c_1, c_2, c_3]$, we can easily compute the local invariants; and vice versa, from a given pair of local invariants G_1 and G_2, we can also find the corresponding triplet. The real space $\mathbb{R}^3$ thus provides a geometric representation of local equivalence classes. However, this representation is not unique, as we can find multiple points in $\mathbb{R}^3$ that correspond to the same local equivalence class. We need to remove the redundancy in this representation. After examining eq. (9), we have the following observations:

(1) Periodicity: $[c_1, c_2, c_3] \sim [c_1+m_1\pi, c_2+m_2\pi, c_3+m_3\pi]$;

(2) Symmetry:

Permutation: $[c_1, c_2, c_3] \sim [c_3, c_2, c_1]$;

Sign flips: $[c_1, c_2, c_3] \sim [-c_1, c_2, -c_3]$.

From periodicity, we can cut down the redundancy and visualize the geometric structure of the two-qubit gates as a cube with side length π in $\mathbb{R}^3$ as shown in Figure 1(a). This provides an equivalent representation of the points on the 3- Torus, since $T^3 \cong \mathbb{R}^3/\mathbb{Z}^3$. Every point in this cube corresponds to a local equivalence class, yet different points in the cube may belong to the same local equivalence class. To further remove the redundancy, we notice that the symmetric points correspond to those points that are reflections of each other with respect to some diagonal planes in the cube. We can cut along these diagonal planes, and the resulting two pieces contain the same local equivalence classes. We can thus discard one of them. After this procedure, we obtain a tetrahedral representation of local equivalence classes of two-qubit as shown in Figure 1(b). Note that for any point $[c_1, c_2, 0]$ on the base of this tetrahedron, its mirror image with respect to the line LA_2, which is $[\pi - c_1, c_2, 0]$, corresponds to the same local equivalence class. Therefore, with the caution that the basal areas LA_2A_1 and LA_2O are identified together, we obtain a geometric representation of local equivalence classes. This is also called Weyl chamber in the Lie group representation theory [12]. In Figure 1(b), the local gates correspond to the points O and A_1, and CNOT is L, SWAP is A_3, and Controlled-U gates are the line OL.

STEERING IN THE TETRAHEDRON

With the geometric representation of local equivalence classes, we can use the steering in the tetrahedron to generate any arbitrary two-qubit operation. The general idea is as follows. For a given quantum system whose dynamics is determined by the Schrodinger equation in eq. (2), control field v will generate a trajectory in the state space U(4).

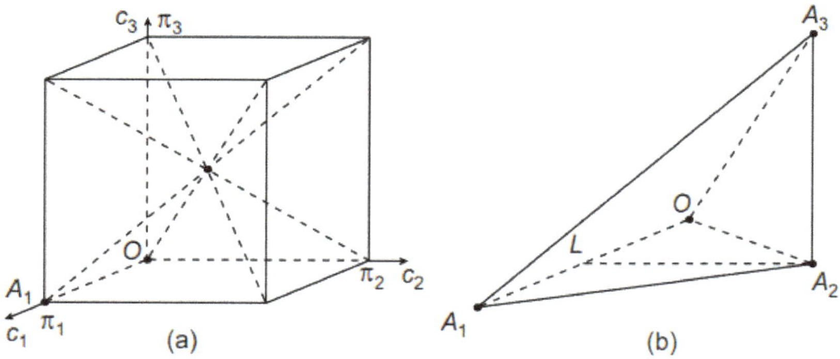

Figure 1: Tetrahedral representation of nonlocal two-qubit operations. (a) Cubic representation. (b) Tetrahedron representation: every point corresponds to a local equivalence class of two-qubit operations.

At any time instant t, U(t) is a point on that trajectory. From Makhlin's procedure, we can calculate local invariants from U(t) as $G_1(U(t))$ and $G_2(U(t))$. Then from eq. (9), we can locate a point $[c_1(t), c_2(t), c_3(t)]$ in the tetrahedron. As time evolves, we can obtain a continuous trajectory. Now if our control target is to generate a quantum operation, say CNOT, we can first find the control field v that drives the trajectory in the tetrahedron to hit the point L. Since we have implemented a gate that is locally equivalent to CNOT, what is left is just to add on two local gates that we assumed can be implemented readily. This is the idea of steering in the tetrahedron. We illustrate the idea by the following two examples.

Example Consider a purely nonlocal Hamiltonian

$$H = -\frac{1}{2}(c_x\sigma_x^1\sigma_x^2 + c_y\sigma_y^1\sigma_y^2 + c_z\sigma_z^1\sigma_z^2). \tag{10}$$

For constant control fields c_x, c_y, and c_z, the quantum operations generated by this Hamiltonian can be written as

$$U(t) = \exp(-iHt) = \exp\frac{i}{2}(c_xt\sigma_x^1\sigma_x^2 + c_yt\sigma_y^1\sigma_y^2 + c_zt\sigma_z^1\sigma_z^2).$$

It thus yields a trajectory $[c_x, c_y, c_z]t$ in the tetrahedron, which is a straight line. We can apply certain gate from a finite group of local operations (dubbed as Weyl group [14]) to change the evolution direction of this trajectory. This indeed corresponds to the reflection with respect to some diagonal plane of the cube in Figure 1(a). Then we can obtain new directions such as $[c_x, -c_y, -c_z]t$, $[-c_x, c_y, -c_z]t$, and $[c_x, c_z, c_y]t$. Piecing together two such trajectories, we can get a quantum operation, e.g. ,

$$e^{\frac{i}{2}\sigma_x^1 \pi} \cdot \exp(-iHt_2) \cdot e^{-\frac{i}{2}\sigma_x^1 \pi} \cdot \exp(-iHt_1),$$

whose endpoint can be written as

$$[c_x, -c_y, -c_z]t_2 + [c_x, c_y, c_z]t_1.$$

We can thus achieve anywhere in the plane OA1A2. If changing direction twice, we can arrive any arbitrary point in the tetrahedron.

We then have the following result [14].

Proposition: For a pure nonlocal Hamiltonian, the following quantum circuit can implement any two-qubit gate.

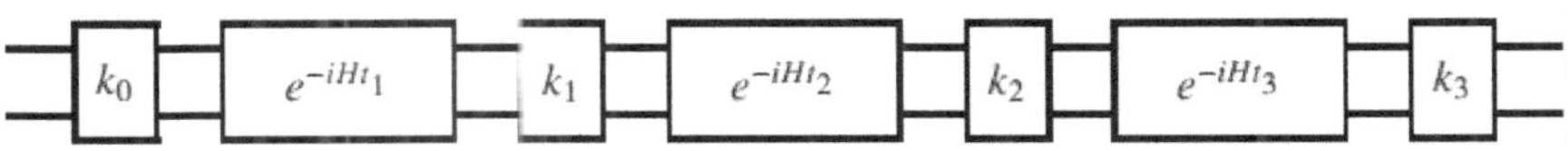

Example: Consider an isotropic exchange with local terms

$$H = (g_1 \cdot \vec{\sigma}) \otimes I + I \otimes (g_2 \cdot \vec{\sigma}) + J(\sigma_x^1 \sigma_x^2 + \sigma_y^1 \sigma_y^2 + \sigma_z^1 \sigma_z^2), \tag{11}$$

Where $g_1, g_2 \in \mathbb{R}^3$, and $\vec{\sigma} = (\sigma_x, \sigma_y, \sigma_z)$. This Hamiltonian can be used to describe, e.g. , spin-coupled quantum dots [15, 16]. We can tune the single-qubit control fields g_1 and g_2 to be parallel, that is, $g_1 = \alpha g_2$, where α characterizes the ratio. In this case, the tetrahedral trajectory is given by

$$c_1(t) = 2Jt,$$

$$c_2(t) = c_3(t) = \left| \sin^{-1}\left(\frac{2J}{\omega} \sin \omega t \right) \right|,$$

Where $\omega = \sqrt{(\|g_1\| - \|g_2\|)^2 + 4J^2}$. We can implement CNOT by choosing $2Jt = \frac{\pi}{2}$ and $\omega t = k\pi$. One particular choice of control parameters can be given as

$$k = 4,\ t = 2.5\pi,\ g_1 = [4, 4, 4],\ g_2 = [3, 3, 3]. \tag{12}$$

The corresponding tetrahedral trajectory is shown in Figure 2.

Quantum circuit construction In this section we consider the problem of generating any two-qubit operation from a given two-qubit gate U_g and local gates. This is closely related to the control problem discussed earlier, and can be another example illustrating the application of tetrahedral representation of nonlocal gates. The motivation is that in many physical situations the Hamiltonian is unknown; or the Hamiltonian is known, but it can be switched on for only a fixed amount of time, e.g. , in encoded quantum systems [17]. Under these circumstances, we are given a specific two-qubit operation, and we want to construct any arbitrary operation from this given gate together with local gates. We now review some results from [18]. The first step in the construction is to implement a Controlled-U gate from two applications of U_g:

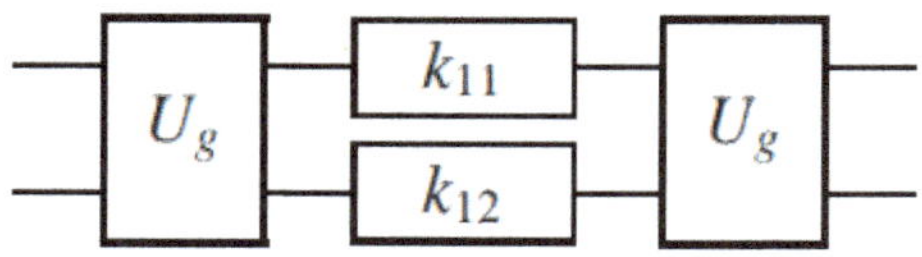

Recall that the Controlled-U gates are located on the line OA_1. We can then use the Controlled-U as a basic building block to generate any desired two-qubit gate. Interested readers may find the detailed analytical procedure in [18].

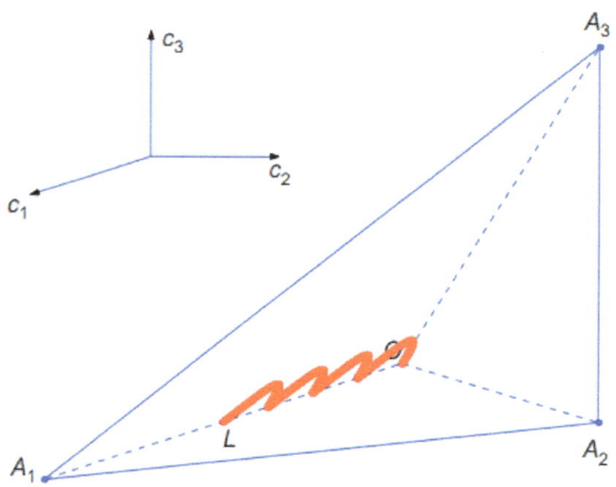

Figure 2: (Color online) Tetrahedral trajectory generated by the Hamiltonian in eq. (11) with parameters (12).

One particularly interesting problem is the efficiency of Controlled-U, that is, for a given Controlled-U gate Ug = $e^{\gamma \frac{i}{2}\sigma_x^1\sigma_x^2}$, how many applications do we need to construct any arbitrary two-qubit operation? By Kostant's convexity theorem, we can prove that the minimum number required is given by

$$n_{\min} = \left\lceil \frac{3\pi}{2\gamma} \right\rceil .$$

In Figure 3, we plot the minimum upper bound for U_g to construct any two-qubit gate as a function of γ. It is clear that the most efficient Controlled-U is when $\gamma = \frac{\pi}{2}$, which is none other than the CNOT gate. Analytical results can be obtained to construct a universal quantum circuit from CNOT by three applications. Probing further, we find that DCNOT given below is equally efficient

$$\exp\left\{\frac{i}{2}\left(\frac{\pi}{2}\sigma_x^1\sigma_x^2 + \frac{\pi}{2}\sigma_y^1\sigma_y^2\right)\right\},$$

as it can generate any two-qubit gate by three applications as well. Note that DCNOT is not a Controlled-U gate, which leads us to wonder what is the most efficient universal gate among all the two-qubit operations. The answer is a gate that we named as the B gate:

$$B = \exp\left\{\frac{i}{2}\left(\frac{\pi}{2}\sigma_x^1\sigma_x^2 + \frac{\pi}{4}\sigma_y^1\sigma_y^2\right)\right\}.$$

The significance of the B gate is that with only two applications it can generate any two-qubit operation:

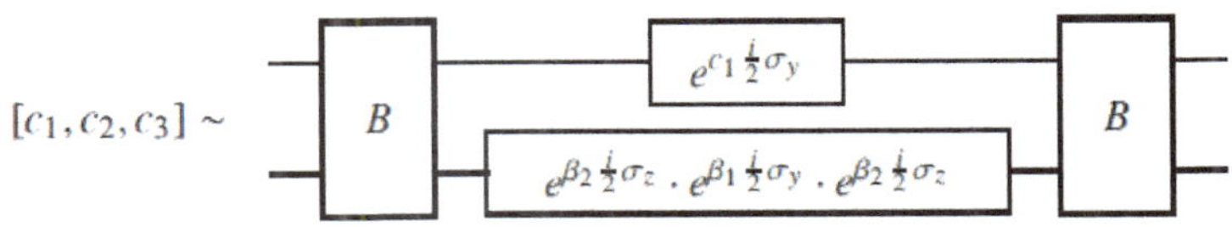

where the parameters β_1 and β_2 are determined by c_2 and c_3 in a closed-form solution.

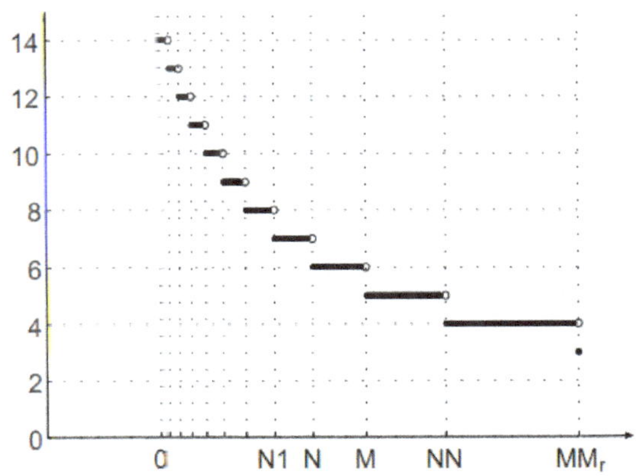

Figure 3: Minimum upper bound of applications required for Ug = $e^{\gamma \frac{i}{2}\sigma_x^1\sigma_x^2}$ to construct an arbitrary two-qubit gate.

SUMMARY

In this paper we reviewed the geometric method in the control of quantum mechanical system. We discussed that a generic open-loop state transfer control is equivalent to the universality problem in quantum information processing. We presented a geometric representation of two-qubit quantum operations, and showed that a control problem on two-qubit system can be converted into a steering problem in a tetrahedron. Two physical examples are given to illustrate this approach. We then studied analytic solutions to generate quantum operations from a generic quantum gate, and investigated the efficiency in the construction of universal quantum circuit.

This work was supported by the Innovation Program of Shanghai Municipal Education Commission (11ZZ20), Shanghai Pujiang Program (11PJ1405800), National Natural Science Foundation of China (61174086) and State Key Laboratory of Advanced Optical Communication Systems and Networks, SJTU, China.

REFERENCES

1. Mabuchi H, Khaneja N. Principles and applications of control in quantum systems. Int J Robust Nonlin Control, 2005, 15: 647–667
2. Tarn T J, Huang G, Clark J W. Modelling of quantum mechanical control systems. Math Model, 1980, 1: 109
3. Huang G M, Tarn T J, Clark J W. On the controllability of quantummechanical systems. J Math Phys, 1983, 24: 2608
4. Peirce A, Dahleh M, Rabitz H. Optimal control of quantum mechanical systems: Existence, numerical approximations, and applications. Phys Rev A, 1988, 37: 4950
5. Rabitz H, Hsieh M, Rosenthal C. Quantum optimally controlled transition landscapes. Science, 2004, 303: 998
6. Shor P W. Algorithms for quantum computation: Discrete logarithms and factoring. In: Goldwasser S, ed. Proceedings of 35nd Annual Symposium on Foundations of Computer Science, 1994. 124–134
7. Braunstein S L, van Loock P. Quantum information with continuous variables. Rev Mod Phys 2005, 77: 513–577
8. Nielsen M, Chuang I. Quantum Computation and Quantum Information. Cambridge: Cambridge University Press, 2000
9. Barenco A, Bennett C H, Cleve R, et al. Elementary gates for quantum computation. Phys Rev A, 1995, 52: 3457
10. Murray R M, Li Z, Sastry S S. A Mathematical Introduction to Robotic Manipulation. Boca Raton: CRC Press, 1994
11. D'Alessandro D, Dahleh M. Optimal control of two-level quantum systems. IEEE Trans Automatic Control, 2001, 46: 866–876
12. Helgason S. Differential Geometry, Lie groups, and Symmetric Spaces Academic. New York: Academic, 1978
13. Makhlin Y. Nonlocal properties of two-qubit gates and mixed states, and the optimization of quantum computations. Quant inf Process, 2002, 1: 243

14. Zhang J, Vala J, Sastry S, et al. Geometric theory of nonlocal two-qubit operations. Phys Rev A, 2003, 67: 042313

15. Loss D, DiVincenzo D P. Quantum computation with quantum dots. Phys Rev A, 1998, 57: 120

16. Ouyang M, Awschalom D D. Coherent spin transfer between molecularly bridged quantum dots. Science. 2003, 301: 1074–1078

17. Kempe J, Bacon D, Lidar D, et al. Theory of decoherence-free faulttolerant universal quantum computation. Phys Rev A, 2001, 63: 042307

18. Zhang J, Vala J, Sastry S, et al. Exact two-qubit universal quantum circuit. Phys Rev Lett, 2003, 91: 027903

5
CHAPTER

QUANTUM CONTROL ROBUST WITH RESPECT TO COUPLING WITH AN EXTERNAL ENVIRONMENT

Łukasz Pawela and Zbigniew Puchała

Keywords: Quantum information, Quantum computation, Control in mathematical physics

INTRODUCTION

The ability to manipulate the dynamics of a given complex quantum system is one of the fundamental issues of the quantum information science. It has been an implicit goal in many fields of science such as quantum physics, chemistry, or implementations of quantum information processing [1–3]. The usage of experimentally controllable quantum systems to perform computational task is a very promising perspective. Such usage is possible only if a system is controllable. Thus, the controllability of a given quantum system is an important issue of the quantum information science, since it concerns whether it is possible to drive a quantum system into a previously fixed state. When manipulating quantum systems, a coherent control strategy is a widely used method. In this case, the application of semiclassical potentials, in a fashion that preserves quantum coherence, is used to manipulate quantum states. If a given system is controllable, it is interesting to obtain control sequence that drives a system to a desired state and simultaneously minimize the value of the disturbance caused by imperfections of practical implementation. In the realistic implementations of quantum control systems, there can be various factors which disturb the evolution [4].

One of the main issues in this context is decoherence—the fact that the systems are very sensitive to the presence of the environment, which often destroys the main feature of the quantum dynamics. Other disturbance can be a result of the restriction on the frequency spectrum of acceptable control parameters [5]. In the case of such

Citation: Łukasz Pawela and Zbigniew Puchała, Quantum control robust with respect to coupling with an external environment, DOI 10.1007/s11128-014-0879-7.

systems, it is not accurate to apply piecewise-constant controls. In an experimental setup that utilizes an external magnetic field [6,7], such restrictions come into play and cannot be neglected. In many situations, the interaction with the control fields causes an undesirable coupling with the environment, which can lead to a destruction of the interesting features of the system [8]. In such situations, it is reasonable to seek a control field with minimal total influence on a system. Depending on a type of interaction with an environment, the influence differs. In this article, we consider an interaction that is proportional to the magnitude of a control field.

To minimize the influence of an environment in such case, when the control field performs the desired evolution, the L1 norm should be minimized. A different dynamical method for beating decoherence in open quantum systems is dynamical decoupling [9–12]. In this case, additional perturbation on a system is added, which protects the evolution against the effects of the environment influence at the same time driving the system to the desired state. In our case, the interaction with the environment is in strict relation to the control strategy, since it emerges only if the control pulses are applied. On the other hand, in a typical dynamical decoupling scheme, the coupling to the environment is constant, given by some Hamiltonian HSE acting on the system and environment. Another approach to robust quantum control is quantum sliding mode control [13]. This model combines unitary control and periodic projective measurements. First, the initial state is driven into a sliding mode and then a periodic projective measurement is performed. Finally, there is risk sensitive quantum control [14,15] that is a robust control method with a feedback loop.

The paper is organized as follows. In Sect. 2, we introduce the model used for simulations. Section 3 describes the simulation setup. In Sect. 4, we show results of numerical simulations, and in Sect. 5, we draw the final conclusions.

OUR MODEL

To demonstrate a method of obtaining piecewise-constant controls, which have minimal energy, we will consider an isotropic Heisenberg spin-1/2 chain of a finite length N. The control will be performed on the first spin only. The total Hamiltonian of the aforementioned quantum control system is given by

$$H(t) = H_0 + H_c(t), \tag{1}$$

Where

$$H_0 = J \sum_{i=1}^{N-1} \left(S_x^i S_x^{i+1} + S_y^i S_y^{i+1} + S_z^i S_z^{i+1} \right), \tag{2}$$

is a drift part given by the Heisenberg Hamiltonian. The control is performed only on a first spin and is Zeeman like, i.e.,

$$H_c(t) = h_x(t)S_x^1 + h_y(t)S_y^1. \tag{3}$$

In the above, S_k^i denotes kth Pauli matrix which acts on the spin i. Time dependent control parameters hx (t) and hy (t) are chosen to be piecewise constant. Furthermore, as opposed to [16], we do not restrict the control fields to be alternating with x and y, i.e., they can be applied simultaneously (see e.g., [17] for similar approach). For notational convenience, we set $\hbar = 1$; and after this, rescaling frequencies and control field amplitudes can be expressed in units of the coupling strength J ; and on the other hand, all times can be expressed in units of 1/J [16]. The system described above is operator controllable, as it was shown in [18] and follows from a controllability condition using a graph infection property introduced in the same article. The controllability of the described system can be also deduced from a more general condition utilizing the notion of hypergraphs [19]. Since the interest here is focused on operator control sequence, a quality of a control will be measured with the use of gate fidelity,

$$F = \frac{1}{2^N}\left|\mathrm{Tr}(U_T^\dagger U(h))\right|, \tag{4}$$

where U_T is the target quantum operation, and U(h) is an operation achieved by control parameters h. We choose gate fidelity as it neglects global phases. In the case of disturbed system, we will measure the quality of the control by a trace distance between Choi–Jamiołkowski states, which gives an estimation of a diamond norm. In many situations, the interaction with the control fields causes an undesirable coupling with the environment, which can lead to a destruction of the interesting features of the system. We will consider a general model described by

$$H(t) = H_0 + H_c(t) + \gamma(|h_x| + |h_y \tag{5}$$

where H_1 denotes a general Hamiltonian responsible for a coupling with an environment, and the interaction is proportional to the magnitude of a control field. To minimize the influence of an environment in this model, we introduce an additional constrain on the control pulses, namely we wish to minimize the L_1 norm of control pulses

$$\|h_k\|_1 = \sum_{i=1}^{n} |h_k^i|, \tag{6}$$

where k ∈ {x, y} and n are the total number of control pulses. In order to make this quantity comparable with fidelity, we impose bounds on the maximal amplitude of the control pulses. To accommodate this, we introduce the following penalty

$$P = \frac{\sum_{i=1}^{n} |h_k^i|}{nb}, \tag{7}$$

where b is the bound on the control pulse amplitude. This leads to the following functional we wish to minimize

$$G = (1 - \mu)P - \mu F, \tag{8}$$

where μ is a weight assigned to fidelity.

To optimize the control pulses, we utilize the BFGS algorithm [20]. In order to use this method effectively, we need to calculate the explicit form of derivatives of Eq. (6). We propose the following functions to be used as the derivative of the absolute value:

- The signum function:

$$\frac{\mathrm{d}|x|}{\mathrm{d}x} = \mathrm{sgn}(x). \tag{9}$$

- A fractional derivative:

$$\frac{\mathrm{d}^{\alpha}|x|}{\mathrm{d}x^{\alpha}} = \pm\frac{\Gamma(2)}{\Gamma(2-\alpha)}x^{1-\alpha}, \tag{10}$$

  where $\Gamma(x) = (x-1)!$ and we set $\alpha = 0.99$.

- Rescaled Fermi–Dirac distribution

$$\frac{\mathrm{d}|x|}{\mathrm{d}x} \approx 2\left(\frac{-1}{\exp\left(\frac{x}{kT}\right)+1} + 0.5\right), \tag{11}$$

where we set kT = 0.01.

The signum function is the natural conclusion when one thinks about the derivative of the L1 norm as it penalizes any nonzero control pulses in the control scheme. To further out studies, we introduce two approximations of the derivative of the L_1 norm. The first one utilizes the idea of fractional derivatives [21]. This allows us to achieve a continuous function, which quickly increases from 0 to 1 for positive values of the argument and decreases from 0 to −1 for negative values. Although continuous, the function has the drawback that control pulses with lower magnitude are less penalized.

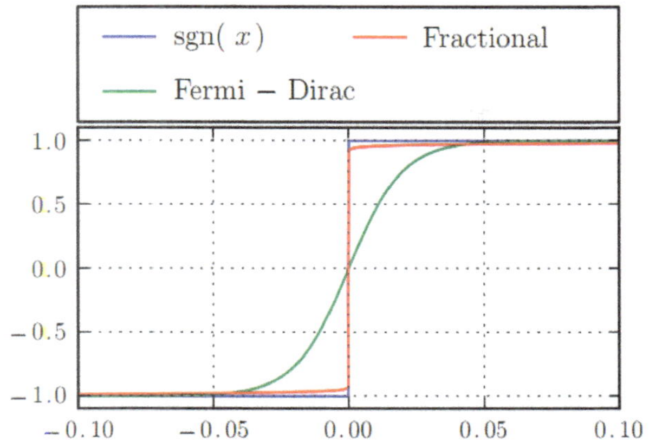

Figure 1: Comparison of different derivative approximations.

The penalty can be adjusted by using the parameter α The last proposed approximation is a rescaled Fermi–Dirac distribution [22]. The distribution is given by

$$f(x) = \frac{1}{\exp\left(\frac{x-E_f}{kT}\right)+1}, \tag{12}$$

where E_f is the Fermi energy level. The usage is justified, as for T = 0 the function is given as

$$f(E) = \begin{cases} 1 & \text{if } E < E_f, \\ 0 & \text{if } E > E_f, \end{cases} \tag{13}$$

We rescale this function in the following manner: First, we set $E_f = 0$. Next, in order to obtain behavior similar to the signum function, we translate and rescale the function, so that f (x) = −1 for x <<0 and f (x) = 1 for x >>0. After these operations we obtain

$$f(x) = 2\left(\frac{-1}{\exp\left(\frac{x}{kT}\right)+1} + 0.5\right), \tag{14}$$

From our point of view, the function has properties similar to the fractional derivative and the penalty for low magnitude pulses can be adjusted by using the "temperature" T . A comparison of these approximations is shown in Fig. 1.

SIMULATION SETUP

To demonstrate the beneficialness of our approach, we study three- and four-qubit spin chains. The control field is applied to the first qubit only. Our target gates are:

$$\mathrm{NOT}_N = \mathbb{1}^{\otimes N-1} \otimes \sigma_x, \tag{15}$$

the negation of the last qubit of the chain, and

$$\mathrm{SWAP}_N = \mathbb{1}^{\otimes N-2} \otimes \mathrm{SWAP}, \tag{16}$$

swapping the states between the last two qubits. This set of gates is universal in quantum computation. We provide an explicit example in which we set the duration of the control pulse to $\Delta t = 0.2$ and the total number of pulses in each direction to $n = 64$ for the threequbit chain and $n = 256$ in the four-qubit case, although the presented method may be applied for arbitrary values of Δt and n. The weight of fidelity in Eq. (8) is set to $\mu = 0.2$ in the three-qubit scenario and to $\mu = 0.4$ in the four-qubit scenario.

RESULTS

We show examples of control sequences obtained by using our method in Figs. 2 and 3. They depict results obtained for the three-qubit NOT gate optimization and four-qubit SWAP gate optimization, respectively. In the three-qubit scenario, we find, as expected, a control sequence that equal to zero most of the time with irregular, high amplitude pulses. A similar case can be made for the swap gate in the fourqubit scenario. The main difference is that in this case, the high amplitude pulses are surrounded by groups of weaker pulses. The results shown here are for the fractional derivative approximation. Simulations for other approximation yield nearly identical results. The fidelity obtained in both cases is $F > 0.99$, and the value of P has the order of 10^{-2}.

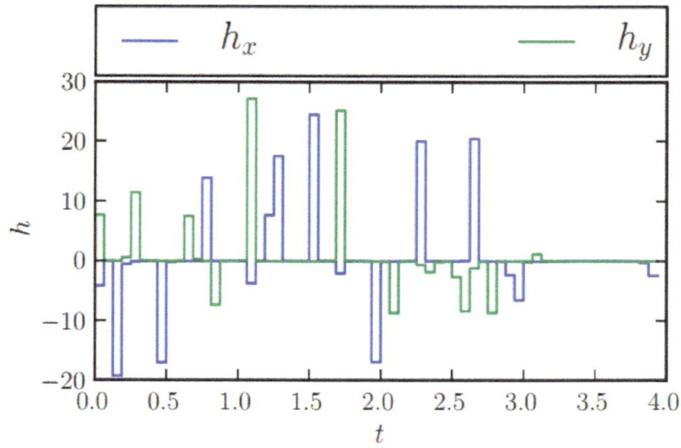

Figure 2: Example control sequences h_x and h_y for the NOT gate in the three-qubit scenario

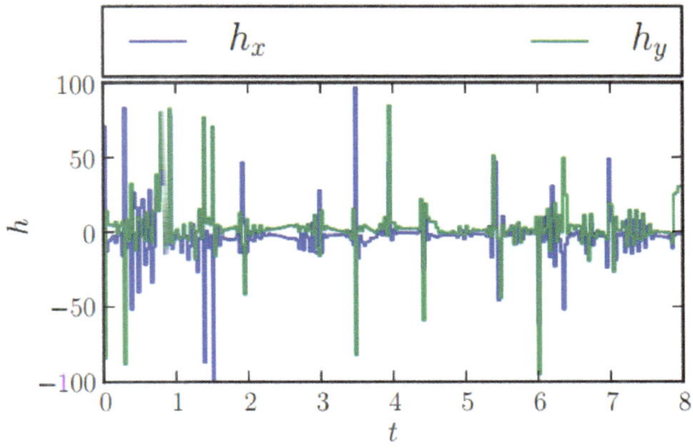

Figure 3: Example control sequences h_x and h_y for the SWAP gate in the four-qubit scenario.

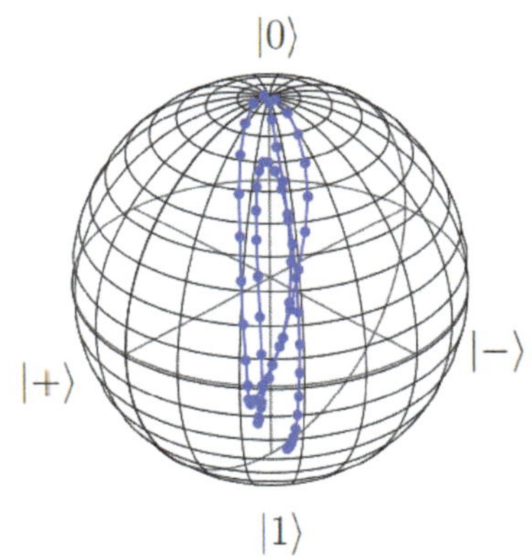

Figure 4: Time evolution of the target qubit of a three-qubit chain from the state $|000\rangle$ to the state $|001\rangle$ under the operator $\mathbb{1} \otimes \mathbb{1} \otimes \sigma_x$ implemented by optimized control sequences.

Finally, we show the evolution of each qubit's state. Let the qubits be in the state $|\psi\rangle_0 = |000\rangle$ in the case of the three-qubit scenario. Figure 4 shows the time evolution of the target qubit in this setup. The final state of the chain is $\psi_f = |001\rangle$. Note that the evolution is smooth, no signs of control pulses are visible in the qubit's trajectory. In the four-qubit scenario the time evolution of the target qubits is shown in Figs. 5 and 6.

Let the initial state of the chain be equal to $|\phi_0\rangle = |0010\rangle$. The final state of the chain is $|\phi\rangle = |0001\rangle$. In this case, the figures show that the transition is performed along the shortest path possible. Also, note that the evolution is quite smooth, and the application of each control pulse is not visible in these figures. In order to demonstrate the advantages of our approach, we perform additional simulations, where we put $\mu = 1$ in Eq. (8). This is the unconstrained problem of finding optimal control pulses. Next, we introduce an interaction with an environment, proportional to $|h_x|+|h_y|$. We model the interaction with the environment by adding a qubit to the chain. The Hamiltonian for this case is

$$H_{\text{graph}}(t) = H_0 + H_c(t) + \gamma(|h_x| + |h_y|)$$
$$\times \sum_{i=1}^{N} \left(S_x^i S_x^{N+1} + S_y^i S_y^{N+1} + S_z^i S_z^{N+1} \right). \tag{17}$$

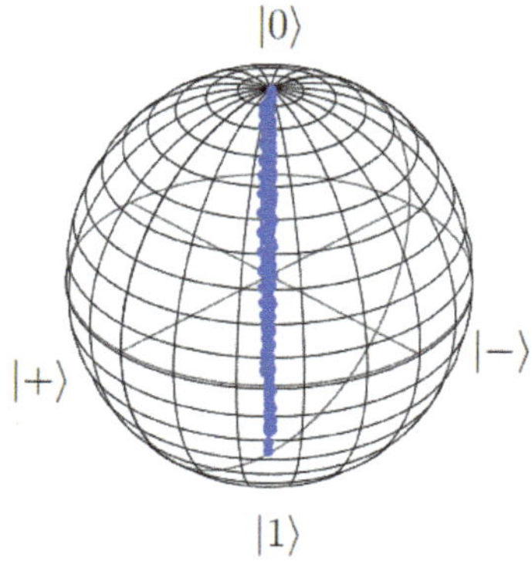

Figure 5: Time evolution of the third qubit of a four-qubit chain from the state $|0010\rangle$ to the state $|0001\rangle$ under the operator 1 ⊗ 1 ⊗ SWAP implemented by optimized control sequences.

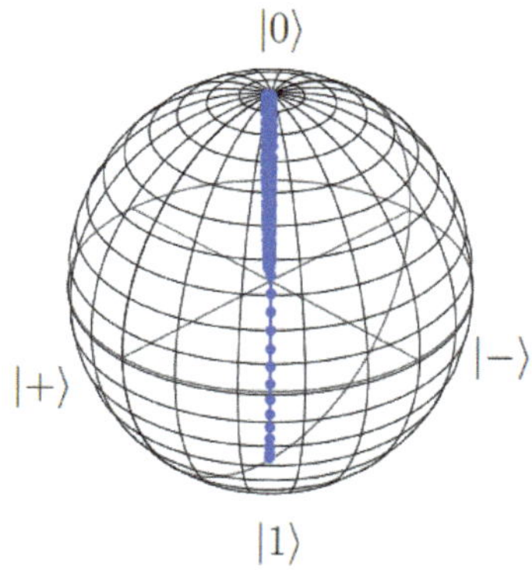

Figure 6: Time evolution of the fourth qubit of a four-qubit chain from the state $|0010\rangle$ to the state $|0001\rangle$under the operator 1 ⊗ 1 ⊗ SWAP implemented by optimized control sequences.

In order to compare the evolution with the additional qubit with a given U_T, we use the following scheme. For a quantum channel Φ, let us write $J(\Phi)$to denote the associated state:

Here, we are assuming that the channel maps n × n complex matrices into m × m complex matrices. The matrix $J(\Phi)$ is sometimes called the Choi–Jamiołkowski representation of Φ. For quantum channels Φ_0 and Φ_1, we may define the "diamond norm distance" between them as

$$\|\Phi_0 - \Phi_1\|_\diamond = \sup_{k,\rho} \|(\Phi_0 \otimes \mathbb{1}_k)(\rho) - (\Phi_1 \otimes \mathbb{1}_k)(\rho)\|_1 \tag{19}$$

where $\mathbb{1}_k$ denotes the identity channel from the set of k ×k complex matrices to itself; $\|\cdot\|_1$ denotes the trace norm; and the supremum is taken over all k ≥ 1 and all density matrices ρ from the set of nk × nk complex matrices. The supremum always happens to be achieved for some choice of k ≤ n and some rank 1 density matrix ρ. A coarse bound for the diamond norm defined in Eq. (19) is known [23]

Table 1: Summary of the value of Eq. (20) for the studied cases

	Without additional qubit		With additional qubit	
	$\mu = 1$	$\mu < 1$	$\mu = 1$	$\mu < 1$
NOT_3	0.0000	0.0000	0.0975	0.0086
NOT_4	0.0000	0.004	0.9788	0.0142
$SWAP_3$	0.0000	0.0001	0.0135	0.0133
$SWAP_4$	0.0000	0.0020	0.0843	0.0064

For μ = 1 we have a control optimization without regarding the L_1 norm of control pulses

Therefore, to compare the target operations with and without the additional qubit, we study the L_1 of the difference of the Jamiołkowski matrices of the respective quantum channels $\|J(\Phi_0) - J(\Phi_1)\|_1$. The results for different target operations are summarized in Table 1. We show results obtained for Fermi–Dirac approximation of the derivative. As stated in the table, the bigger the system under consideration is the greater is the gain from using our method.

CONCLUSIONS

In this work, we introduced a method of obtaining a piecewise-constant control field for a quantum system with an additional constrain of minimizing the L_1 norm. To demonstrate the beneficialness of our approach, we have shown results obtained for a spin chain, on which we implemented two quantum operations: negation of the last qubit of the chain and swapping the states of the two last qubits of the chain. Our results show that it is possible to obtain control fields which have minimal energy and still give a high fidelity of the quantum operation. Our method may be used in situations where the interaction with the control field causes additional coupling to the environment. As our method allows one to minimize the number of control pulses, it also minimizes the amount of coupling to the environment. It is important to note that our model differs from known in the literature dynamical decoupling, in which additional perturbation on a system is added, which protects the evolution against the effects of the environment influence. In our case the interaction with the environment is related to the control strategy, and it emerges only if the control is applied. Our model allows to optimize high fidelity control pulses for the cases with and without external environment, as shown in Table 1 as long as the coupling is induced by the control pulses themselves. Other possible usage of our method includes systems, in which it is possible to use rare, but high value of control pulses, for example, superconducting magnets with high impulse current.

ACKNOWLEDGMENTS

Ł. Pawela was supported by the Polish National Science Centre under the Grant Number N N514 513340. Z. Puchała was supported by the Polish Ministry of Science and Higher Education under the Project Number IP2011 044271.

REFERENCES

1. d'Alessandro, D.: Introduction to Quantum Control and Dynamics. Chapman & Hall, London (2008)
2. Albertini, F., D'Alessandro, D.: The Lie algebra structure and controllability of spin systems. Linear Algebra Appl. 350(1), 213–235. http://www.sciencedirect.com/science/article/pii/S0024379502002902 (2002)
3. Werschnik, J., Gross, E.: Quantum optimal control theory. J. Phys. B At. Mol. Opt. Phys. 40, R175–R211 (2007)
4. Pawela, Ł., Sadowski, P.: arXiv preprint arXiv:1310.2109 (2013) (under review)
5. Pawela, Ł., Puchała, Z.: Quantum control with spectral constraints. Quantum Inf. Process. 13, 227–237 (2014)
6. Chaudhury, S., Merkel, S., Herr, T., Silberfarb, A., Deutsch, I.H., Jessen, P.S.: Quantum control of the hyperfine spin of a Cs atom ensemble. Phys. Rev. Lett. 99, 163002 (2007). doi:10.1103/PhysRevLett.99.163002
7. Jami, S., Amerian, Z., Ahmadi, F., Motevalizadeh, L.: Effects of external magnetic field on thermal entanglement in a spin-one chain with three particles. Indian J. Phys. 87(4), 367–372 (2013)
8. Zhang, S., Jie, Q., Wang, Q.: Revival and decay of entanglement in a two-qubit system coupled to a kicked top. Indian J. Phys. 86(5), 387–393 (2012)
9. Viola, L., Knill, E., Lloyd, S.: Dynamical decoupling of open quantum systems. Phys. Rev. Lett. 82, 2417 (1999a)
10. Viola, L., Lloyd, S., Knill, E.: Universal control of decoupled quantum systems. Phys. Rev. Lett. 83, 4888 (1999b)
11. Viola, L., Knill, E.: Robust dynamical decoupling of quantum systems with bounded controls. Phys. Rev. Lett. 90, 037901 (2003)

12. Dahleh, M., Peirce, A., Rabitz, H.: Optimal control of uncertain quantum systems. Phys. Rev. A 42, 1065 (1990)
13. Dong, D., Petersen, I.R.: Sliding mode control of quantum systems. New J. Phys. 11, 105033 (2009)
14. James, M.: Risk-sensitive optimal control of quantum systems. Phys. Rev. A 69, 032108 (2004)
15. D'Helon, C., Doherty, A., James, M., Wilson, S.: In: 2006 45th IEEE Conference on Decision and Control (IEEE, 2006), pp. 3132–3137 (2006)
16. Heule, R., Bruder, C., Burgarth, D., Stojanović, V.M.: Local quantum control of Heisenberg spin chains. Phys. Rev. A 82, 052333 (2010). doi:10.1103/PhysRevA.82.052333
17. Khaneja, N., Reiss, T., Kehlet, C., Schulte-Herbrüggen, T., Glaser, S.J.: Optimal control of coupled spin dynamics: design of NMR pulse sequences by gradient ascent algorithms. J. Magn. Reson. 172, 296–305. http://www.sciencedirect.com/science/article/pii/S1090780704003696 (2005)
18. Burgarth, D., Bose, S., Bruder, C., Giovannetti, V.: Local controllability of quantum networks. Phys. Rev. A 79, 060305 (2009). doi:10.1103/PhysRevA.79.060305
19. Puchała, Z.: Local controllability of quantum systems. Quantum Inf. Process. 12, 459–466 (2013). doi:10.1007/s11128-012-0391-x
20. Press, W., Flannery, B., Teukolsky, S., Vetterling, W.: Numerical Recipes in FORTRAN 77: Volume 1, Volume 1 of Fortran Numerical Recipes: The Art of Scientific Computing, vol. 1. Cambridge university press, Cambridge (1992)
21. Miller, K.S., Ross, B.: An Introduction to the Fractional Calculus and Fractional Differential Equations. Wiley, New York (1993)
22. Smirnov, B.M.: Fermi–Dirac Distribution, pp. 57–73. Wiley-VCH Verlag GmbH & Co. KGaA (2007). doi:10.1002/9783527608089.ch4
23. Kitaev, A.Y., Shen, A.H., Vyalyi, M.N.: Classical and Quantum Computation, vol. 47 of Graduate Studies in Mathematics. American Mathematical Society, Providence (2002)

6
CHAPTER

HYBRID OPTIMIZATION SCHEMES FOR QUANTUM CONTROL

Michael H Goerz , K Birgitta Whaley and Christiane P Koch

Keywords: quantum control, optimization methods, quantum information

INTRODUCTION

Coherent quantum control has long been a key component in the effort towards future quantum technologies. It relies on the interference between multiple pathways to steer the quantum system in some desired way [1, 2]. Originally conceived for the control of chemical reactions [3–5], it has since been extended to a wide variety of applications, see Ref. [6] for a review. In this context, numerical optimal control theory (OCT) is a particularly powerful tool. OCT follows an iterative approach, improving the control field in each iteration to steer the dynamics to the optimization target. Generally, the fastest converging algorithms are those that take into account information about the gradient of the optimization functional with respect to variations in the control. Gradient-based methods assume an open-loop setup, where the entire optimization procedure is performed based on the knowledge of the system dynamics. Two widely used methods are Krotov's method [7–9] and gradient ascent pulse engineering (GRAPE) [10]. Krotov's method guarantees monotonic convergence for time-continuous control fields. The Limited memory Broyden-Fletcher-Goldfarb-Shannon (LBFGS) method [11] can be used to extend both GRAPE [12] and Krotov's method [13], considering not only the gradient but also an estimate for the Hessian, i.e., the second order derivative. This has been demonstrated to improve the convergence in some instances, in particular when close to the optimum [14]. In experimental setups that only allow for limited control and knowledge of the dynamics, closed-loop control schemes have often been preferred [15]. There, the controls are updated based only on a measurement of the figure of

Citation: Michael H Goerz, K Birgitta Whaley, Christiane P Koch, Hybrid optimization schemes for quantum control, DOI: 10.1140/epjqt/s40507-015-0034-0.

merit, e.g. using genetic algorithms [16] or other gradient-free optimization methods. More recently, gradient-free optimization methods have also been employed in an open-loop context, prompted by the observation that evaluation of the gradient in many-body systems is often numerically infeasible. The chopped random basis (CRAB) method [17, 18] has been formulated for such applications. It expands the control in a relatively small number of randomly chosen spectral components and then applies a Nelder-Mead simplex optimization to the expansion coefficients. In principle, gradient-free methods are applicable if the control can reasonably be described by only a few free parameters and the optimization landscape has no local minima in the vicinity of the initial 'guess'.

Optimal control theory is particularly relevant for quantum information processing. Both Krotov's method and GRAPE have been extensively used to obtain high-fidelity quantum gates [19–24]. Short gate durations are crucial, in order to minimize detrimental effects of decoherence. With OCT, this is achieved by systematically decreasing the gate duration until no solution can be found [25, 26], thus operating at the quantum speed limit [27, 28]. Moreover, OCT may be used to actively minimize the effects of decoherence [29, 30], and to increase robustness with respect to classical noise [31]. Robustness is a requirement that is generally difficult to fulfill with analytical approaches.

Here, we explore the possibility of combining gradient-free and gradient-based methods at different stages in the optimization, exploiting the benefits of each method. The application of a simplex optimization to a guess pulse, described by only a few free parameters, efficiently yields a comparably simple first optimized pulse of moderate fidelity. This pre-optimized pulse then provides a good starting point for further optimization using a gradient-based method. The second optimization stage relaxes the restrictions on the search space implied by a simple parametrization and may then quickly converge towards a high fidelity. The simplex pre-optimization addresses the observation that typically in direct gradient-based optimizations, due to the large size of the search space, the majority of the numerical effort is spent in 'getting off the ground'. Pre-optimization thus allows to locate a region of the search space in which the gradient is large enough to provide meaningful information. Such good guess pulses may sometimes be designed by hand, but this requires a very good intuition of the underlying control mechanisms. We propose here instead to simply prune the search space for the initial phase of the optimization by reducing the complexity of the control. The second optimization stage can then more easily identify high fidelity solutions.

We illustrate the use of such hybrid optimization schemes by optimizing a quantum gate on superconducting qubits, using an example inspired by the recently proposed resonator-induced phase gate [32]. Superconducting circuits are a prime candidate for the implementation of quantum computing, due to the flexibility in qubit parameters, their inherent controllability, and the promise of scalability. Moreover, with recent advances in the transmon architecture, decoherence times are approaching 0.1 ms [33], allowing to reach fault-tolerance with sufficiently fast gates. However, the flexibility and large number of different gate mechanisms [34] also imply a challenge from a control perspective, as it is not immediately obvious what are good qubit parameters, or good guess pulses. This makes superconducting circuits especially well-suited for combining a coarse search using simplex methods, which may then be refined with a more powerful gradient-based method.

The paper is organized as follows: In Section 2, we review the realization of a geometric phase gate on a system of two transmon qubits coupled with a shared transmission line resonator, using all microwave control. In Section 3, we first introduce a functional targeting the geometric phase gate, i.e., any diagonal perfect entangler, and then show the results of a direct optimization using the gradient-based Krotov method. In Section 4, we apply the gradient-free Nelder-Mead simplex optimization to obtain pre-optimized guess pulses, which then become the starting point for an optimization with Krotov's method. We compare the control pulses obtained by the gradient-based, gradient-free, and hybrid schemes, and the dynamics they induce, as well as the numerical effort necessary to obtain converged results in each scheme. Section 5 concludes.

MODEL AND GATE MECHANISM

We consider a system of two transmon qubits, coupled via a shared transmission line resonator ('cavity')35,36]. The Hamiltonian reads

$$\hat{\mathbf{H}} = \sum_{q=1,2}\left[\omega_q \hat{\mathbf{b}}_q^\dagger \hat{\mathbf{b}}_q + \frac{\alpha_q}{2}\hat{\mathbf{b}}_q^\dagger \hat{\mathbf{b}}_q^\dagger \hat{\mathbf{b}}_q \hat{\mathbf{b}}_q + g_q\left(\hat{\mathbf{b}}_q^\dagger \hat{\mathbf{a}} + \hat{\mathbf{b}}_q \hat{\mathbf{a}}^\dagger\right)\right] + \omega_c \hat{\mathbf{a}}^\dagger \hat{\mathbf{a}} + \epsilon^*(t)\hat{\mathbf{a}} + \epsilon(t)\hat{\mathbf{a}}^\dagger, \tag{1}$$

where ω_c, ω_1, ω_2 are the frequency of the cavity and the first and second qubit, respectively; α_1, α_2 are the qubit anharmonicities, and g_1, g_2 are the coupling between each qubit and the cavity. The operators $\hat{\mathbf{a}}, \hat{\mathbf{b}}_1, \text{and } \hat{\mathbf{b}}_2$ are the standard annihilation operators for the cavity $(\hat{\mathbf{a}})$ and the two qubits $(\hat{\mathbf{b}}_1, \text{and } \hat{\mathbf{b}}_2)$, respectively. For numerical purposes, the Hilbert spaces for the qubit and cavity are truncated after 6, respectively 70, levels. It has been verified that the inclusion of additional levels yields no significant change in the results of the subsequent sections. The parameters take the values listed in Table 1. The system is driven by the microwave field $\epsilon(t)$, with a pulse duration T. An off-resonant pulse results in a state-dependent shift of the resonator frequency [32]. For a slowly-varying pulse shape with $\epsilon(0) = \epsilon(T) = 0$, such that the level shifts occur adiabatically, the dynamics result in a geometric phase on each of the qubit levels. That is, the resulting gate takes the diagonal form

$$\hat{\mathbf{U}} = \mathrm{diag}\left[e^{i\phi_{00}}, e^{i\phi_{01}}, e^{i\phi_{10}}, e^{i\phi_{11}}\right]. \tag{2}$$

The maximal reachable concurrence of such a diagonal gate is

$$C(\gamma) = \left|\sin\left(\frac{\gamma}{2}\right)\right|, \quad \gamma \equiv \phi_{00} - \phi_{10} - \phi_{01} + \phi_{11}, \tag{3}$$

where γ defines the non-local two-qubit phase [26]. The concurrence is obtained from the theory of local invariants for two-qubit gates [37, 38]. The local invariants for a diagonal gate evaluate to $G_1=\cos^2(\gamma/2)$ and $G_2=1+2\cos^2(\gamma/2)$. From these, the Weyl chamber coordinates may be calculated as $c_1=\gamma/2$, $c_2=c_3=0$ [38]. Following Ref. [39], the concurrence is evaluated as a function of the Weyl chamber coordinates to yield Eq. (3). The gate is a perfect entangler for $\gamma=\pi$.

Table 1: Parameters for two transmon qubits coupled via a shared transmission line resonator.

Cavity frequency	ω_c	8.10	GHz
Left qubit frequency	ω_1	6.85	GHz
Right qubit frequency	ω_2	7.25	GHz
Drive frequency	ω_d	8.14	GHz
Qubit anharmonicity	α_1, α_2	−300	MHz
Qubit-cavity coupling	g_1, g_2	70	MHz

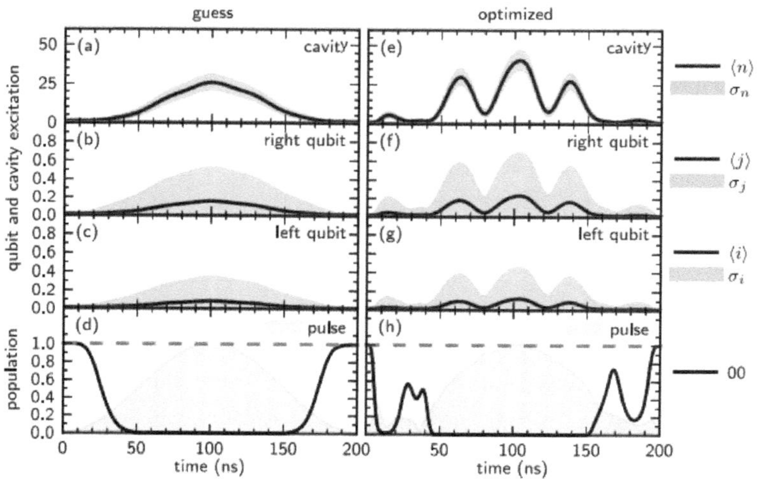

Population dynamics under guess and optimized pulse. The figure shows the population dynamics of the initial state $|\Psi\rangle(t=0)=|00\rangle$ for the guess pulse, panels (a)-(d), and the pulse obtained from direct optimization of the J_T^{geo} functional, panels (e)-(h). In panels (a), (e), expectation value $\langle n\rangle$ of the cavity excitation, plus-minus the standard deviation σ_n. In panels (b), (f), and (c), (g), expectation values and standard deviations for the excitation of the right and left qubit, respectively. In panels (d), (h), population in the state $|00\rangle$. The pulse shape (normalized by the peak amplitude E_0=300 MHz, cf. Figure 2) is shown in the background of panels (d), (h). The guess pulse implements a geometric phase gate with a gate error of $\varepsilon_{avg}=8.3\times10^{-2}$, with concurrence error $\varepsilon_C=1.9\times10^{-1}$ and population loss from the logical subspace $\varepsilon_{pop}=5.9\times10^{-3}$. The optimized pulse decreases the gate error to $\varepsilon_{avg}=1.4\times10^{-4}$, with $\varepsilon_C=1.8\times10^{-6}$ and $\varepsilon_{pop}=1.4\times10^{-4}$, see Table 2.

We consider pulse shapes of the form

$$\epsilon(t) = E_0 \sin^2\left(\pi \frac{t}{T}\right) \cos(\omega_d t) \qquad (4)$$

with a fixed driving frequency ω_d given in Table 1. For simplicity, we neglect the dephasing induced by high cavity populations which allows solution of wave-packet dynamics with the time-dependent Schrödinger equation including microwave control fields. For an arbitrarily chosen guess pulse of duration T=200 ns and peak amplitude E_0=300 MHz, the population dynamics resulting from the initial condition $|\Psi(t=0)\rangle=|00\rangle$ are shown in panels (a)-(d) of Figure 1.

If the pulse induces adiabatic dynamics, it shifts the qubit and cavity levels proportionally to $\epsilon(t)^2/\Delta$ where Δ is the detuning from the respective level [32]. In the original field-free frame, this is equivalent to shifting the initial wave packet proportionally to the square of the pulse. Thus, the excitation and population dynamics should smoothly follow the pulse shape. Specifically, the condition for adiabaticity is that if the pulse shape were to be stretched in time, the population dynamics would simply stretch correspondingly. For a given peak amplitude, the larger the detuning of the drive ω_d from the frequencies ω_c, ω_2, ω_1, the smaller the excitation in the respective Hilbert space. Since the drive is detuned by only 40 MHz from the cavity, the cavity excitation, panel (a), reaches a large value $\langle n\rangle\approx 30$. The far-detuned right qubit, panel (b), and even farther detuned left qubit, panel (c), only show a small excitation. As indicated by the standard deviations shown as shaded areas in panels (a)-(c), the excitation remains relatively localized in energy. It is noteworthy, however, that the excitation curves (specifically of the cavity) show some imperfections. These 'wobbles' can be interpreted as deviations from the expected adiabatic dynamics, e.g. jumping over an avoided crossing between highly excited cavity states. This ultimately results in a small loss of population from the logical subspace, as the system does not perfectly return to its original state. The loss of population for the given example is $\varepsilon_{pop}=5.9\times 10^{-3}$. While such a small deviation is not discernible in the plot of the population in panel (d), it is nonetheless a significant error in the objective of obtaining high fidelity gates below the quantum error correction limit, typically resulting in gate errors below 10^{-3}. In principle, cavity population can be suppressed by tuning the qubit parameters and using non-trivial pulse shapes [32]. The dynamics for the remaining two-qubit basis states are similar to those of the $|00\rangle$ state in Figure 1. The concurrence of the gate implemented by the propagated guess pulse, according to Eq. (3), yields a value of $C\approx 0.8$. This implements the closest diagonal perfect entangler with an average gate error of 8.3×10^{-2}. Both the loss of population from the logical subspace and the small value of the generated entanglement imply that the chosen pulse parameters are sub-optimal with respect to the desired geometric phase gate. We therefore turn to numerical optimal control to obtain a high fidelity gate.

DIRECT OPTIMIZATION WITH KROTOV'S METHOD

Optimization Functionals and Method

The standard approach for implementing a specific quantum gate $\hat{O}$ using optimal control theory is to maximize the overlap between the time evolution $\hat{U}(T,0;\epsilon(t))$ under the control $\epsilon(t)$ and the target gate [19]. For a two-qubit gate, this is commonly expressed in the final time functional [40]

$$J_T^{\mathrm{sm}} = 1 - \frac{1}{16}\left|\sum_{k=1}^{4}\langle k|\hat{\mathbf{O}}^{\dagger}\hat{\mathbf{U}}(T,0;\epsilon(t))|k\rangle\right|^2, \quad |k\rangle \in \{|00\rangle, |01\rangle, |10\rangle, |11\rangle\}, \tag{5}$$

which goes to zero as the target gate $\hat{\mathbf{O}}$ is implemented, up to a global phase. Krotov's method allows to iteratively improve the control field, changing the control $\epsilon^{(i)}(t)$ to the updated control $\epsilon^{(i+1)}(t)$ in the ith iteration. The final time functional J_T, given e.g. by Eq. (5), is augmented with a running cost to result in the total functional

$$J[\epsilon^{(i)}(t)] = J_T(\{\phi_k^{(i)}(T)\}) + \int_0^T g_a[\epsilon^{(i)}(t)]\,dt, \qquad |\phi_k^{(i)}(t)\rangle = \hat{\mathbf{U}}(t,0;\epsilon^{(i)}(t))|k\rangle. \tag{6}$$

Monotonic convergence is ensured by the choice [9,40]

$$g_a[\epsilon^{(i)}(t)] = \frac{\lambda_a}{S(t)}(\Delta\epsilon(t))^2, \quad \Delta\epsilon(t) \equiv \epsilon^{(i+1)}(t) - \epsilon^{(i)}(t), \tag{7}$$

where λ_a is an arbitrary scaling parameter and $S(t) \in [0, 1]$ is a shape function that ensures smooth switch-on and switch-off.

The iterative update scheme is then given in terms of three coupled equations [9],

$$\Delta\epsilon(t) = \frac{S(t)}{\lambda_a}\,\mathrm{Im}\left[\sum_{k=1}^{4}\left(\langle\chi_k^{(i)}(t)|\left(\frac{\partial\hat{\mathbf{H}}}{\partial\epsilon}\Big|_{\substack{\phi^{(i+1)}(t)\\ \epsilon^{(i+1)}(t)}}\right)|\phi_k^{(i+1)}(t)\rangle\right.\right.$$

$$\left.\left. + \frac{1}{2}\sigma(t)\langle\Delta\phi_k(t)|\left(\frac{\partial\hat{\mathbf{H}}}{\partial\epsilon}\Big|_{\substack{\phi^{(i+1)}(t)\\ \epsilon^{(i+1)}(t)}}\right)|\phi_k^{(i+1)(t)}\rangle\right)\right], \tag{8a}$$

$$\frac{\partial}{\partial t}|\phi_k^{(i+1)}(t)\rangle = -\frac{i}{\hbar}\hat{\mathbf{H}}^{(i+1)}|\phi_k^{(i+1)}(t)\rangle, \qquad |\phi_k^{(i+1)}(0)\rangle = |k\rangle, \tag{8b}$$

$$\frac{\partial}{\partial t}|\chi_k^{(i)}(t)\rangle = -\frac{i}{\hbar}\hat{\mathbf{H}}^{\dagger(i)}|\chi_k^{(i)}(t)\rangle, \qquad |\chi_k^{(i)}(T)\rangle = -\frac{\partial J_T}{\partial\langle\phi_k|}\Big|_{\phi_k^{(i)}(T)}, \tag{8c}$$

With $|\Delta\phi_k(t)\rangle \equiv |\phi_k^{(i+1)}(t)\rangle - |\phi_k^{(i)}(t)\rangle$. The second order contribution to the update, with the prefactor $\sigma(t)$, is required for certain types of functionals [9], as we will see below. For the choice of J_T^{sm} we may set $\sigma(t)=0$[9].

For the gate mechanism outlined in Section 2, we obtain a phase on each of the four logical basis states. These phases should combine to produce a perfect entangler, with

$\gamma = \pi$ according to Eq. (3). The individual phases φ_{00}, φ_{01}, φ_{10} and φ_{11} depend delicately on the shape, amplitude, and duration of the pulse. It is therefore not known a priori for a given guess pulse which exact geometric phase gate will or can be reached, and thus what should be the target gate $\hat{\mathbf{O}}$ of the optimization. For a gate $\hat{\mathbf{U}}_0$ induced by a guess pulse, we may construct the closest diagonal perfect entangler by numerically evaluating

$$\hat{\mathbf{O}} = \underset{\phi_{00},\phi_{01},\phi_{10}}{\arg\min} \left\| \hat{\mathbf{O}}_{\text{diag}}(\phi_{00},\phi_{01},\phi_{10}) - \hat{\mathbf{U}}_0 \right\|, \tag{9}$$

With

$$\hat{\mathbf{O}}_{\text{diag}}(\phi_{00},\phi_{01},\phi_{10}) = \text{diag}\left[e^{i\phi_{00}}, e^{i\phi_{01}}, e^{i\phi_{10}}, e^{i(\pi+\phi_{01}+\phi_{10}-\phi_{00})}\right], \tag{10}$$

which includes the condition $\gamma=\pi$ to make the gate a perfect entangler. Using this gate as a target fully determines the optimization problem, with two caveats. First, the target gate depends on $\hat{\mathbf{U}}_0$ induced by the (arbitrary) guess pulse, and second, the construction of the closest diagonal perfect entangler does not take into account the topology of the optimization landscape; the 'closest' gate is by no means guaranteed to be the one that is easiest to reach.

An approach that addresses these issues is to go beyond the standard functional of Eq. (5) and formulate a functional that targets the properties of the geometric phase gate specifically, without stipulating the phases on all of the logical states. We split the functional into two terms,

$$J_T^{\text{geo}} = \frac{1}{8}(J_{\text{diag}} + J_\gamma), \tag{11}$$

where J_{diag} goes to zero if and only if the gate is diagonal (with arbitrary phases) and J_γ goes to zero if and only if the gate is also a perfect entangler, $\gamma = \pi$. Thus, the functional is conceptually similar to a recently proposed functional targeting an arbitrary perfect entangler [41]. However, the additional restriction to enforce a diagonal gate is important, as the Hamiltonian also allows for non-diagonal gates, but only through undesired nonadiabatic effects. The two terms take the form

$$J_{\text{diag}} = 4 - \tau_{00}\tau_{00}^* - \tau_{01}\tau_{01}^* - \tau_{10}\tau_{10}^* - \tau_{11}\tau_{11}^*, \tag{12a}$$

$$J_\gamma = 2 + \tau_{00}\tau_{01}^*\tau_{10}^*\tau_{11} + \tau_{00}^*\tau_{01}\tau_{10}\tau_{11}^*, \tag{12b}$$

With

$$\tau_{00} \equiv \langle 00|\hat{\mathrm{U}}|00\rangle, \qquad \tau_{01} \equiv \langle 01|\hat{\mathrm{U}}|01\rangle,$$

$$\tau_{10} \equiv \langle 10|\hat{\mathrm{U}}|10\rangle, \qquad \tau_{11} \equiv \langle 11|\hat{\mathrm{U}}|11\rangle. \tag{12c}$$

The construction of J_γ is based on the observation that

$$\gamma = \pi \quad \Longleftrightarrow \quad 2 + e^{i\gamma} + e^{-i\gamma} = 0,$$

which becomes Eq. (12b) by associating τ_k with $e^{i\varphi k}$ and using the definition of γ in Eq. (3). Both J_{diag} and J_γ take values $\in [0, 4]$, hence the normalization factor 1/8 in Eq. (11) to bring the value of the functional closer to that of J_T^{sm}. In contrast to J_T^{sm}, the functional J_T^{geo} is not convex, since the states enter in higher than quadratic order. The Krotov update equation (8a) must then include the second order contribution, where $\sigma(t)$ can be determined numerically in each iteration as [9]

$$\sigma(t) = -\max(\epsilon_A, 2A + \epsilon_A), \quad A = \frac{2\sum_{k=1}^{4}\mathrm{Re}[\langle\chi_k(T)|\Delta\phi_k(T)\rangle] + \Delta J_T}{\sum_{k=1}^{4}|\Delta\phi_k(T)|^2}, \tag{13}$$

with a small non-negative number ϵ_A, and $\Delta J_T \equiv J_T(\{\phi_k^{(i+1)}(T)\}) - J_T(\{\phi_k^{(i)}(T)\})$. The boundary condition for the backward propagated states in Eq. (8c) yields

$$|\chi_{00}(T)\rangle = (\tau_{00} - \tau_{01}\tau_{10}\tau_{11}^*)|00\rangle, \tag{14a}$$

$$|\chi_{01}(T)\rangle = (\tau_{01} - \tau_{00}\tau_{10}^*\tau_{11})|01\rangle, \tag{14b}$$

$$|\chi_{10}(T)\rangle = (\tau_{10} - \tau_{00}\tau_{01}^*\tau_{11})|10\rangle, \tag{14c}$$

$$|\chi_{11}(T)\rangle = (\tau_{11} - \tau_{00}^*\tau_{01}\tau_{10})|11\rangle. \tag{14d}$$

Both J_T^{sm} and J_T^{geo} are only loosely connected to the average gate fidelity that is accessible to experimental measurement. In the case of a two-qubit gate and non-dissipative dynamics this can be evaluated as [42]

$$F_{\mathrm{avg}} = \int |\langle\Psi|\hat{\mathrm{O}}^\dagger\hat{\mathrm{U}}|\Psi\rangle|^2 d\Psi = \frac{1}{20}\left(|\mathrm{tr}[\hat{\mathrm{O}}^\dagger\hat{\mathrm{U}}]|^2 + \mathrm{tr}[\hat{\mathrm{O}}^\dagger\hat{\mathrm{U}}\hat{\mathrm{U}}^\dagger\hat{\mathrm{O}}]\right). \tag{15}$$

Thus, F_{avg}, respectively the gate error $\varepsilon_{avg} \equiv 1 - F_{avg}$, provides a well-defined measure of the optimization success independent of the choice of optimization functional. For an optimization with J_T^{geo}, we may evaluate ε_{avg} with respect to the closest geometric phase gate resulting from propagation with the optimized pulse, according to Eq. (9).

Table 2: Optimization success for different optimization schemes

	T [ns]	prop.	ε_C	ε_{pop}	ε_{avg}
Guess	200	0	1.92×10^{-1}	5.94×10^{-3}	8.25×10^{-2}
Direct s.m.	200	20,000	2.23×10^{-2}	4.13×10^{-3}	1.45×10^{-2}
Direct geo.	200	11,032	1.83×10^{-6}	1.42×10^{-4}	1.43×10^{-4}
Simplex	185	116	1.95×10^{-5}	1.40×10^{-2}	1.40×10^{-2}
Pre-opt. s.m.	185	518	5.07×10^{-5}	1.11×10^{-5}	3.36×10^{-5}
Pre-opt. geo.	185	300	5.24×10^{-5}	1.40×10^{-5}	3.50×10^{-5}

For each scheme, we give the gate duration T, the total number of propagations, the concurrence error $\varepsilon_C \equiv 1 - C$ by which the gate differs from a perfect entangler, the loss of population ε_{pop} from the logical subspace, and the gate error $\varepsilon_{avg} \equiv 1 - F_{avg}$ with respect to a geometric phase gate. The number of propagations for 'pre-opt. s.m.' and 'pre-opt. geo.' include both the 116 propagations of the first stage simplex optimization and the propagations from the second-stage optimization using Krotov's method, with 201 respectively 92 iterations and two propagations per iteration. The reported number of propagations is thus proportional to the total CPU and wall-clock time required to obtain the result starting from the original guess pulse.

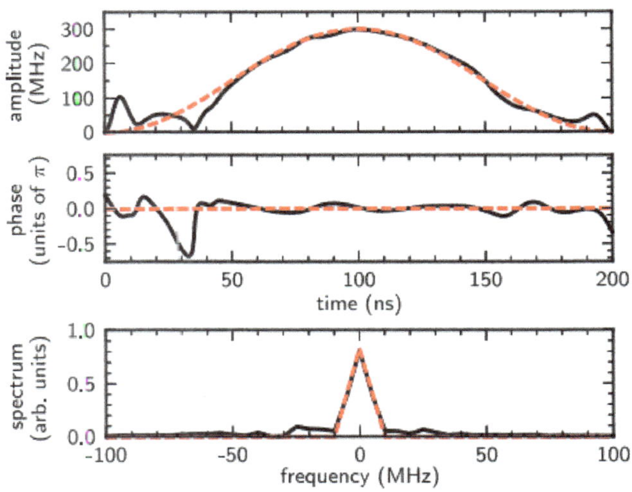

Figure 2: Optimized pulse resulting from direct optimization with Krotov's method using the J_T^{geo} functional. In the panels from top to bottom, absolute value of the pulse shape, complex phase, and spectrum of the optimized pulse (solid black lines) and of the guess pulse (dashed red/gray lines).

Optimization results The optimization starts from the guess pulse described by Eq. (4), with T = 200 ns and E_0 = 300 MHz, as discussed in Section 2, with the dynamics shown in panels (a)-(d) of Figure 1. The gate error with respect to the closest geometric phase gate for this guess is $\varepsilon_{avg} = 8.3 \times 10^{-2}$, with a loss of population from the logical subspace of $\varepsilon_{pop} = 5.9 \times 10^{-3}$. The concurrence error, defined as $\varepsilon_C \equiv 1 - C$, takes the value 1.9×10^{-1}. Optimization using Krotov's method and J_T^{geo} as the optimization functional converges within 5,516 iterations of the algorithm. Convergence is assumed when the relative change of the functional $\Delta J_T / J_T$ falls below 10^{-4}, such that no significant further improvement is to be expected. The gate error is reduced to $\varepsilon_{avg} = 1.4 \times 10^{-4}$. It is dominated by the remaining loss of population from the logical subspace, $\varepsilon_{pop} \approx \varepsilon_{avg}$, as the concurrence error is only $\varepsilon_C = 1.8 \times 10^{-6}$, see Table 2. The resulting optimized pulse is shown in Figure 2, with the guess pulse indicated by the dashed line. For the center 100 ns, there are only small deviations from the guess pulse (both in shape and phase). Significant deviations occur only at the very beginning and end, most notably the variation in the complex phase (center panel) between 20 and 40 ns. The spectral width of the pulse (bottom panel) remains well within a bandwidth of ±50 MHz. The dynamics of the $|00\rangle$ state under the optimized pulse are shown in panels (e)-(h) of Figure 1. The difference to the dynamics under the guess pulse, panels (a)-(d), is striking; the excitations no longer smoothly follow the pulse shape, but show strong oscillations on top of the expected behavior. The features at the beginning of the optimized pulse provide a kick to the system, inducing oscillations in the populations, with a counter-kick near the end of the pulse. These kicks are very visible in the population of the $|00\rangle$ state in panel (h), compared to the smooth dynamics for guess pulse in panel (d).

It is worth noting that the gate obtained with the optimized pulse is not the closest geometric phase gate to the gate implemented by the guess pulse, as in Eq. (9). This illustrates the benefit of using J_T^{geo} over J_T^{sm}. The latter optimizes towards a specific, pre-determined gate, according to Eq. (9), while J_T^{geo} can dynamically adjust which specific geometric phase gate is easiest to reach, allowing it to fulfill the objective much more easily. Table 2 shows that optimization with J_T^{geo} requires significantly less propagations (which are directly proportional to CPU time) than optimization with J_T^{sm}, for both direct and pre-optimized strategies.

While the optimization yields a gate error well below the quantum error correction threshold, it deviates significantly from the simple geometric phase gate scheme, resulting in complex dynamics. The numerical effort required to obtain a high fidelity solution is considerable, with several thousand iterations (each iteration requiring two full propagations of four logical basis states). We have discussed here only the optimization for a fixed gate duration of T=200 ns. Generally, significantly faster quantum gates could potentially be implemented using other mechanisms, e.g. [24]. The geometric phase gate, however, relies on adiabatic shifts of the energy levels such that loss of population from the logical subspace inhibits realization of high fidelities when pushing the gate duration significantly below 200 ns.

The complexity of the optimized pulse is typical for Krotov's method or other gradient-based optimization methods. For the present example, this clashes with the gate mechanism of the geometric phase gate that intends to use simple and smooth pulse shapes.

HYBRID OPTIMIZATION SCHEME

Simplex Optimization

The gate error of the guess pulse is dominated by the insufficient amount of entanglement that is generated. A natural approach is to maintain the analytical pulse shape of Eq. (4) for the time being, and to vary the free parameters E_0 and T in order to maximize the figure of merit. Such an optimization of a pulse determined by only a handful of parameters (in this case two) is easily performed using a gradient-free method such as Nelder-Mead simplex. This has the additional benefit that there are no restrictions on the choice of optimization functional. Specifically, there is no need to formulate it in such a way that derivatives can be calculated analytically. Thus, we can include the objective of minimizing the gate duration in the functional and modify Eq. (11) to read

$$J_T^{\text{splx}} = J_{\text{diag}} + J_\gamma + \frac{T}{T_0}, \qquad T_0 = 200 \text{ ns}. \tag{16}$$

Note that with the addition of penalizing the gate duration, the functional is no longer a distance measure that approaches zero as the target is reached; the simplex optimization will find a local minimum of Eq. (16) at a value of $J_T^{\text{splx}} \lesssim 1$.

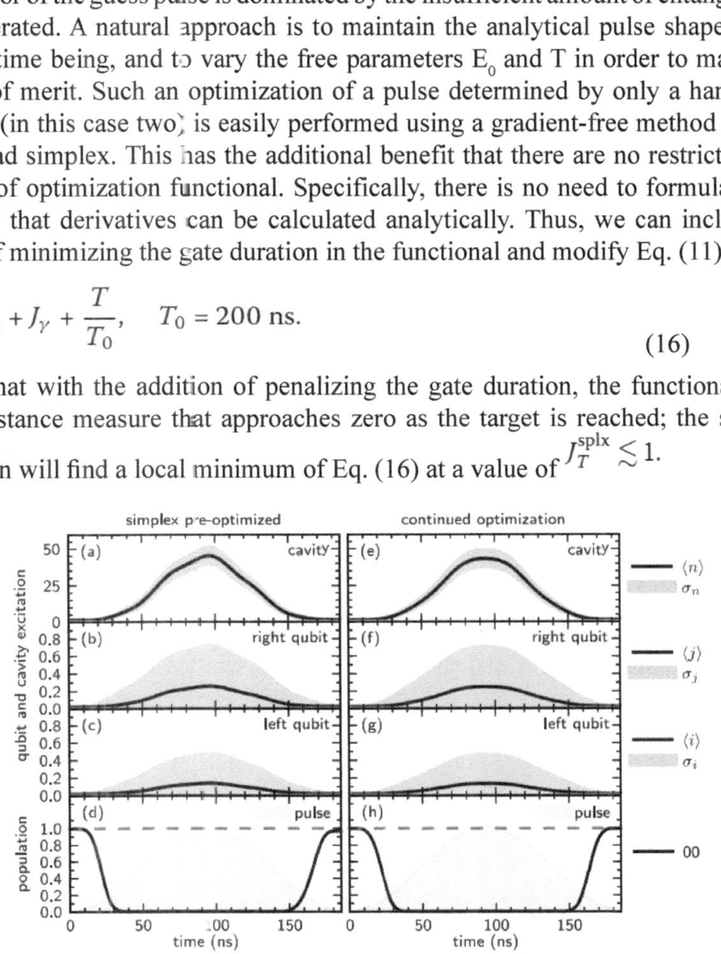

Figure 3: Population dynamics under (pre-optimized) guess and optimized pulse. The figure follows the conventions of Figure 1. Panels (a)-(d) show the dynamics resulting from a simplex optimization, see text for details. The resulting pulse is the starting point for a continued optimization using Krotov's method with the J_T^{sm} functional. The optimized dynamics are shown in panels (e)-(h). The pulse amplitude indicated in the background of panels (d), (h) is normalized to the peak amplitude of $E_0 \approx 400$ MHz. The simplex-optimized pulse implements a geometric phase gate with a gate error of $\varepsilon_{\text{avg}} = 1.4 \times 10^{-2}$, with $\varepsilon_C = 2.0 \times 10^{-2}$ and $\varepsilon_{\text{pop}} = 1.4 \times 10^{-2}$. The continued optimization decreases the gate error to $\varepsilon_{\text{avg}} = 3.4 \times 10^{-5}$, with $\varepsilon_C = 5.1 \times 10^{-5}$ and $\varepsilon_{\text{pop}} = 1.1 \times 10^{-5}$, see Table 2.

Using only 116 propagations, the algorithm converges to a solution that reduces the gate duration from 200 to 185 ns, while bringing the entanglement error down to $\varepsilon_C=2.0\times10^{-5}$, see Table 2. The resulting dynamics are shown in panels (a)-(d) of Figure 3. They are similar to those of the original guess pulse, cf. panel (a)-(d) of Figure 1. The shorter pulse duration and larger pulse intensity (from 300 MHz in the original guess to ≈400 MHz) results in a significantly larger cavity excitation. It also leads to more non-adiabatic defects (wobbles in the cavity excitation). Consequently, the loss of population from the logical subspace is increased by about a factor of two to $\varepsilon_{pop}=1.4\times10^{-2}$, and limits the total gate error to $\varepsilon_{avg}=1.4\times10^{-2}$. Thus, while the simplex search yields a dramatic improvement over the original guess pulse, it does not reach a sufficiently high fidelity to approach the quantum error correction limit. To remedy this, we turn to a hybrid approach, combining simplex and gradient-based optimization.

Continued Optimization with Krotov's Method

For the present example, we use the final pulse of the previous section as the starting point of an optimization with Krotov's method. Since now the guess pulse already has a relatively high fidelity, both the J_T^{sm} and J_T^{geo} functionals may be used interchangeably, as we are only searching in a very small vicinity of the starting point. Both methods converge rapidly to $\Delta J_T/J_T<10^{-4}$ in under 200 iterations. The dynamics resulting from the propagation of the $|00\rangle$ state under the pulse obtained from optimization with the JsmTJTsm functional is shown in panels (e)-(h) of Figure 3. The comparison to the dynamics of the pre-optimized guess pulse in panels (a)-(d) is striking: the excitations now follow the pulse shape smoothly. The non-adiabatic defects, i.e., the wobbles especially in the cavity excitation in panel (a), have been corrected. Consequently, the loss of population from the logical subspace is now reduced to a value of $\varepsilon_{pop}=1.1\times10^{-5}$. Together with only a slight increase in the concurrence error to $\varepsilon_C=5.1\times10^{-5}$, the overall gate error of the optimized pulse is $\varepsilon_{avg}=3.4\times10^{-5}$. This is an improvement of half an order of magnitude compared to the direct optimization in Section 3.2. Moreover, the result has been obtained at a small fraction of the numerical cost. The pre-optimized guess and post-optimized pulse shape, indicated as the blue shaded area in panels (d) and (h), appear visually indistinguishable.

The correction to the pre-optimized guess pulse is shown in Figure 4. Indeed, the corrections are on the order of 1 MHz, much smaller than the peak amplitude of $E_0\approx400$ MHz. The corrections follow a regular pattern, both in the shape (top panel) and the complex phase (center panel), as indicated by the existence of sharp peaks in the spectrum (bottom panel). It appears that the non-adiabatic defects in the pre-optimized guess pulse can be corrected by a series of small kicks in amplitude and phase at regular intervals. This results in an optimized pulse that is conceptually simpler and yields a higher fidelity than the direct application of Krotov's method. The comparison illustrates the power of a hybrid approach to steer the physical characteristics of the optimized control. In our example, the parametrization used in the first optimization stage enforces the desired pulse shape. Without this restriction, pulses of undesirable complexity are obtained. On the other hand, the restriction must ultimately be relaxed to allow for the representation of the higher-frequency features that correct non-adiabatic defects. Thus, the role of the two-stage optimization is also to influence the physical features of the control, in addition to its numerical benefits.

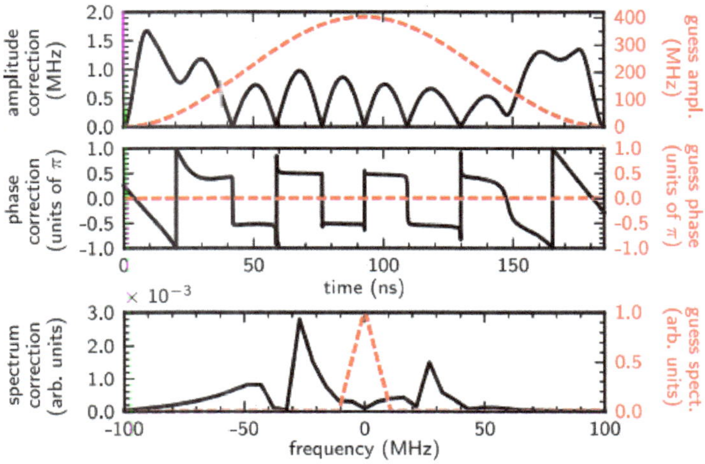

Figure 4: Pulse corrections obtained with Krotov's method. The figure summarizes the differences between the optimized pulse, cf. panel (h) in Figure 3, and the (pre-optimized) guess pulse, cf. panel (d) in Figure 3, also indicated by the dashed red/gray line in each panel. The panels from top to bottom show the corrections to absolute value, complex phase, and spectrum of the pulse, respectively (solid black lines). The amplitude, phase, and spectrum of the guess pulse (dashed red/gray lines) are shown using the alternative axis scaling in red/gray.

The striking numerical efficiency of the hybrid optimization scheme can be seen by comparing its convergence to that of the direct optimization. This is shown in Figure 5. The direct optimization shows an extended plateau for at least the first 100 iterations, before slowly converging. This behavior is typical for ill-chosen guess pulses [43]. For a direct optimization with the J_T^{sm} functional, the plateau extends for several thousand iterations, and does not yield convergence within 10,000 iterations. In contrast, the optimization starting from a pre-optimized pulse has no plateau (note the log-scale of the x-axis); both J_T^{sm} and J_T^{geo} converge at roughly the same rate. The improvement by the simplex (pre-) optimization compared to the original guess pulse can be seen from the difference in the y-intercept between the 'direct' and 'pre-optimized' curves in Figure 5; in addition, there is also a reduction in the gate duration from 200 to 185 ns at no additional numerical cost.

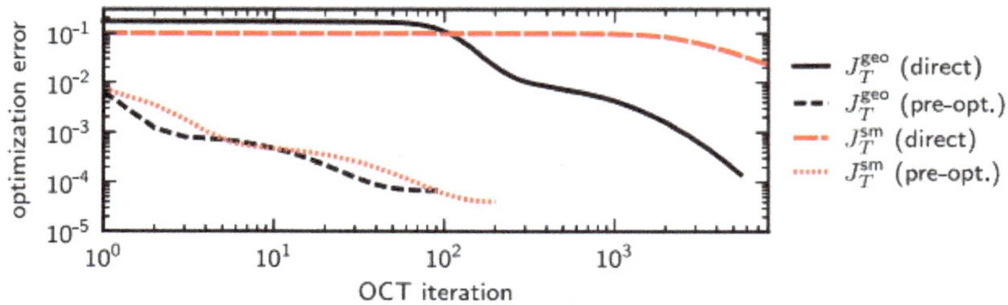

Figure 5: Convergence of optimization towards a geometric phase gate. Value of the final-time optimization functional J_T^{geo}, respectively J_T^{sm}, over the number of iterations using Krotov's method. Each iteration requires two full propagations; the number of propagations are proportional to the required CPU time. The direct optimization starts from an arbitrary guess pulse, see Section 3.2 for details. In the pre-optimized case, the guess pulse was the result of a simplex optimization, see Section 4.2 for details.

The gate duration for the geometric phase gate mechanism is limited by the requirement of adiabaticity. We have also performed the hybrid optimization scheme for a gate duration of ≈100 ns. In this case, the simplex search yields significant non-adiabatic defects, with a loss of population of 1.2×10^{-1}. The concurrence error is 2.3×10^{-3} and the total gate error, $\varepsilon_{avg}=1.2\times10^{-1}$, is dominated by the population loss. Post-optimization using Krotov's method significantly reduces the total gate error to $\varepsilon_{avg}=1.4\times10^{-2}$. The post-optimization result is obtained at a numerical cost very close to that for the T=200 ns gate, and yields pulse corrections very similar to those shown in Figure 4. This results in a correction of the non-adiabatic defects and thereby lowers the population loss to only 6.5×10^{-3}. The total gate error is now dominated instead by the increased concurrence error of 1.7×10^{-2}, which is insufficient for a high quality phase gate. The observation that an overall improvement in gate performance at T=100 ns from hybrid optimization is only possible by increasing adiabaticity at the cost of reduced entanglement, indicates that a quantum speed limit has been reached for the specific gate mechanism. These results show that a hybrid optimization scheme may be used successfully even when operating close to the quantum speed limit.

While it is not the aim of this work to characterize the quantum speed limit for the coupled transmon system, we note that this could be quantified by systematically scanning over the gate duration [26]. Under the constraints of the specific gate mechanism employed here, a numerical approach appears necessary to extract the speed limit. However, for the Strauch gate of Ref. [44], an analytic estimate may be made from summing the minimal precession periods required for each component of the gate, i.e., two iSWAP and one controlled-Z gates. With the qubit-cavity interaction strength of g=70 MHz employed here, this would suggest that gate durations as short as a few tens of nanoseconds might be attainable for this system.

We may also compare the total number of propagations necessary to obtain the optimized pulse, included in Table 2 together with measures of success: concurrence error, loss of population from the logical subspace, and gate error. Each iteration using

Krotov's method requires two propagations. For the hybrid optimization schemes ('pre-opt s.m.', 'pre-opt hol.'), the number of propagations includes simplex as well as the additional propagations due to Krotov's method. The gate error is measured with respect to the exact target gate when using the square-modulus functional, and with respect to the closest geometric phase gate to the optimized dynamics in all other cases. In terms of numerical efficiency, the hybrid schemes outperform the direct optimization by nearly two orders of magnitude while resulting in a significantly better gate fidelity.

CONCLUSIONS

For the example of a geometric phase gate on a system of two transmon qubits with a shared transmission line resonator, we have considered the application of gradient-based optimization methods, specifically Krotov's method, and a gradient-free optimization method, Nelder-Mead simplex. The objectives of the geometric phase gate can be formulated in a specialized optimization functional J_T^{geo} that reaches its optimal value for any diagonal perfect entangler. We have shown that for a direct optimization, this functional vastly outperforms the 'standard' square-modulus functional for gate optimization. Convergence is aided by the use of a functional that formulates the objective as general as possible. This is in agreement with recent results for optimization using a functional targeting arbitrary perfect entanglers [41, 45].

The direct optimization using Krotov's method can in principle find controls that implement the desired gate with high fidelity. However, the resulting dynamics are complex and the numerical effort is dominated by an extended plateau in the initial phase of the optimization; for short gate durations, optimization becomes increasingly harder and the fidelity is limited by loss from the logical subspace. The numerical effort to reach high fidelities quickly becomes unfeasible in this case. In contrast, parametrizing the pulse by its duration and peak amplitude only, and applying a simplex optimization on those parameters, we are able to find a simple analytic pulse that implements the desired gate with moderate fidelity, still one order of magnitude above the quantum error correction limit. Thus, for the example presented here, neither the application of a gradient-based algorithm nor the simplex optimization alone yield satisfactory results; only the combination of both methods into a hybrid optimization scheme is able to obtain controls with a clear mechanism that implement a geometric phase gate to high fidelity, with a minimum of numerical effort.

These results prompt the recommendation to generally adapt hybrid optimization schemes, i.e., obtain guess pulses for gradient-based optimization from a gradient-free pre-optimization, when there is insufficient knowledge to design good guess pulses by hand. There is great flexibility in the choice of parametrization. Here, we have taken the two free parameters, peak amplitude and gate duration, in a simple fixed analytical formula for the pulse shape. Generally, one might use slightly more sophisticated parametrizations, following e.g. the CRAB approach [18]. The small number of free parameters and relatively high quality of the original guess pulse with a fidelity of already >90% results in particularly fast convergence of the simplex method. For a more sophisticated example, at least several hundred propagations would probably be required in the simplex stage. However, even in that case, this numerical effort is by far outweighed by the large number of iterations required to leave the initial plateau in the optimization landscape for a bad guess pulse in a direct gradient-based optimization. Moreover, any figure of merit is suitable for optimization with the simplex method, as there is no need to derive the gradient for the optimization functional. For example, the gate duration can be included in the figure of merit, something that generally is not straightforward in gradient-based methods [46]. The hybrid scheme is aimed at providing optimal solutions in an open-loop context, and still leaves open the possibility of further combining open-loop and closed-loop optimization methods when targeting a specific experimental setup [47].

In principle, the approach could thus be extended to multiple stages, where in each stage, a different parameterization and a suitable optimization algorithm is used. There is no requirement for a specific method such as Nelder-Mead simplex or Krotov's method to be employed. For example, a two-stage optimization using a genetic algorithm in the first stage, and a gradient algorithm in the second stage has been used for the optimization of quantum gates in strongly coupled two-level systems [48, 49]. There, starting with a simple parametrization, and then moving to a less restricted search space in the second optimization stage was also found to benefit the overall optimization performance, in agreement with the results shown here.

Adding an additional stage might also be beneficial when the optimization landscape is non-trivial and contains traps or saddle points. This may happen e.g. when the optimization is performed with limited resources [50]. Repeating the simplex search from a different starting point - either by systematic variation or by random search - may then find solutions that do not get stuck in traps of the optimization landscape. Of course, the truncation of the search space due to the low-dimensional parametrization may itself introduce additional traps in the landscape. However, these traps would disappear when returning to the full search space in the final optimization stage.

For the specific example of the geometric phase gate considered here, we have found the post-optimization to introduce small corrections to non-adiabatic defects. Due to the small relative strength of the correction, the experimental realization of the geometric phase gate would require extraordinary precision. Moreover, the large excitation of the cavity would limit the fidelity when dissipative effects, specifically spontaneous decay of the cavity are taken into account. However, we stress that the method of employing hybrid optimization schemes presented here is entirely general and aimed at reducing the numerical effort in obtaining high fidelity solutions to arbitrary quantum control problems, not limited to quantum information processing. Moreover, robustness with respect to both fluctuations in the control parameters and dissipation can be achieved using complimentary advanced control techniques, such as a description in Liouville space and ensemble optimization [31, 43]. These approaches are numerically even more demanding, such that the reduction of the optimization cost achieved by the hybrid scheme discussed here may become imperative.

ACKNOWLEDGEMENTS

We would like to thank F. Motzoi for fruitful discussions. Financial support from DAAD under grant PPP USA 54367416 is gratefully acknowledged.

REFERENCES

1. Brumer P, Shapiro M. Principles and applications of the quantum control of molecular processes. New York: Wiley; 2003.
2. Rice SA, Zhao M. Optical control of molecular dynamics. New York: Wiley; 2000.
3. Tannor DJ, Rice SA. Control of selectivity of chemical reaction via control of wave packet evolution. J Chem Phys. 1985;83:5013.
4. Tannor DJ, Kosloff R, Rice SA. Coherent pulse sequence induced control of selectivity of reactions: exact quantum mechanical calculations. J Chem Phys. 1986;85:5805.
5. Levin L, Skomorowski W, Rybak L, Kosloff R, Koch CP, Amitay Z. Coherent control of bond making. Phys Rev Lett. 2015;114:233003.
6. Brif C, Chakrabarti R, Rabitz H. Control of quantum phenomena: past, present and future. New J Phys. 2010;12:075008.
7. Konnov AI, Krotov VF. On global methods of successive improvement of controlled processes. Autom Remote Control. 1999;60:1427.
8. Sklarz SE, Tannor DJ. Loading a Bose-Einstein condensate onto an optical lattice: an application of optimal control theory to the nonlinear Schrödinger equation. Phys Rev A. 2002;66(5):053619.
9. Reich DM, Ndong M, Koch CP. Monotonically convergent optimization in quantum control using Krotov's method. J Chem Phys. 2012;136:104103.
10. Khaneja N, Reiss T, Kehlet C, Schulte-Herbrüggen T, Glaser SJ. Optimal control of coupled spin dynamics: design of NMR pulse sequences by gradient ascent algorithms. J Magn Res. 2005;172:296.
11. Byrd RH, Lu P, Nocedal J, Zhu C. A limited memory algorithm for bound constrained optimization. SIAM J Sci Comput. 1995;16(5):1190.
12. de Fouquières P, Schirmer SG, Glaser SJ, Kuprov I. Second order gradient ascent pulse engineering. J Magn Res. 2011;212:412.
13. Eitan R, Mundt M, Tannor DJ. Optimal control with accelerated convergence: combining the Krotov and quasi-Newton methods. Phys Rev A. 2011;83(5):053426.
14. Machnes S, Sander U, Glaser S, de Fouquières P, Gruslys A, Schirmer S, Schulte-Herbrüggen T. Comparing, optimizing, and benchmarking quantum-control algorithms in a unifying programming framework. Phys Rev A. 2011;84:022305.
15. Judson RS, Rabitz H. Teaching lasers to control molecules. Phys Rev Lett. 1992;68:1500.
16. Goldberg DE. Genetic algorithms in search, optimization and machine learning. Boston: Addison-Wesley; 1989.
17. Doria P, Calarco T, Montangero S. Optimal control technique for many-body quantum dynamics. Phys Rev Lett. 2011;106:190501.
18. Caneva T, Calarco T, Montangero S. Chopped random-basis quantum optimization. Phys Rev A. 2011;84:022326.

19. Palao JP, Kosloff R. Quantum computing by an optimal control algorithm for unitary transformations. Phys Rev Lett. 2002;89:188301.
20. Tesch CM, de Vivie-Riedle R. Quantum computation with vibrationally excited molecules. Phys Rev Lett. 2002;89:157901.
21. Treutlein P, Hänsch TW, Reichel J, Negretti A, Cirone MA, Calarco T. Microwave potentials and optimal control for robust quantum gates on an atom chip. Phys Rev A. 2006;74:022312.
22. Schirmer S. Implementation of quantum gates via optimal control. J Mod Opt. 2009;56:831.
23. Müller MM, Reich DM, Murphy M, Yuan H, Vala J, Whaley KB, Calarco T, Koch CP. Optimizing entangling quantum gates for physical systems. Phys Rev A. 2011;84:042315.
24. Egger DJ, Wilhelm FK. Optimized controlled-z gates for two superconducting qubits coupled through a resonator. Supercond Sci Technol. 2014;27:014001.
25. Caneva T, Murphy M, Calarco T, Fazio R, Montangero S, Giovannetti V, Santoro GE. Optimal control at the quantum speed limit. Phys Rev Lett. 2009;103:240501.
26. Goerz MH, Calarco T, Koch CP. The quantum speed limit of optimal controlled phasegates for trapped neutral atoms. J Phys B. 2011;44:154011.
27. Bhattacharyya K. Quantum decay and the Mandelstam-Tamm-energy inequality. J Phys A. 1983;16:2993.
28. Margolus N, Levitin LB. The maximum speed of dynamical evolution. Physica D. 1998;120:188.
29. Kallush S, Kosloff R. Quantum governor: automatic quantum control and reduction of the influence of noise without measuring. Phys Rev A. 2006;73:032324.
30. Schulte-Herbrüggen T, Spörl A, Khaneja N, Glaser SJ. Optimal control for generating quantum gates in open dissipative systems. J Phys B. 2011;44:154013.
31. Goerz MH, Halperin EJ, Aytac JM, Koch CP, Whaley KB. Robustness of high-fidelity Rydberg gates with single-site addressability. Phys Rev A. 2014;90:032329.
32. Cross AW, Gambetta JM. Optimized pulse shapes for a resonator-induced phase gate. Phys Rev A. 2015;91:032325.
33. Rigetti C, Gambetta JM, Poletto S, Plourde BLT, Chow JM, Córcoles AD, Smolin JA, Merkel ST, Rozen JR, Keefe GA, Rothwell MB, Ketchen MB, Steffen M. Superconducting qubit in a waveguide cavity with a coherence time approaching 0.1 ms. Phys Rev B. 2012;86:100506.
34. Chow JM, Gambetta JM, Cross AW, Merkel ST, Rigetti C, Steffen M. Microwave-activated conditional-phase gate for superconducting qubits. New J Phys. 2013;15:115012.
35. Koch J, Yu TM, Gambetta J, Houck AA, Schuster DI, Majer J, Blais A, Devoret MH, Girvin SM, Schoelkopf RJ. Charge-insensitive qubit design derived from the Cooper pair box. Phys Rev A. 2007;76:042319.
36. Majer J, Chow JM, Gambetta JM, Koch J, Johnson BR, Schreier JA, Frunzio L,

Schuster DI, Houck AA, Wallraff A, Blais A, Devoret MH, Girvin SM, Schoelkopf RJ. Coupling superconducting qubits via a cavity bus. Nature. 2007;449:443.

37. Makhlin Y. Nonlocal properties of two-qubit gates and mixed states, and the optimization of quantum computations. Quantum Inf Process. 2002;1(4):243.
38. Zhang J, Vala J, Sastry S, Whaley KB. Geometric theory of nonlocal two-qubit operations. Phys Rev A. 2003;67(4):042313.
39. Kraus B, Cirac JI. Optimal creation of entanglement using a two-qubit gate. Phys Rev A. 2001;63(6):062309.
40. Palao JP, Kosloff R. Optimal control theory for unitary transformations. Phys Rev A. 2003;68:062308.
41. Watts P, Vala J, Müller MM, Calarco T, Whaley KB, Reich DM, Goerz MH, Koch CP. Optimizing for an arbitrary perfect entangler: I. Functionals. Phys Rev A. 2015;91:062306.
42. Pedersen LH, Møller NM, Mølmer K. Fidelity of quantum operations. Phys Lett A. 2007;367:47.
43. Goerz MH, Reich DM, Koch CP. Optimal control theory for a unitary operation under dissipative evolution. New J Phys. 2014;16:055012.
44. Strauch FW, Johnson PR, Dragt AJ, Lobb CJ, Anderson JR, Wellstood FC. Quantum logic gates for coupled superconducting phase qubits. Phys Rev Lett. 2003;91:167005.
45. Goerz MH, Gualdi G, Reich DM, Koch CP, Motzoi F, Whaley KB, Vala J, Müller MM, Montangero S, Calarco T. Optimizing for an arbitrary perfect entangler. II. Application. Phys Rev A. 2015;91:062307.
46. Ndong M, Koch CP, Sugny D. Time optimization and state-dependent constraints in the quantum optimal control of molecular orientation. J Mod Opt. 2014;61(10):857.
47. Egger DJ, Wilhelm FK. Adaptive hybrid optimal quantum control for imprecisely characterized systems. Phys Rev Lett. 2014;112:240503.
48. Grace M, Brif C, Rabitz H, Walmsley IA, Kosut RL, Lidar DA. Optimal control of quantum gates and suppression of decoherence in a system of interacting two-level particles. J Phys B. 2007;40(9):103.
49. Grace MD, Brif C, Rabitz H, Lidar DA, Walmsley IA, Kosut RL. Fidelity of optimally controlled quantum gates with randomly coupled multiparticle environments. J Mod Opt. 2007;54:2339.
50. Moore KW, Rabitz H. Exploring constrained quantum control landscapes. J Chem Phys. 2012;137:134113.

7

CHAPTER

CONTROLLING QUANTUM INFORMATION PROCESSING IN HYBRID SYSTEMS ON CHIPS

Guy Bensky, Robert Amsüss, Johannes Majer, David Petrosyan, Jörg Schmiedmayer, Gershon Kurizki

Keywords: Quantum information, Hybrid quantum systems, Atom chip

INTRODUCTION

Each quantum system considered for quantum information processing has its strengths and weaknesses. Some systems are better suited for processing, others for storage, and there are those best suited for writing-in/reading-out or transferring quantum information. Realistically, to build robust quantum information systems, one will need to combine different systems to pool their strengths in order to overcome their weaknesses. This has prompted the suggestion of hybrid, composite quantum systems [1– 9] that would combine the advantages of their different subsystems. An important requirement of such hybrids is to faithfully transfer qubits between the subsystems. Such hybrid quantum devices may be envisioned to integrate very different subsystems, or "blocks" on a single chip. As a general example, but without loss of generality, we will discuss here systems that are constructed around superconducting qubits. On a superconducting chip we envision three main components: (1) A quantum-processor block, (2) a quantum memory block and (3) a quantum "bus" that acts as an interface:

- The envisioned quantum-processor is based on superconducting qubits [10–12] that can rapidly and efficiently perform quantum gate operations. Such qubits are vulnerable to decoherence due to their strong coupling to the environment and the noise of the external controls. For the purpose of this paper we can view the superconducting qubit as an artificial atom with a strong anharmonicity and a very strong coupling to the electromagnetic field [13]. The anharmonicity of the potential defining the qubit states ensures that one creates in a single

Citation: Guy Bensky, Robert Amsüss, Johannes Majer, David Petrosyan, Jörg Schmiedmayer, Gershon Kurizki Controlling Quantum Information Processing in Hybrid Systems On Chips DOI: 10.1007/s11128-011-0302-6.

qubit only a single excitation like in a single atom. The strong coupling to the electromagnetic field allows efficient coupling to the quantum "bus" discussed below

- The quantum-memory block is weakly coupled to the environment and the external controls, and is robust against decoherence and therefore suitable for implementing a long lived storage of quantum information. This may be a single isolated system like a trapped ion [14], or a collective qubit encoded in, for example, a spin ensemble [15]. For the purpose of this paper we concentrate on collective states encoded in spin ensembles as the memory qubit. These can be very robust and can have coherence properties similar to a single spin [16,17]. The two systems we consider here are (1) an atomic physics implementation where the qubit is encoded in the hyper- fine ground states of (trapped) atoms or molecules [9,18–20]; (2) a solid state implementation where the qubit is encoded in the electron or nuclear spin states of nitrogen-vacancy (NV) color centers in diamond [21–24]
- The interface or quantum "bus", mediates among the different blocks or qubits and allows therefore coupling to the quantum memory. It should be designed to provide strong coupling and high-fidelity transfer of qubits to the quantum memory. Here we consider as the bus a transmission line microwave cavity [25]. Such cavities have very small mode volume (typically $10^{-4}\lambda^3$), which results in a large field per photon and consequently in a strong electromagentic interaction allowing to wire up quantum systems[4]. In addition, the interaction with the different qubits can quickly switched on and off by, e.g., tuning them in and out of resonance.

Even though the mode volume of chip-based microwave cavities is very small, their coupling to the long lived spin states is very weak (typically 100 Hz). This limitation can be overcome by using polar molecules [1–3] or Rydberg atoms [5,6,26,27] which couple strongly to the cavity field to assist the transfer of quantum information (QI). Alternatively, we may employ ensembles with $N>>1$ spins (atoms),whose $\sqrt{N}$ scaling enhances the coupling to collective states of the ensemble [8,9,28]. Recent experimental demonstrations of trapping atoms on superconducting atom chips [29–34], decelerating and trapping of polar molecules on a chip [35], and strong coupling between spin ensembles and microwave photons in superconducting cavities [36–40] are important steps towards implementing functional, hybrid quantum devices of this kind. Implementations using trapped atoms or molecules on a chip [41] have the advantage that the ensembles are homogeneous, but small and require the large overhead of cooling and trapping to be implemented in a cryogenic environment at mK temperatures. On the other hand, implementations using solid-state based spin ensembles are simpler to prepare and handle at mK temperatures, and will allow large ensembles with $N \sim 10^{12}$ but are prone to inhomogeneous spectral broadening. This broadening can both reduce the fidelity of the QI transfer and limit its coherent storage time [42–44]. Although spin-echo methods [45] can in principle prolong the storage time, they are restricted by the requirement to be faster than the inverse spectral width of the spin ensemble [46,47]. In addition, these methods cannot compensate for the fidelity loss during the QI transfer/retrieval. In this work we discuss possible avenues for realizing such integrated hybrid quantum devices. QI transfer between the cavity and the ensemble-based quantum memory is typically the main experimental bottleneck. The coupling between the processing (superconducting) qubit and the cavity is, by contrast, very strong and can be quickly switched on and off at will [25,48–51]. It is interesting to note that at present, high-Q cavities are the best media for quantum-state storage (see, for example, Ref. [52]). Hence, the quality benchmark of integrated hybrid devices should be such that as to allow longer or higher-fidelity storage than such cavities. We discuss the prospects for achieving this benchmark. The paper is organized as follows: We first give an overview of the superconducting qubit–cavity coupling, Sect. 2, and then concentrate on the coupling and transfer to the quantum memory, Sects. 3–6. We discuss how the fidelity of this transfer, and of the entire hybrid QI device, can be significantly improved by combining several methods of dynamical control: (i) In Sect. 3 we examine a strategy for maximizing the average fidelity of quantum state-transfer in such hybrids, from a noisy subspace (fragile under decoherence), used for QI writing / reading, to a quiet (robust) subspace, used for memory / storage, by choosing an appropriate dynamical control field. (ii) In Sect. 4 we analyze the effect 'of optimal spectral filtering of the inhomogeneous atomic or spin ensemble on the total fidelity of transfer, memory and retrieval. In Sects. 5 and 6 we then discuss these strategies in the context of three different implementations, and point out in Sect. 7 ways to achieve the quality benchmark discussed above.

Superconducting Qubit–Cavity Coupling

To have in mind a concrete scenario, we consider a solid state qubit that is represented by a SC Cooper pair box coupled to a SC electrode via two tunnel junctions at the rate E_J (charge qubit) in the SQUID configuration [12]. At the charge degeneracy point, the energy separation $\hbar\omega_{10} = 2E_J\cos(\pi\Phi/\Phi_0)$ between the qubit states $|0\rangle$ and $|1\rangle$ can be dynamically controlled via an external magnetic field B⊥ that induces flux $\Phi = AB_\perp$ through the SQUID area A (A ($\Phi_0 = hc/2e$ is the flux quantum). Typical values for $E_J/\hbar$ are the microwave range (10–20 GHz). The dipole moment $\wp_{01}$ for the transition $|0\rangle \leftrightarrow |1\rangle$ is typically very large, $\wp_{01} \simeq 10^4 a_0 e$. Thus, many fast quantum logic gates can be implemented within the qubit dephasing time $1/\gamma_q \gtrsim 1\,\mu s$ [53]. Charge qubits can be embedded in near-resonant SC transmission line resonators, such as a coplanar waveguide (CPW) cavity of Fig. 1a [4,25,48–50], having a high quality factor Q up to 106. The tight confinement of the cavity field in a small volume (see below) yields very large field per photon ε_c and strong coupling (vacuum Rabi frequency) $\eta_{qc} = (\wp_{01}/\hbar)\varepsilon_c u(\mathbf{r}) \sim 2\pi \times 50$ MHz between the cavity field and the qubit located at position r near the field antinode where the cavity mode function u(r) 1. In the frame rotating with the cavity field frequency ω_c, the Hamiltonian has the form

$$H_{qc} = \hbar\delta_{qc}\hat{\sigma}^+\hat{\sigma}^- - \hbar\eta_{qc}(\hat{\sigma}^+\hat{c} + \hat{c}^\dagger\hat{\sigma}^-), \tag{1}$$

where $\delta_{qc} = \omega_{10} - \omega_c$ is the externally controlled (via B⊥) detuning, $\hat{\sigma}^-$ $(\hat{\sigma}^+)$ is the qubit lowering (rising) operator, and $\hat{c}$ $(\hat{c}^\dagger)$ is the cavity photon annihilation (creation) operator. One can incorporate many SC qubits in the same CPW cavity, each qubit located near the cavity field antinode (Fig. 1a). The cavity can then mediate long-range controlled interactions between pairs of resonant qubits [25,48–50,4], realizing, e.g., the two-qubit √swap gate, which together with the single qubit rotations form the universal set of logic gates in such a quantum computer. However, due to rapid dephasing and relaxation, neither SC qubits nor the cavity mode can carry out reliable long term storage of QI. In what follows, we show that this task can be accomplished by dynamical control of the QI transfer to the quantum memory ensemble positioned near or at the surface of an integrated chip that incorporates the CPW cavity and SC qubits.

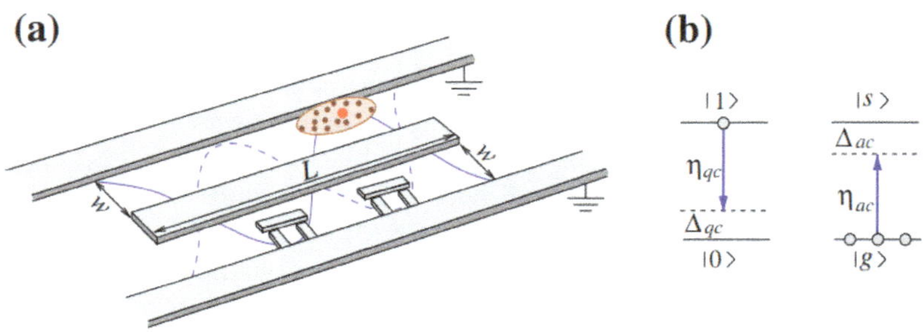

Figure 1: a CPW cavity with strip-line length L and electrode distance w. SC qubits are placed at the antinodes of the standing wave field and ensembles of ultracold atoms or NV centers are placed near the CPW surface. b SC qubit (left) and ensemble qubit (right) can couple to a common mode of the CPW cavity.

As we will show in Sect. 6, to achieve high fidelity gates using CPW interface it is crucial to suppress the cavity temperature corresponding to less than 0.01 thermal photons.

OPTIMIZED TRANSFER BY DYNAMICAL CONTROL

We use dynamical control to optimize a reliable transfer of a quantum state from the fragile (noisy) SC qubit to the robust (quiet) memory qubit via the cavity. We choose to focus on the case of two resonant qubits with temporally controlled coupling strength as an example, since it is simple enough to be solvable analytically yet still holds surprising results. However, the same method can be numerically applied to more complex systems. Our goal here is to devise the best solution when the transfer is limited by the homogeneous broadening (decoherence) of the SC qubit and the cavity, rather than by the inhomogeneous broadening of the memory (to be discussed in Sect. 4). The free Hamiltonian of a system of two qubits with controlled coupling is

$$\begin{aligned}\hat{H}(t) &= \hat{H}_S(t) + \hat{H}_I + \hat{H}_B \\ \hat{H}_S(t) &= \frac{\omega_0}{2}\left(\hat{\sigma}_z^{(1)} + \hat{\sigma}_z^{(2)}\right) + H_c(t), \\ \hat{H}_c(t) &= V(t)\hat{\sigma}_x^{(1)} \otimes \hat{\sigma}_x^{(2)}.\end{aligned} \tag{2a}$$

Here HB is the Hamiltonian for the environment (bath), HI is the coupling operator between the system and the environment and $\hat{H}_c(t)$ is the Hamiltonian (or effective Hamiltonian in some cases) for the controlled interaction between the qubits, V(t), describing the adjustable amplitude of the interaction (see Fig. 2 for examples of how such adjustable coupling amplitudes can be achieved in practice). Since our goal is to study a system with decoherence only in the noisy qubit 1, the system-bath interaction Hamiltonian H_I is taken to be

$$\hat{H}_I = \hat{S} \otimes \hat{B} = \hat{\sigma}_z^{(1)} \otimes \hat{B}, \qquad (2b)$$

where $\hat{B}$ is a bath operator. There is no restriction on the choice of either B or H_B. This model represents a single noise source of proper dephasing in the source qubit 1 due to the bath operator $\hat{B}$, whereas the target qubit 2 is robust against decoherence. This model can be generalized to any degree of asymmetry between the decoherence properties of the two qubits. The accumulated phase

$$\phi(t) = \int_0^t V(t')dt' \qquad (3)$$

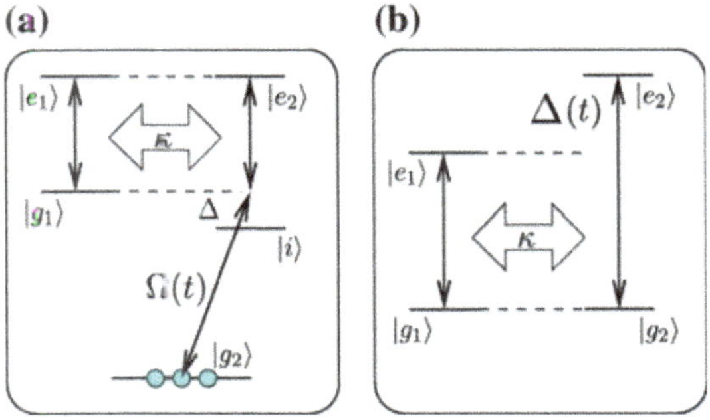

Figure 2: Schemes of dynamically controllable coupling between "noisy" source qubit 1 and quiet target qubit 2: a 2-photon transfer off-resonantly through $|i\rangle$ gives effective $\sigma_x^1 \otimes \sigma_x^2$ coupling with a controllable strength $V(t) = \frac{\kappa\Omega(t)}{\Delta}$

Where $\Omega(t)$ is the Rabi frequency of an external laser field. This allows for a smooth control, but limits the coupling to $V(t) \lesssim \kappa/10$. **b** Controlled detuning of the source qubit from the target qubit (t). This is limited to an "on/off" control, but allows a coupling of up to V(t) = κ, where κ is the natural (non-controlled) coupling between the qubits. is our control function. In the ideal case, without decoherence or leakage, the state transfer from qubit 1 to qubit 2 can be perfectly realized if at the final time, t f, the phase φ(t) satisfies $\phi(t_f) = \frac{\pi}{2}$, whence any initial state of qubit 1 is mapped onto that of qubit 2 (initially in the ground state)

$$(\alpha|g_1\rangle + \beta|e_1\rangle)\,|g_2\rangle \rightarrow |g_1\rangle\,(\alpha|g_2\rangle - i\beta|e_2\rangle)\,, \tag{4}$$

for any normalized α, β. Here the states $|g_1\rangle$ ($|g_2\rangle$) and $|e_1\rangle$ ($|e_2\rangle$) are respectively the ground and the excited states of the source qubit 1 and the target qubit 2. There are two conflicting noise (error) considerations for the transfer, each affecting a different subsystem: (i) In the presence of interaction between the source qubit 1 and the bath, the longer the information stays in qubit 1 the lower the fidelity of the transfer. (ii) On the other hand, if we make the transfer extremely fast, it may result in population from $|g_1\rangle\,|g_2\rangle$ leaking into $|e_1\rangle\,|e_2\rangle$, thus lowering the fidelity of transfer. Such leakage [54–60] signifies the violation of the rotating wave approximation (RWA). Namely, fast modulation V(t) may incur unwanted, off-resonant, transitions if the transfer rate is comparable to the energy difference (level distance) of the qubits, ω_0.

We here focus on bath-related errors (i), assuming that the RWA is valid, i.e., there is no leakage because of the RWA violation. This may be the case if the transfer time is much slower than the energy separation ω_0. The control Hamiltonian $H_c(t)$ then has the RWA form [61–65]:

$$\hat{H}_c(t) \equiv V(t)H_c; \quad H_c = |e_1 g_2\rangle\langle g_1 e_2| + |g_1 e_2\rangle\langle e_1 g_2| \tag{5}$$

The average fidelity of the transfer has an involved dependence on the modulation V(t) and the transfer time t_f . The problem at hand is to find the optimal transfer that minimizes the average error at time t_f, $1 - \overline{\mathcal{F}(t_f)}$, $\overline{\mathcal{F}}$ being the fidelity averaged over all possible initial states of qubit 1, with equal distribution over the Bloch sphere. Obviously, zero error is obtainable for infinitely fast (zero-time) transfer, if we allow infinitely strong control field. Since this is unphysical, we add a constraint on the total energy E of the field that causes the transfer process

$$\int_0^{t_f} dt\,(V(t))^2 = \int_0^{t_f} dt \left(\frac{d\phi(t)}{dt}\right)^2 = E. \tag{6}$$

Although other constraints may be considered, this constraint is both simple and physical: if we use the scheme described in Fig. 2a the energy constraint translates to a limit on the average laser intensity. In addition, using this constraint can prevent leakage to levels out of the operational qubit subspace [54–60]. The constraint defines the minimum possible time for the transfer $t_{\min} = \frac{\pi^2}{4E}$. Choosing an exact value of the energy might seem arbitrary since one usually has some freedom as to the amount of energy the system can handle. However, the exact value we choose would only matter if a small change in the energy could result in a large change in the fidelity. Yet, we have shown numerically and analytically that the error and optimal modulation change smoothly with the energy constraint and are nearly linear in the characteristic transfer time E^{-1}. Hence, the chosen energy constraint is simply a typical value allowed for the process. Consider in what follows a typical non-Markovian Lorentzian bath spectrum, i.e., the correlation function $\Phi(t)$ of the bath operator B is exponentially decaying $\Phi(t) = \frac{\gamma}{t_c} e^{-|t|/t_c}$, t_c being the correlation (memory) time. One might expect that for such a simple bath the best strategy is the fastest possible transfer under the energy constraint, i.e. when the modulation is given by $V(0 \leq t \leq t_{\min}) = 2E/\pi$. Surprisingly, a slower transfer ($t_f > t_{min}$) under an appropriate modulation φ(t) (detailed below) can improve the average fidelity even for a purely Markovian bath, with negligible correlation (memory) time $t_c/t_{min} \rightarrow 0$, and more so for baths with memory times longer than the transfer time, $t_c \gtrsim t_{\min}$.

a. Markovian limit— When the bath is memoryless, i.e., Markovian, this improvement is limited, as shown in Fig. 4, to about 12%. By comparing the "best" solution to the "fastest" one (Fig. 3), one can see the the "best" solution starts off faster and then slows down, being overtaken by the "fastest" solution only at $t \approx 0.9t_{min}$. This illustrates what we believe to be the source of the Markovian-noise error reduction: the "best" solution starts off faster, so as to transfer more of the information while it is still nearly untainted by the bath. Obviously, towards the end it must slow down so as to comply with the energy constraint, thus resulting in total transfer time t f that is longer than the fastest time t_{min} for the given energy. This system is analytically solvable in the Markovian limit, under the assumption that the effect of the bath is weak during the entire transfer. In the frame rotating with $H_S(t)$, and using the RWA from Eq. (5), the Lindblad master equation of the system is [66]

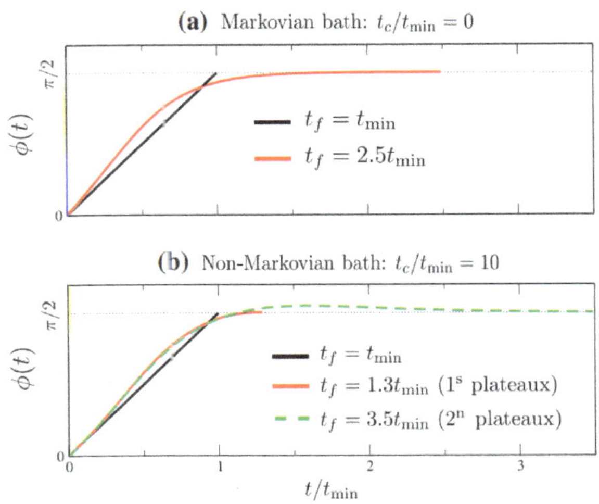

Figure 3: The transfer phase φ(t) versus t/tmin. The fastest modulation (black, dotted) with the Markovian optimal modulation (red, solid) and the non-Markovian optimal modulation (green, dashed), in Markovian (a) and non-Markovian (b) baths. In the Markovian bath the optimal modulation transfer starts off faster than the "fastest" transfer (when the information is still "fresh"), and slows down subsequently. For the non-Markovian bath, optimal modulation achieves full transfer (φ(t) = π/2) well within the modulation, but then "overshoots" (φ(t) > π/2), and eventually returns to φ(t) = π/2 (Color figure online)

$$\dot{\rho}(t) = \gamma[2S(t)\rho(t)S(t) - S(t)S(t)\rho(t) - \rho(t)S(t)S(t)]$$
$$S(t) = e^{\imath H_c \phi(t)} |e_1 g_2\rangle\langle e_1 g_2| e^{-\imath H_c \phi(t)}$$
$$\left[e^{\imath H_c \phi(t)} |e_1 g_2\rangle = \cos\phi |e_1 g_2\rangle + \imath \sin\phi |g_1 e_2\rangle\right] \quad (7)$$

Where $\phi(t) = \int_0^t V(t')dt'$ istheaccumulatedactiononthesystemand$\gamma = \mathrm{Tr}\langle B B \rho_B \rangle$ is the Markovian decoherence rate. We assume that the effect of the bath is weak and that in the frame rotating with $H_s(t)$ the density matrix ρ(t) does not change much during the transfer $\rho(t) \approx \rho_0$. We can then find an expression for the fidelity of the system f (t) at time t given an initial state of $|\psi_0\rangle = \alpha |e_1 g_2\rangle + \beta |g_1 g_2\rangle, (\rho_0 = |\psi_0\rangle\langle\psi_0|)$:

$$\rho(t) = \rho_0 + \int_0^t \dot{\rho}(t')dt'$$
$$f(t) = \langle\psi_0| \rho(t) |\psi_0\rangle$$
$$\approx \langle\psi_0| \rho_0 |\psi_0\rangle$$
$$+\gamma \int_0^t dt' \langle\psi_0| 2S(t')\rho_0 S(t') |\psi_0\rangle \quad (8)$$
$$-\gamma \int_0^t dt' \langle\psi_0| S(t')S(t')\rho_0 + \rho_0 S(t')S(t') |\psi_0\rangle$$
$$= 1 + 2\gamma \int_0^t dt' \left(|\alpha|^4 \cos^4\phi(t') - |\alpha|^2 \cos^2\phi(t')\right)$$

This is the fidelity for a specific initial state $\alpha |e\rangle + \beta |g\rangle$. By averaging this result over the entire Bloch sphere (all initial states) we find the average fidelity:

$$\overline{f(t)} \approx 1 + 2\gamma \int_0^t dt' \left(\overline{|\alpha|^4} \cos^4\phi(t') - \overline{|\alpha|^2} \cos^2\phi(t')\right)$$
$$= 1 + 2\gamma \int_0^t dt' \left(\frac{1}{3}\cos^4\phi(t') - \frac{1}{2}\cos^2\phi(t')\right)$$
$$= 1 - \gamma \int_0^t dt' \left(\frac{1}{3}\cos^4\phi(t') + \frac{1}{4}\sin^2 2\phi(t')\right) \quad (9)$$

We can use the Euler-Lagrange variational method to find an analytical solution for the optimal modulation phase φ(t), given a Markovian ($t_c \rightarrow 0$) bath, at long transfer times $t_f \rightarrow \infty$. The energy constraint gives

$$E = \int_0^\infty (\dot{\phi}(t))^2 dt \Rightarrow \frac{\partial E}{\partial \phi(t)} = -2\ddot{\phi}(t). \quad (10)$$

The error $(1 - \overline{f(t_f)})$ we wish to minimize can be expressed, for $t_f \to \infty$, as:

$$error \approx \frac{\gamma}{2} \int_0^\infty \left(\frac{\sin^2(2\phi(t))}{2} + 2\frac{\cos^4(\phi(t))}{3} \right) dt$$
$$\Rightarrow \frac{\partial(error)}{\partial\phi(t)} \approx \frac{\gamma}{2} \left(\sin(4\phi(t)) + 8\frac{\cos^3(\phi(t))\sin(\phi(t))}{3} \right) \tag{11}$$

With the Euler-Lagrange equation (using $-2\lambda/\gamma$ as Lagrange multiplier), we obtain

$$\frac{\partial E}{\partial\phi(t)} = -\frac{2\lambda}{\gamma}\frac{\partial(error)}{\partial\phi(t)}$$
$$\Rightarrow \frac{\partial E}{\partial\phi(t)}\dot{\phi}(t) = -\frac{2\lambda}{\gamma}\frac{\partial(error)}{\partial\phi(t)}\dot{\phi}(t)$$
$$\frac{d}{dt}(\dot{\phi}(t))^2 = \lambda\frac{d}{dt}\left(\frac{\sin^2(2\phi(t))}{2} + 2\frac{\cos^4(\phi(t))}{3} \right) \tag{12}$$

$$(\dot{\phi}(t))^2 = \lambda \left(\frac{\sin^2(2\phi(t))}{2} + 2\frac{\cos^4(\phi(t))}{3} \right)$$
$$\phi(t) = \sqrt{\lambda} \int_0^t dt' \sqrt{\frac{\sin^2(2\phi(t'))}{2} + 2\frac{\cos^4(\phi(t'))}{3}}$$

where we used the condition that $\phi(t) \xrightarrow[t\to\infty]{} \pi/2$ and $\dot{\phi}(t) \xrightarrow[t\to\infty]{} 0$ to find the integration constant. For $\lambda = 1$, the Markovian phase $\varphi_M(x)$ as a function of the dimensionless argument x satisfies

$$\frac{d\phi_M(x)}{dx} = \sqrt{\frac{\sin^2(2\phi_M(x))}{2} + 2\frac{\cos^4(\phi_M(x))}{3}}, \tag{13}$$

with $\varphi_M(0) = 0$. Equation (13) can be solved numerically to determine the shape of $\varphi_M(x)$ and its energy $e_M = \int_0^\infty |\phi'_M(x)|^2 dx = 1.038\ldots$ (where both x and eM are dimensionless). By rescalling $\varphi_M(x)$ we find the general Markovian optimal modulation at infitite time for any energy E, $\phi(t) = \phi_M(\frac{E}{e_M}t)$, with an error (1−fidelity) of $\frac{\gamma}{E}e_M^2 = \frac{\gamma}{E}1.077\ldots$, γ being the dephasing rate of the source qubit 1. The fastest modulation with energy E has an error of $\frac{\gamma}{E}\frac{\pi^2}{8} = \frac{\gamma}{E}1.233\ldots$. This means that the optimal modulation has about 12% less error than the fastest modulation for the same energy. This improvement is not large: using this optimal solution will increase the fidelity from, e.g., 0.9 to 0.912. Yet this is for a Markovian, memoryless, bath, where using current dynamical decoupling methods [56–60,67–71] no improvement was possible. For non-Markovian baths that cause the noisy-qubit decoherence the decrease in error (and hence increase in fidelity) can be much more significant, as shown below.

b. Non-Markovian bath effects— When the memory-time t_c of the bath is comparable to or larger than the characteristic transfer time $t_c \gtrsim t_{\min}$, a much larger improvement can be achieved (see Fig. 4). Remarkably, the best solution actually performs a full transfer, $\varphi(t) = \pi/2$, well within the modulation time, but rather than stopping at $\varphi = \pi/2$ it then "overshoots" the transfer, so that $\varphi(t) > \pi/2$, and then returns slowly to $\pi/2$ (see Fig. 3). This can explain the source of the noise reduction—when "overshooting", the information partially returns from the target (storage) qubit to the source (noisy) qubit, but with a negative sign. Hence, similarly to the "echo" method, the noise now operates in the reverse direction, correcting itself, i.e., the nonMarkovian bath effect is undone. This requires transfer times significantly larger than the minimal transfer time tmin, ranging from 3tmin to 10tmin or more, yet the fidelity increases substantially (up to 50% in Fig. 4). Two kinds of non-Markovian baths may play a role in the noisy SC qubit decoherence: (a) noisy controls with slow drift (memory) time [61–65]; (b) qubit relaxation in a cavity whose memory time is determined by its Q-factor [72].

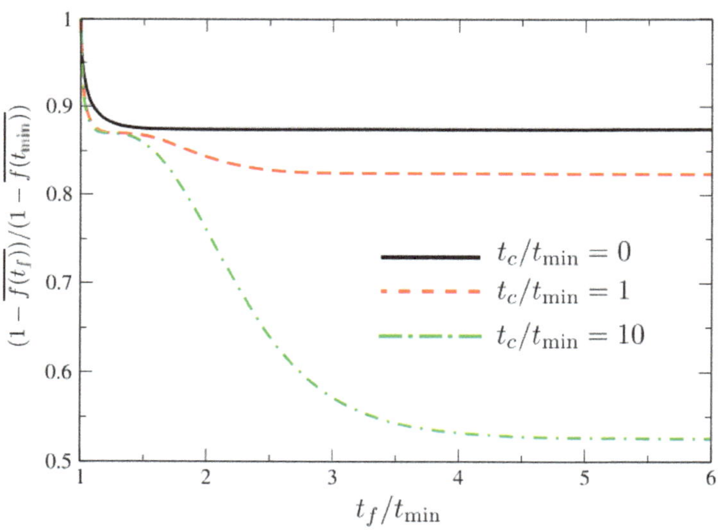

Figure 4: Dependence of the lowest achievable average error $\overline{(1 - \text{fidelity})}$ (averaged over all possible initial states of the source qubit) on the transfer time t f normalized to the error of the fastest transfer (over time tmin) at a given transfer energy (Eq. 6). This function is plotted for various bath memory times: (black, solid) $t_c = 0$ (Markovian);(red, dash) $t_c = t_{\min}$; (green, dash-dot) $t_c = 10t_{\min}$. Even for Markovian baths $(t_c \ll t_{\min})$ the best solution is not the fastest one. For non-Markovian baths (tc tmin) two plateaux (regions of insensitivity to t_f) can be seen. The first plateau is independent of the memory time, and matches the Markovian plateau. The second plateau is lower the longer the memory of the bath (Color figure online)

OPTIMIZING STORAGE AND TRANSFER FIDELITIES BY SPECTRAL FILTERING OF INHOMOGENEOUS SPIN ENSEMBLES

When the coupling of a single spin or atom to the cavity field is weak, we have to encode a qubit in collective states of an ensemble of spins whose ground state is $|\psi_0\rangle = |g_1, g_2, \ldots, g_N\rangle$ and the fully symmetrized, collective, single-excitation (Dicke) state is $|\psi_1\rangle = N^{-1/2}\sum_j |j\rangle$, where $|j\rangle = |g_1, g_2, \ldots, e_j, \ldots, g_N\rangle$ denotes a state with only spin j excited. If all the spins had the same resonant frequency ω_0, the single excitation state would evolve at time t = τ > 0 to $t = \tau > 0$ to $|\psi_1(\tau)\rangle = N^{-1/2}\sum_j e^{-i\omega_0\tau}|j\rangle$ and coherence is preserved if the position of the spins stays fixed. In general spins in a solid state environment have different resonant frequencies ω_j which results in inhomogeneous broadening. Then, even if the symmetric state $|\psi_1\rangle$ is prepared at t = 0, it would evolve into $|\tilde{\psi}_1(\tau)\rangle = N^{-1/2}\sum_j e^{-i\omega_j\tau}|j\rangle$, while the ground (or vacuum) state $|\psi_0\rangle$ remains unchanged. We thus define the "storage fidelity" as the squared overlap at time τ of state $|\psi_1(\tau)\rangle$ with its in homogeneously broadened counterpart:

$$\mathcal{F}(\tau) \equiv \left|\langle\psi_1(\tau)|\tilde{\psi}_1(\tau)\rangle\right|^2 = \left|\frac{1}{N}\int n(\omega)e^{-i(\omega-\omega_0)\tau}d\omega\right|^2. \tag{14}$$

where n(ω)is the spectral density normalized to the total number of atoms, $\int n(\omega)d\omega = N$. Our goal is to modify the spectrum n(ω) of the atomic (spin) ensemble so as to maximize the fidelity of transfer, storage and retrieval. To this end, we employ a third

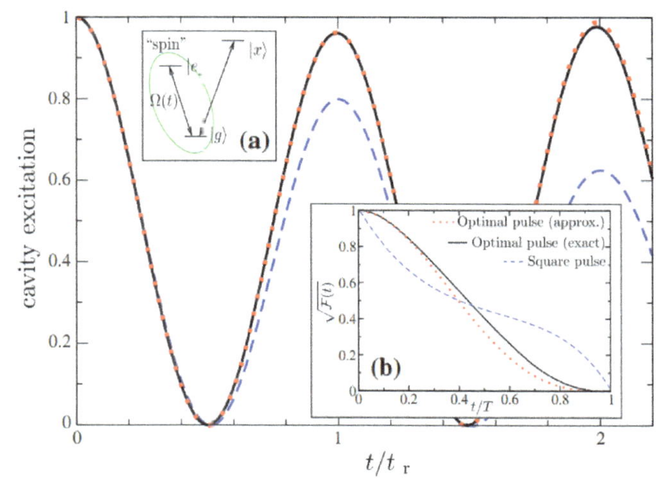

Figure 5: Inset a atomic level scheme. Main panel: Rabi oscillations of a single excitation between the cavity and the atomic ensemble for the optimal pulse (exact solution: black solid line; approximate analytical solution: red dotted line), and the square pulse (blue dashed line). Inset b the corresponding $\sqrt{\mathcal{F}(t)}$ for preparation time T = 10t_{tr} (Color figure online)

(auxiliary), long-lived state $|x\rangle$, outside the spin- 1/2 space of the ground $|g\rangle$ and excited $|e\rangle$ states used for QI storage (see Fig. 5a). The preparation of the ensemble proceeds in three steps: (1) Starting with the entire ensemble in the ground state $|g\rangle$ apply an external pulse of Rabi frequency $\Omega(t)$ exciting the atoms to the state $|e\rangle$. The duration T of the pulse should be long enough in order to select only the atoms with transition frequencies ω within a desired range of $\delta \sim 2\pi/T$ around ω_0, while the shape of the pulse $\Omega(t)$ is designed to optimize the resulting frequency spectrum (see below). (2) Transfer all those atoms remaining in state $|g\rangle$ to the auxiliary, long lived state $|x\rangle$ by another strong pulse, using, e.g., an adiabatic sweep across the $|g\rangle \rightarrow |x\rangle$ transition [73,74]. (3) Return the atoms selected in stage (i) from $|e\rangle$ to $|g\rangle$ by, e.g., the adiabatic transfer. The chosen subensemble is now ready to use. Its spectrum is $n(\omega)$ $P(\omega-\omega_0)$, where

$$P(\omega - \omega_0) = \left| \langle e| \, T_+ \, e^{-i \int_0^T H(t)dt} \, |g\rangle \right|^2, \tag{15}$$

With $H(t) = \frac{1}{2}(\omega - \omega_0)\sigma_z + \Omega(t)\sigma_x$. Equation (15) is the probability for the preparation pulse with frequency ω_0 and envelope $\Omega(t)$ in step 1) to excite an atom (spin) with resonant frequency ω. If $P(\omega)$ is much narrower than the initial spectrum of the atoms $n(\omega)$, we may replace $n(\omega) \rightarrow n(\omega_0)$, obtaining

$$\mathcal{F}(\tau) \approx \frac{4\pi^2 n^2(\omega_0)}{N^2} \left| \int_0^T dt \, \Omega(t + \tau)\Omega(t) \right|^2, \tag{16}$$

Where $N = 2\pi n(\omega_0) \int_0^T dt \, \Omega^2(t)$. Hence, for $\tau > T$ the fidelity vanishes and one cannot store the QI for a time longer than the preparation time.

To determine the optimal preparation pulse-shape $\Omega(t)$ we use the Euler-Lagrange method to maximize the fidelity (Eq. 16) for a specific storage time τ, while fixing the number of atoms N and the preparation time T.

For $\tau \ll T$ we find

$$\Omega(t) = \Omega_0 \sin(\pi t/T), \qquad N = \pi n(\omega_0)\Omega_0^2 T. \tag{17}$$

The resulting storage fidelity of the atomic ensemble is given by

$$\mathcal{F}(\tau) = \left| \frac{(T - \tau)\cos(\pi\tau/T)}{T} + \frac{\sin(\pi\tau/T)}{\pi} \right|^2 \approx 1 - \frac{\pi^2\tau^2}{T^2}. \tag{18}$$

Hence, using the preparation pulse of Eq. (17), the fidelity loss for small times $\tau << T$ becomes quadratic in τ, which should be contrasted with linear fidelity loss for the square preparation pulse (Fig. 5b). This can greatly improve the ensemble coherence time. We illustrate these results for NV-centers in Sect. 5.2

MAGNETIC COUPLING OF THE CAVITY FIELD TO SPIN ENSEMBLES

The magnetic dipole coupling to a microwave field is much weaker than the electric dipole coupling to optical-fields. Nevertheless one can achieve strong coupling to collective states of spin ensembles with currently available coplanar waveguide resonators (CPWR) due to their highly localized magnetic field. The high concentration of field energy near the surface results from a dramatic reduction of the effective volume $V_{\rm eff} \sim \frac{\pi}{2}\lambda l^2$ of the CPWR mode, where l is the decay length of the field which is of the order of the gap between the central conductor and the ground wire of the CPWR. For example, for the ^{87}Rb microwave transition at 6.83 GHz (wavelength $\lambda \sim 3$ cm) and a typical decay length $l \sim 3\mu$m one expects an enhancement of the atom-photon coupling strength of $(\lambda/l) \sim 10,000$. A full calculation of the local electromagnetic field [9] normalized to a single photon results in the field shown in Fig. 6b–d. We obtain at a distance of 1 μm above the surface a field of >40μG for a single photon, which confirms the simple estimate given above. Below we will illustrate two different implementations of magnetic coupling to spin ensembles: (i) trapped ultracold atoms [9] and (ii) electron spins in diamond [36–38].

Implementation in an Atomic Ensemble

Atom chips [75–77] are capable of trapping, positioning and manipulating large ensembles of ultracold atomic ensembles a few μm above a gap [78,79]. The advantage of atoms is that they do not have inhomogeneous broadening. Their disadvantage is that they are difficult to handle in a mK environment and cannot attain very large numbers N: their $\sqrt{N}$ enhancement is currently limited to $\sim 10^3$.

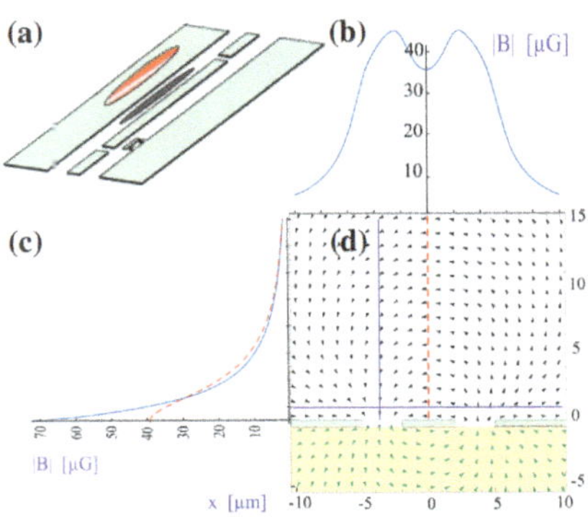

Figure 6: a Schematic of the CPWR including a solid-state qubit and a cloud of ultracold atoms trapped above one of the gaps. b The magnetic field strength of a single photon as a function of

lateral distance 1μm from the chip surface and c as a function of distance to the chip surface at the gap (full line) and at the central conductor (dashed line). d Vector plot of the magnetic field in the resonator.

We envision [9] placing an ensemble of 10^6 ultracold ^{87}Rb atoms with the ground state hyperfine splitting $\omega_{sg}/2\pi = 6.83$ GHz between $|F = 1\rangle \equiv |g\rangle$ and $|F = 2\rangle \equiv |s\rangle$ in the near field zone of the superconducting CPW cavity. Currently we are building up such an experiment [80,81]. Let us choose the frequency of the CPW cavity to be near-resonant with that of the atomic transition $|g\rangle \leftrightarrow |s\rangle$ (Fig. 1b). When the atomic ensemble is near the field antinode, with the spacial dimension of the cloud being small compared to the mode wavelength, all the atoms couple symmetrically to the cavity field. The corresponding Hamiltonian can be expressed as

$$H_{ac} = \hbar\delta_{ac}\hat{s}^{\dagger}\hat{s} + \hbar\eta_{ac}(\hat{s}^{\dagger}\hat{g}\,\hat{c} + \hat{c}^{\dagger}\hat{g}^{\dagger}\hat{s}), \tag{19}$$

where $\delta_{ac} = \omega_{sg} - \omega_c$ is the detuning and $\eta_{ac} = i(\wp_{sg}/\hbar)\varepsilon_c u(\mathbf{r})$ is the coupling rate between the cavity field and a (single) atom at position r. Since $|F = 1\rangle \leftrightarrow |F = 2\rangle$ is a magnetic dipole transition, the corresponding matrix element is small, $\wp_{sg} \simeq \frac{1}{2}\alpha a_0 e$ with $\alpha = 1/137$, which yields $\eta_{ac} \sim 2\pi \times 40$ Hz [for u(r) 1] [9]. The operators $\hat{g}$ ($\hat{g}^{\dagger}$) and $\hat{s}$ ($\hat{s}^{\dagger}$)annihilate (create) an atom in the corresponding state $|g\rangle$ and $|s\rangle$; these essentially bosonic operators live in a space of completely symmetrized states $|n_g, n_s\rangle$ with n_g atoms in state $|g\rangle$ and n_s atoms in state $|s\rangle$while $n_g + n_s = N$. With 106 trapped ultracold ^{87}Rb atoms and a CPWR with a quality factor $Q \sim 10^6$ one can achieve strong coupling between a microwave photon in the CPWR and a collective hyperfine qubit state in the atomic ensemble with $g_{\text{eff}}/2\pi \sim 40$ kHz larger than the cavity line width of $\kappa/2\pi \sim 7$ kHz [9]. This will allow to perform magnetic coupling cavity QED experiments with MW photons and collective state qubits stored in the atomic ensemble (see Fig. 7).

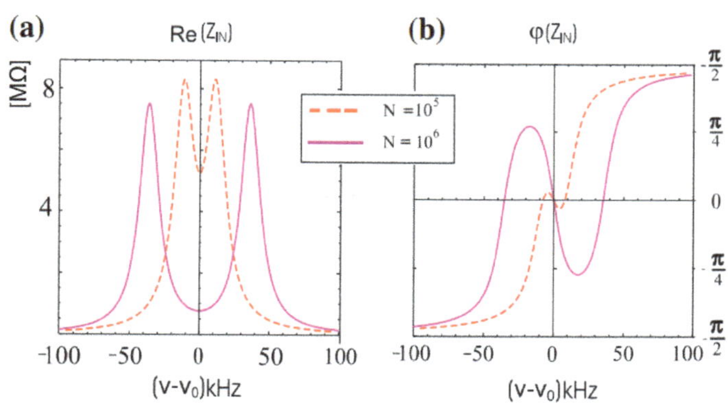

Figure 7: Response of the atom–cavity system as illustrated by its complex impedance Z: a the real part of Z is related to the spectrum and b the phase of Z directly illustrates the phase shift of

the transmitted microwave radiation. The calculations are for $Q = 10^6$, $N = 10^6$ and $N = 10^5$ and plotted versus incident field detuning from the resonance frequency.

The most direct approach to state transfer from the SC qubit to the atomic ensemble would be to prepare all the atoms in state $|g\rangle$, the cavity field in vacuum $|0_c\rangle$, and choose the frequencies of the cavity mode and the atomic hyperfine transition to be the same, $\delta_{ac} = 0$. Then, by tuning the SC qubit frequency to resonance with the cavity, $\delta_{qc} = 0$, during time τ_{qc} such that $2\eta_{qc}\tau_{qc} = \pi$, an arbitrary quantum state $|\psi\rangle = \alpha|0\rangle + \beta|1\rangle$ will be transferred from the SC qubit to the CPW cavity field (cf. Eq. 1). Next, it follows from Eq. (19) that the collective coupling rate of the cavity field and the atomic ensemble via the transition $|n_g = N, n_s = 0; 1_c\rangle \rightarrow |n_g = N-1, n_s = 1; 0_c\rangle$ is given by $\sqrt{N}\eta_{ac} \sim 2\pi \times 40\,\mathrm{KHz}$. Thus, during time $\tau_{sg} = \pi/(2\sqrt{N}\eta_{ac}) \sim 6\,\mu s$, corresponding to half a Rabi cycle, the cavity photon will be absorbed by the atoms and we will have achieved our goal. The time τsg is, however, comparable to the photon lifetime in the CPW cavity, $\kappa^{-1} = Q/\omega_c \sim 22\,\mu s$. Thus the photon will be lost with probability $P_{\mathrm{loss}} \approx \kappa\tau_{sg} \sim 0.27$ before being coherently absorbed by the atoms. It is therefore necessary to improve the CPW cavity by increasing its quality factor Q and thereby decreasing the photon decay rate κ. In an alternative setup, the SC qubit and the atoms are tuned to be resonant with each other, $\delta_{qc,ac} \simeq \delta$, but detuned from the cavity mode frequency. For large detuning $\delta \gg \eta_{qc}$, the adiabatic elimination of the cavity mode yields an effective photon decay rate $\kappa_{\mathrm{eff}} = \kappa\eta_{qc}^2/\delta^2$, while the corresponding second-order interaction Hamiltonian, $V_{qa}^{(2)} = \hbar\eta_{\mathrm{eff}}(\hat{s}^\dagger\hat{g}\hat{\sigma}^- + \hat{\sigma}^+\hat{g}^\dagger\hat{s})$, with $\eta_{\mathrm{eff}} = \eta_{qc}\eta_{ac}/\delta$, describes an effective swap of an excitation between the SC qubit and atomic ensemble with the rate $\sqrt{N}\eta_{\mathrm{eff}}$, mediated by virtual photon exchange in the cavity. Thus the effective coupling is reduced by a factor of δ/η_{qc}, while the decoherence rate caused by the cavity is reduced by a factor of $(\delta/\eta_{qc})^2$. For $\delta = 10\eta_{qc}$ we then have $\sqrt{N}\eta_{\mathrm{eff}} \sim 2\pi \times 4\,\mathrm{KHz}$ while $\kappa_{\mathrm{eff}} \simeq 2\pi \times 100\,\mathrm{Hz}$, which yields a low probability of photon decay during the excitation swap, $P_{\mathrm{loss}} \approx \kappa_{\mathrm{eff}}\pi/(2\sqrt{N}\eta_{\mathrm{eff}}) = 0.04$.

Recall, however, that the SC qubit decoherence (dephasing) $\gamma_q \lesssim 1\,\mathrm{MHz}$ is much larger than $\sqrt{N}\eta_{\mathrm{eff}}$, hampering the detuned scheme above. In turn, γ_q is much smaller than the coupling η_{qc}, hence this decoherence does not pose a problem during the resonant excitation exchange between the SC qubit and CPW cavity. This rapid stage may therefore be accomplished with high fidelity, as opposed to the much slower stage of excitation transfer from the cavity to the atoms with weak magnetic dipole transition. The dynamical control by modulation discussed in Sect. 3 can be used to suppress the SC qubit decoherence effects. It should be possible to improve both off- and on-resonant transfer fidelity, i.e., reduce the error by 50% (Fig. 4).

Implementation in A NV Center Ensemble

Magnetic coupling cavity QED can also be implemented using an ensemble of electron spins in diamond [36–38]. The ground $|g\rangle$, excited $|e\rangle$ and auxiliary $|x\rangle$states of the NV color center in diamond correspond, respectively, to the m = 0, m = −1 and m = 1 Zeeman sublevels of the ground electronic (spin-triplet) state of the NV. The $|g\rangle \rightarrow |e\rangle, |x\rangle$ transition frequencies are around 2.88 GHz. The quantization axis of the electron spin orientation is given by the orientation in the crystal. A static magnetic field does not (to first order) change the orientation but can be used to tune the transition | $|g\rangle \rightarrow |e\rangle$ in and out of resonance with the CPWR mode (see Fig. 8c). With a magnetic moment of $2\mu_B$ the resulting coupling is similar to the coupling of the Rb ground state spins, but the ensemble shows strong inhomogeneous broadening of a few MHz. In a typical experimental setup, Fig. 8a [38], a diamond is glued or pressed on top of the SC resonator chip. By placing the diamond sample at the center of the λ/2 cavity one ensures that a large fraction of the magnetic mode volume is occupied by spins.

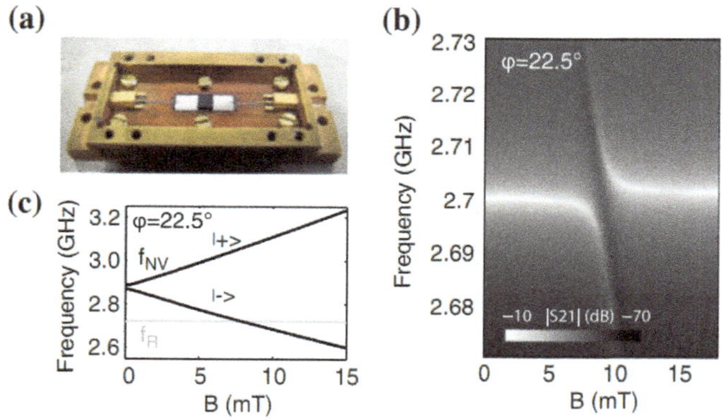

Figure 8 a A (001) diamond sample with a high NV center density is placed on top of a SC–CPW resonator where the oscillating magnetic field possesses an anti-node. b NV center transition frequencies $|g\rangle \rightarrow |-1\rangle$ ($|g\rangle \rightarrow |+1\rangle$) as a function of an externally applied magnetic field. Since the resonance frequency of the CPWR is not changed by a magnetic field in the plane of the resonator, a static magnetic field can be used to tune the NV ensemble into resonance with the cavity. c Measured MW transmission $|S_{21}|^2$ versus applied magnetic field. A large Rabi splitting of 2×8.7 MHz is observed when ensemble and resonator are brought into resonance, an indication for strong magnetic coupling.

A high concentration of NV centers can for example be obtained by irradiating a high-temperature high-pressure diamond of an initial nitrogen impurity concentration $\gtrsim 100$ ppm with neutrons and finally annealing the sample at 900 °C. An estimation of the coupling strength for N ≈ 10^{12} NV center spins uniformly coupled to the same electromagnetic field mode results in $\sqrt{N}\eta_{ac}/2\pi \sim 10$ MHz. Figure 8b shows the experimentally observed avoided crossing originating from the magnetic spin-cavity coupling. In this measurement the cavity transmission is probed with a vector network analyzer while an external magnetic field is used to tune the spin ensemble. When both, NV center spins and cavity, have the same frequency, a large Rabi splitting of 2 × 8.7 MHz is observed. Apart from coupling to electron spin ensembles, the hyperfine interaction between the NV centers electron spin and a nuclear spin in a neighboring lattice site can mediate the coupling to a nuclear spin ensemble. A demonstration of the coupling to ^{13}C nuclear spins in diamond with I = 1/2 was shown in [38]. Although strong coupling was not yet observed, this is a first step towards a nuclear spin memory. We will now illustrate the results of the optimized filtering discussed in Sect. 4 for the case of the in-homogeneously broadened NV ensembles. The degeneracy of levels $|e\rangle$ and $|x\rangle$ are lifted by stress in the crystal. The transitions $|g\rangle \rightarrow |e\rangle, |x\rangle$ can be selectively addressed by the external σ±-polarized microwave fields. We assume the inhomogeneous spectrum of the ensemble of NV centers has the total width of $\Delta/2\pi \sim 7$ MHz, composed of three partially overlapping Lorentzians of widths ~ 2.6 MHz split by ~ 2.2 MHz due to the hyperfine coupling to the I = 1 nuclear spin of the ^{14}N atom.

For an ensemble–cavity coupling strength $\sqrt{N\bar{\eta^2}} \simeq 2\pi \times 13$ MHz, the excitation transfer time is $t_{tr} = \pi/(\sqrt{N\bar{\eta^2}}) \approx 40$ ns. As an example, assume that one can achieve a CPWR with a quality factor of Q = 10^6. The photon lifetime in the cavity, $\kappa^{-1} \approx 55$ μs, is much longer than the transfer time. This allows a preparation that reduces the ensemble spectral width Δ, and the number of active atoms N, by $5 \cdot 10^3$, almost 4 orders of magnitude, while still keeping the transfer time $t_{tr} \approx 2.8\,\mu s = 0.05\kappa^{-1} \ll \kappa^{-1}$, well within the cavity lifetime. The new ensemble, created by the optimal preparation pulse $\Omega(t)$ of $T = 0.7$ ms duration, has the storage fidelity $F(t) \approx 1 - (t/0.22\ \mathrm{ms})^2$. This ensemble hardly loses any fidelity during the transfer. Furthermore, it allows us to store QI for $t \simeq 50$ μs with 95% fidelity, compared to 3 ns storage with the original ensemble and 2.8 μs storage in the cavity. Hence, we can outperform cavity storage even when Q is high (106). By contrast, had we used the square preparation pulse, the new ensemble would have had a storage fidelity of $F(t) \approx 1 - t/0.17$ ms, which can store QI for $t \simeq 8.5$μs with 95% fidelity—more than 5 times worse than with the optimal preparation.

ATOM–CAVITY ELECTRIC-DIPOLE COUPLING VIA RYDBERG STATES

Very strong atom–cavity field coupling can be achieved at microwave frequencies for electric-dipole transitions between highly-excited Rydberg states [26,27]. This allows the atoms to be placed much further away from the chip surface, and still achieve strong collective coupling.

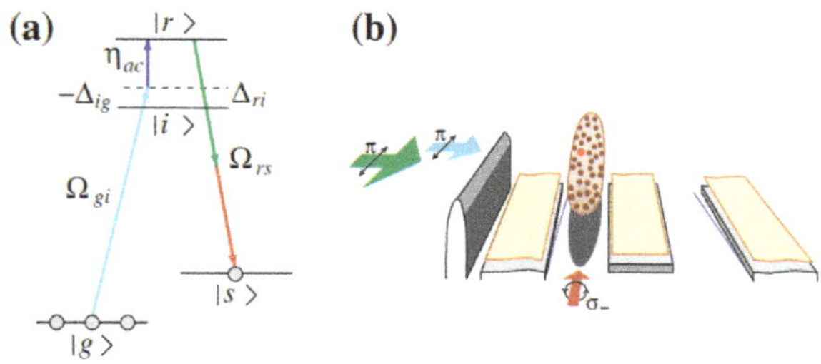

Figure 9: a Atomic Rydberg states $|i\rangle$ and $|r\rangle$ and relevant couplings for the excitation transfer from the CPW cavity to the atomic storage state $|s\rangle$. b Propagation geometry of the corresponding optical fields and the beam-stop and metallic mirrors on top of the SC electrodes of the CPW cavity.

We would like to note here that we use the Rydberg states to mediate the coupling between the MW photons and the stored collective spin state of the atomic ensemble. Since the qubits give single MW photons at a time, we do not have to employ Rydberg blockade [82] to ensure the writing of single quantum states.

Let us select a pair of Rydberg states $|i\rangle$ and $|r\rangle$such that the frequency ωri of transition $|i\rangle \leftrightarrow |r\rangle$ is close to the cavity mode frequency ωc, the corresponding detuning being $\delta_{ri} = \omega_{ri} - \omega_c$. We envision a level scheme sketched in Fig. 9(a), where the transition $|g\rangle \leftrightarrow |i\rangle$ is driven by an external optical field with Rabi frequency Ω_{gi} and detuning δ_{ig}. The Hamiltonian reads

$$
\begin{aligned}
H_{ac} &= \hbar\delta_{ig}\hat{i}^\dagger\hat{i} + \hbar(\delta_{ig} + \delta_{ri})\hat{r}^\dagger\hat{r} \\
&\quad -\hbar(\Omega_{gi}\,\hat{i}^\dagger\hat{g} + \eta_{ac}\,\hat{r}^\dagger\hat{i}\,\hat{c} + \text{H.c.}),
\end{aligned}
\tag{20}
$$

Where $\eta_{ac} = (\wp_{ir}/\hbar)\varepsilon_c u(\mathbf{r})$ is the atom-cavity field coupling rate, $\wp_{ir}$ being the corresponding dipole matrix element, while operators $\hat{i}\ (\hat{i}^\dagger)$ and $\hat{r}\ (\hat{r}^\dagger)$ annihilate (create) an atom in state $|i\rangle$ and $|r\rangle$, respectively.

We set the detunings as $\delta_{ri} \simeq -\delta_{ig} = \delta$. Then, given a photon in the cavity, the external field Ω_{gi} and the cavity field induce a two-photon transition from the ground state $|g\rangle$ to the Rydberg state $|r\rangle$ via non-resonant intermediate Rydberg state $|i\rangle$. If $\delta \gg \eta_{ac}, \sqrt{N}\Omega_{gi}$, state $|i\rangle$ is never populated, and we obtain an effective interaction Hamiltonian $V_{ac}^{(2)} = \hbar\eta_{\text{eff}}(\hat{r}^\dagger\hat{g}\,\hat{c} + \hat{c}^\dagger\hat{g}^\dagger\hat{r})$, with $\eta_{\text{eff}} = \Omega_{gi}\eta_{ac}/\delta$. Thus, starting from the initial state of the system $|n_g = N, n_{i,r,s} = 0; 1_c\rangle$, by pulsing Ω_{gi} for time $\tau_{gr} = \pi/(2\sqrt{N}\eta_{\text{eff}})$, the cavity photon will be coherently absorbed and a single atom from the ensemble will be excited to the Rydberg state $|r\rangle$. Next, another (bichromatic) external field with Rabi frequency Ω_{rs} pulsed for a time $\tau_{rs} = \pi/(2\Omega_{rs})$ can resonantly transfer the single collective Rydberg excitation of the atomic ensemble to the storage state $|s\rangle$: this process is described by $H_{rs} = -\hbar\Omega_{rs}\hat{s}^{\dagger}\hat{r} + \text{H.c.}$ At a later time, when required, the reverse process can add a single photon in the cavity while all the atoms will end up in state $|g\rangle$. This single photonic excitation can then be quickly transferred to the SC qubit, as described above.

In what follows, we survey the relevant experimental parameters. An elongated trapping volume $V_a \sim d \times d \times l$, with $d \simeq 5\,\mu$m and $l \simeq 1$ mm, contains $N \simeq 10^6$ atoms at density $\rho_a \sim 4 \times 10^{13}\,\text{cm}^{-3}$ The atomic lower states $|g\rangle$ and $|s\rangle$ correspond to the $|F = 1, M_F = -1\rangle$ and $|F = 2, M_F = 1\rangle$ sublevels of the ground electronic state $5s_{1/2}$ of ^{87}Rb. We choose the Rydberg states $|i\rangle \equiv |np_{1/2}, F = 2, M_F = -1\rangle$ and $|r\rangle \equiv |(n+1)s_{1/2}, F = 1, M_F = 0\rangle$ with n = 68 the principal quantum number. The quantum defects for the $s_{1/2}$ and $p_{1/2}$ Rydberg states of Rb are $\delta_s = 3.131$ and $\delta_p = 2.6545$ [26], with which the corresponding transition frequency is $\omega_{ri} = 2\pi \times 12.2$ GHz. Calculation of the relevant transition dipole matrix element involving the radial and angular parts gives $\wp_{ir} \simeq 1520 a_0 e$.

With the strip-line length $L \simeq 1$ cm and effective dielectric constant $\epsilon_r \sim 6$, the frequency of the mth standing-wave mode of the cavity is $\omega_c = \pi mc/L\sqrt{\epsilon_r}$ (Fig. 1a). The grounded SC electrodes at distance $w \simeq 10$ µm confine the cavity field within the effective volume $V_c = \int d^3r|u(\mathbf{r})|^2 \simeq \frac{\pi}{2}w^2L$ L yielding $\varepsilon_c = \sqrt{\hbar\omega_c/2\epsilon_0 V_c} \gtrsim 0.5$ V/m. Taking the full-wavelength (m = 2) linearly polarized cavity mode with $\omega_c/2\pi \simeq 12.16$ GHz, we estimate [9] that at the position of atomic cloud about 10 µm above the CPW surface the mode function $u(\mathbf{r}) \simeq e^{-1}$ which yields the vacuum Rabi frequency $\eta_{ac} = (\wp_{ir}/\hbar)\varepsilon_c u(\mathbf{r}) \simeq 2\pi \times 3.85$ MHz and appropriately large detuning $\delta_{ri} \simeq 10\eta_{ac}$.

The transition $|g\rangle \rightarrow |i\rangle$ is driven by linearly π-polarized UV field with wavelength $\lambda_{ig} \simeq 297$nm and detuning $\delta_{ig} = -\delta_{ri}$. To optimize the transition rate, its Rabi frequency is chosen as $\sqrt{N}\Omega_{gi} \simeq \eta_{ac}$, with which the transfer time is $\tau_{gr} \simeq 0.65$ µs. The required UV field intensity at the atomic cloud is $I_{gi} = 0.46$ W cm^{-2}. Next $|r\rangle \rightarrow |s\rangle$ is a two-photon transition via non-resonant intermediate state $|5p_{1/2}, F = 2, M_F = 0\rangle = |e\rangle$. The wavelengths are $\lambda_{re} \simeq 474$nm (linearly π-polarized field) and $\lambda_{es} \simeq 795$nm (circularly σ−-polarized field). With the corresponding intensities $I_{re} = 440$ W cm^{-2} and $I_{es} = 2.25$ mW cm^{-2} and intermediate detuning $\delta_{es} = 2\pi \times 25$ MHz, the two-photon Rabi frequency is $\Omega_{rs} = 2\pi \times 250\,\text{KHz}$ leading to the transfer time of τ_{rs} 1µs. Note that τgr and τ_{rs} are short compared to the lifetimes of cavity photon $1/\kappa \sim 10$ µs and Rydberg states $1/\gamma R \sim 100$ µs [26].

Figure 10 shows the results of numerical integration of the master equation [83,84] for density operator $\hat{\rho}(t)$ whose evolution is governed by Hamiltonians H_{qc}, H_{ac} of (20) and Hrs with the addition of the cavity decay and SC qubit dephasing times as discussed in Sect. 2. As seen, in the case of $\langle\hat{n}_c(0)\rangle = 0$, the state transfer is nearly ideal, with the small final error probability $P_{\mathrm{err}} \lesssim 0.04$ due to relaxation of the qubit, the cavity field and the atoms. However, in the case of finite temperature, $\langle \hat{n}_c(0) \rangle = 0.5$, during the transfer, as expected, the cavity field and the collective atomic state occupation numbers $\langle \hat{n}_c \rangle$ and $\langle \hat{n}_{r,s} \rangle$ exceed unity, and the resulting error probability $P_{\mathrm{err}} \simeq 0.3$ is large. Hence, suppressing the cavity temperature is crucial for achieving high transfer fidelity. We have perfomed simulations of the dynamics of the SC qubit–cavity transfer process with temperature-dependent initial population $\langle \hat{n}_c(0) \rangle$ of the CPW cavity photon field, for typical frequencies of few GHz [4,25,48–50]. We characterize the transfer process for a given initial state $|\psi\rangle_q$ of the SC qubit by the conditional fidelity $F_\psi = \mathrm{Tr}(\hat{\rho}\, |\psi_f\rangle\langle\psi_f|)$, where $|\psi_f\rangle$ denotes the final state stored in the memory for an ideal transfer.

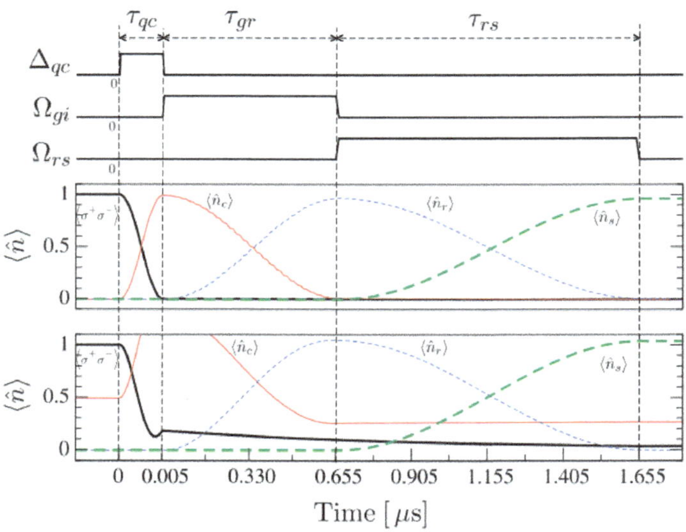

Figure 10: Numerical simulations of the three-step excitation transfer from the SC qubit to the ultracold atomic ensemble. The cavity is taken to have a Q-factor of 10^6 and the SC qubit is taken to have a dephasing time of $1/\gamma_q \gtrsim 1\,\mu\mathrm{s}$. Top panel illustrates the sequence of (π-)pulses: first the SQ qubit is brought to resonance with the CPW cavity, by pulsing $\delta_{qc}(t) = 0$ for time τ_{qc}; next, the $\Omega_{gi}(t)$ field is pulsed for time τ_{gr}; finally, the $\Omega_{rs}(t)$ field is pulsed for time τ_{rs}. Central and lower panels show the dynamics of occupation numbers for the qubit exited state $\langle \hat{\sigma}^+\hat{\sigma}^- \rangle$, the cavity field $\langle \hat{n}_c \rangle \equiv \langle \hat{c}^\dagger \hat{c} \rangle$ and the collective atomic states $\langle \hat{n}_r \rangle \equiv \langle \hat{r}^\dagger \hat{r} \rangle$ and $\langle \hat{n}_s \rangle \equiv \langle \hat{s}^\dagger \hat{s} \rangle$. Initially, the mean thermal photon number is $\langle \hat{n}_c(0) \rangle = 0$ in the central panel, and $\langle \hat{n}_c(0) \rangle = 0.5$ in the lower panel.

The mean transfer fidelity $\bar{F}$ is obtained by averaging F_ψ over all possible $|\psi\rangle$. The dependence of F_ψ and $\bar{F}$ on the CPW cavity temperature T, or the mean thermal photon number $\langle \hat{n}_c \rangle = (e^{\hbar\omega_c/k_B T} - 1)^{-1}$ is shown in Fig. 11. Below $k_B T/\hbar\omega_c \simeq 0.2$, corresponding to $\langle \hat{n}_c \rangle \lesssim 0.01$, the transfer fidelity is fairly high, $\bar{F} > 98\%$, but then it quickly degrades due to the detrimental effect of even a small number of thermal photons. For the above parameters this critical temperature is $T \sim 0.1$ K necessitating cryogenic conditions. Hence the importance of keeping the cavity temperature very low (see Sect. 2).

CONCLUSIONS

We have discussed here two general strategies for quantum state-transfer and storage optimization within hybrid open systems, and their consequences for experimental implementation on a SC chip. A: Our analysis of dynamically controlled transfer from a "noisy" qubit to its "quiet" counterpart, has revealed an intriguing interplay between the ability to avoid bathinduced errors that profoundly depend on the bath-memory time and the limitations imposed by leakage out of the operational subspace. Counterintuitively, under no circumstances is the fastest transfer optimal (for a given transfer energy). An important issue that merits discussion in this context is that of cavity relaxation effects. The situation studied here pertains to qubits where the main source of decoherence is dephasing noise. Yet, since the noise suppression attained here requires longer transfer time than that of the fastest transfer, one may wonder whether the (competing) cavity relaxation effects may not increase as well. However, although the total transfer time is increased, the main part of the transfer takes actually less time than in the fastest case. Hence, a reduction in cavity relaxation may occur, similar to the dephasing reduction: this is indeed confirmed by our simulation results.

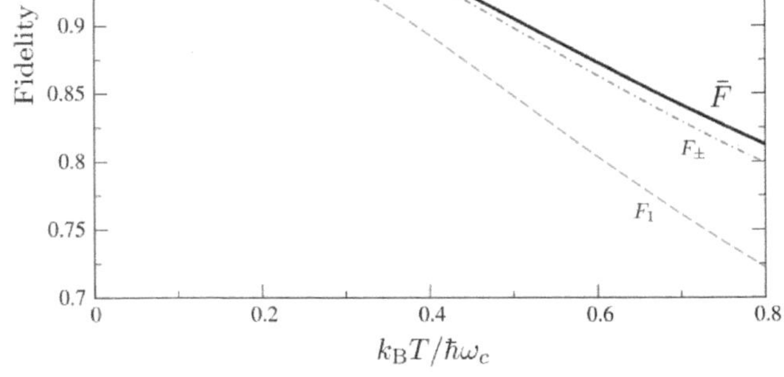

Figure 11: Conditional F_ψ and mean $\bar{F}$ transfer fidelities vs temperature k_BT (lower horizontal axis) or mean thermal photon number $\langle \hat{n}_c \rangle$ (upper horizontal axis).In F_ψ, $\psi = 0, 1, \pm$ correspond to states $|0\rangle$, $|1\rangle$ and $\frac{1}{\sqrt{2}}[\,|0\rangle \pm (i)\,|1\rangle]$

B: We have proposed a method that alleviates the shortcomings of the spin-ensemble inhomogeneous broadening, assuming that the cavity lifetime is much longer than the transfer time. This method allows us to optimize the memory, given the cavity, the spectral density of the ensemble and their mean coupling strength. This protocol is compatible with dynamical control [61–65], especially with its optimal version [54,55,85] which can help mitigate the qubit decoherence, while the cavity temperature effects can be partially tolerated [6,28] or suppressed [86]. When the spin-ensemble temperature is an issue, dynamical initialization may be useful [87–89]. Three possible experimental implementations of the above methods have been discussed: (i) magnetic dipole coupling of CPWR to ultracold atom ensembles; (ii) magnetic dipole coupling of CPWR to NV-center ensembles; (iii) electric dipole coupling to Rydberg atom ensembles. Overall, we have developed a promising approach for realizing efficient quantum state transfer between superconducting charge qubits and mesoscopic atomic or spin ensembles coupled to a microwave coplanar waveguide cavity. Our estimates show that this hybrid system may allow longer and higher-fidelity storage than even a high-Q cavity.

ACKNOWLEDGMENTS

This research was supported by the EU through MIDAS, by DIP, the Humboldt-Meitner Award (G.K.), the Humboldt Foundation (D.P.) and the Wittgenstein Prize (J.S.).

REFERENCES

1. Andre A. et al.: Polar molecules near superconducting resonators: a coherent, all-electrical atom-mesocopic interface. Nat. Phys. 2, 636 (2006)
2. Rabl P. et al.: Hybrid quantum processors: molecular ensembles as quantum memory for solid state circuits. Phys. Rev. Lett. 97(3), 033003 (2006)
3. Tordrup Karl, Mølmer Klaus: Quantum computing with a single molecular ensemble and a cooper-pair box. Phys. Rev. A 77(2), 020301 (2008)
4. Schoelkopf R.J., Girvin S.M.: Wiring up quantum systems. Nature 451, 664 (2008)
5. Petrosyan D., Fleischhauer M.: Quantum information processing with single photons and atomic ensembles in microwave coplanar waveguide resonators. Phys. Rev. Lett. 100, 170501 (2008)
6. Petrosyan D. et al.: Reversible state transfer between superconducting qubits and atomic ensembles. Phys. Rev. A 79(4), 040304(R) (2009)
7. Wallquist M. et al.: Hybrid quantum devices and quantum engineering. Phys. Scr. t137, 014001 (2009)
8. Imamoglu A.: Cavity qed based on collective magnetic dipole coupling: spin ensembles as hybrid two-level systems. Phys. Rev. Lett. 102(8), 083602 (2009)
9. Verdú J. et al.: Strong magnetic coupling of an ultracold gas to a superconducting waveguide cavity. Phys. Rev. Lett. 103(4), 043603 (2009)
10. Devoret M.H., Martinis J.M.: Implementing qubits with superconducting

integrated circuits. Quantum Inf. Process. 3, 163–203 (2004)

11. You J.Q., Nori F.: Superconducting circuits and quantum information. Phys. Today 58(11), 42–47 (2005)
12. Makhlin Y., Schön G., Shnirman A.: Quantum-state engineering with Josephson-junction devices. Rev. Mod. Phys. 73(2), 357–400 (2001)
13. You J.Q., Nori F.: Atomic physics and quantum optics using superconducting circuits. Nature 474, 589 (2011)
14. Duan L.-M., Monroe C.: Colloquium: quantum networks with trapped ions. Rev. Mod. Phys. 82(2), 1209–1224 (2010)
15. Taylor J.M., Marcus C.M., Lukin M.D.: Long-lived memory for mesoscopic quantum bits. Phys. Rev. Lett. 90(20), 206803 (2003)
16. Fleischhauer M., Lukin M.D.: Quantum memory for photons: dark-state polaritons. Phys. Rev. A 65(2), 022314 (2002)
17. Lukin M.D.: Colloquium: trapping and manipulating photon states in atomic ensembles. Rev. Mod. Phys. 75(2), 457–472 (2003)
18. Zhao B. et al.: A millisecond quantum memory for scalable quantum networks. Nat. Phys. 5, 95 (2009)
19. Dudin Y.O., Zhao R., Kennedy T.A.B., Kuzmich A.: Light storage in a magnetically dressed optical lattice. Phys. Rev. A 81(4), 041805 (2010)
20. Deutsch C. et al.: Spin self-rephasing and very long coherence times in a trapped atomic ensemble. Phys. Rev. Lett. 105(2), 020401 (2010)
21. Gurudev Dutt M.V. et al.: Quantum register based on individual electronic and nuclear spin qubits in diamond. Science 316(5829), 1312–1316 (2007)
22. Taylor J.M. et al.: High-sensitivity diamond magnetometer with nanoscale resolution. Nat. Phys. 4, 810–816 (2008)
23. Acosta V.M. et al.: Diamonds with a high density of nitrogen-vacancy centers for magnetometry applications. Phys. Rev. B 80(11), 115202 (2009)
24. Stanwix P.L. et al.: Coherence of nitrogen-vacancy electronic spin ensembles in diamond. Phys. Rev. B 82(20), 201201 (2010)
25. Blais A. et al.: Cavity quantum electrodynamics for superconducting electrical circuits: an architecture for quantum computation. Phys. Rev. A 69(6), 062320 (2004)
26. Gallagher T.F.: Rydberg Atoms. Cambridge University Press, Cambridge (1994)
27. Brune M., Raimond J.-M., Haroche S.: Manipulating quantum entanglement with atoms and photons in a cavity. Rev. Mod. Phys. 73, 565 (2001)
28. Henschel K., Majer J., Schmiedmayer J., Ritsch H.: Cavity qed with an ultracold ensemble on a chip: prospects for strong magnetic coupling at finite temperatures. Phys. Rev. A 82(3), 033810 (2010)
29. Nirrengarten T. et al.: Realization of a superconducting atom chip. Phys. Rev. Lett. 97(20), 200405 (2006)

30. Mukai T. et al.: Persistent supercurrent atom chip. Phys. Rev. Lett. 98(26), 260407 (2007)
31. Roux C. et al.: Bose-Einstein condensation on a superconducting atom chip. Europhys. Lett. 81, 56004 (2008)
32. Kasch B. et al.: Cold atoms near superconductors: atomic spin coherence beyond the Johnson noise limit. New J. Phys. 12, 065024 (2010)
33. Hufnagel Ch., Mukai T., Shimizu F.: Stability of a superconductive atom chip with presistent current. Phys. Rev. A 79, 053641 (2009)
34. Emmert A. et al.: Measurement of the trapping lifetime close to a cold metallic surface on a cryogenic atom–chip. Eur. Phys. J. D 51, 173 (2009)
35. Meek S.A., Conrad H., Meijer G.: Trapping molecules on a chip. Science 324, 1699–1702 (2009)
36. Schuster D.I. et al.: High-cooperativity coupling of electron-spin ensembles to superconducting cavities. Phys. Rev. Lett. 105(14), 140501 (2010)
37. Kubo Y. et al.: Strong coupling of a spin ensemble to a superconducting resonator. Phys. Rev. Lett. 105(14), 140502 (2010)
38. Amsüss R. et al.: Cavity qed with magnetically coupled collective spin states. Phys. Rev. Lett. 107(6), 060502 (2011)
39. Wu H. et al.: Storage of multiple coherent microwave excitations in an electron spin ensemble. Phys. Rev. Lett. 105(14), 140503 (2010)
40. Bushev P. et al.: Ultralow-power spectroscopy of a rare-earth spin ensemble using a superconducting resonator. Phys. Rev. B 84(6), 060501 (2011)
41. Reichel, J., Vuletic, V. (eds): Atom Chips. Wiley, VCH (2011)
42. Roos I., Mølmer K.: Quantum computing with an inhomogeneously broadened ensemble of ions: Suppression of errors from detuning variations by specially adapted pulses and coherent population trapping. Phys. Rev. A 69(2), 022321 (2004)
43. Tordrup K., Mølmer K.: Quantum-state reconstruction with imperfect rotations on an inhomogeneously broadened ensemble of qubits. Phys. Rev. A 75(4), 042318 (2007)
44. Wesenberg J.H., Kurucz Z., Mølmer K.: Dynamics of the collective modes of an inhomogeneous spin ensemble in a cavity. Phys. Rev. A 83(2), 023826 (2011)
45. Hahn E.L.: Spin echoes. Phys. Rev. 80(8), 580–594 (1950)
46. Nilsson M., Kröll S.: Solid state quantum memory using complete absorption and re-emission of photons by tailored and externally controlled inhomogeneous absorption profiles. Opt. Commun. 247(4–6), 393–403 (2005)
47. Alexander A.L., Longdell J.J., Sellars M.J., Manson N.B.: Photon echoes produced by switching electric fields. Phys. Rev. Lett. 96(4), 043602 (2006)
48. Wallraff A. et al.: Strong coupling of a single photon to a superconducting qubit using circuit quantum electrodynamics. Nature 431, 162–167 (2004)
49. Majer J. et al.: Coupling superconducting qubits via a cavity bus. Nature 449,

443–447 (2007)

50. Sillanpaa M.A., Park J.I., Simmonds R.W.: Coherent quantum state storage and transfer between two phase qubits via a resonant cavity. Nature 449, 438–442 (2007)
51. Brennecke F. et al.: Cavity qed with a Bose-Einstein condensate. Nature 450, 268–271 (2007)
52. Mariantoni M. et al.: Photon shell game in three-resonator circuit quantum electrodynamics. Nat. Phys. 7, 287 (2011)
53. Houck A.A. et al.: Controlling the spontaneous emission of a superconducting transmon qubit. Phys. Rev. Lett. 101, 080502 (2008)
54. Clausen J., Bensky G., Kurizki G.: Bath-optimized minimal-energy protection of quantum operations from decoherence. Phys. Rev. Lett. 104(4), 040401 (2010)
55. Gordon G., Kurizki G., Lidar D.A.: Optimal dynamical decoherence control of a qubit. Phys. Rev. Lett. 101(1), 010403 (2008)
56. Wu L.A., Kurizki G., Brumer P.: Master equation and control of an open quantum system with leakage. Phys. Rev. Lett. 102, 080405 (2009)
57. Wu L.-A., Byrd M.S., Lidar D.A.: Polynomial-time simulation of pairing models on a quantum computer. Phys. Rev. Lett. 89, 057904 (2002)
58. Byrd M.S., Wu L-A., Lidar D.A.: Overview of quantum error prevention and leakage elimination. J. Mod. Opt. 51, 2449 (2004)
59. Byrd M.S., Lidar D.A., Wu L-A., Zanardi P.: Universal leakage elimination. Phys. Rev. A 71, 052301 (2005)
60. Lidar D.A.: Towards fault tolerant adiabatic quantum computation. Phys. Rev. Lett. 100, 160506 (2008)
61. Kofman A.G., Kurizki G.: Unified theory of dynamically suppressed qubit decoherence in thermal baths. Phys. Rev. Lett. 93(13), 130406 (2004)
62. Kofman A.G., Kurizki G.: Universal dynamical control of quantum mechanical decay: Modulation of the coupling to the continuum. Phys. Rev. Lett. 87, 270405 (2001)
63. Kofman A.G., Kurizki G.: Acceleration of quantum decay processes by frequent observations. Nature (London) 405, 546 (2000)
64. Gordon G., Erez N., Kurizki G.: Universal dynamical decoherence control of noisy single- and multi-qubit systems. J. Phys. B 40(9), S75 (2007)
65. Gordon G., Kurizki G.: Universal dephasing control during quantum computation. Phys. Rev. A 76(4), 042310 (2007)
66. Breuer H.P., Petruccione F.: The Theory of Open Quantum Systems. Oxford University Press, Oxford (2002)
67. Viola L., Knill E., Lloyd S.: Dynamical generation of noiseless quantum subsystems. Phys. Rev. Lett. 85(16), 3520–3523 (2000)
68. Viola L., Knill E., Lloyd S.: Dynamical decoupling of open quantum systems. Phys. Rev. Lett. 82(12), 2417–2421 (1999)

69. Viola L., Lloyd S.: Dynamical suppression of decoherence in two-state quantum systems. Phys. Rev. A 58(4), 2733–2744 (1998)

70. Vitali D., Tombesi P.: Heating and decoherence suppression using decoupling techniques. Phys. Rev. A 65(1), 012305 (2001)

71. Uhrig G.S.: Keeping a quantum bit alive by optimized π-pulse sequences. Phys. Rev. Lett. 98(10), 100504 (2007)

72. Kofman A.G., Kurizki G.: Quantum Zeno effect on atomic excitation decay in resonators. Phys. Rev. A 54(5), R3750–R3753 (1996)

73. Abragam A.: The Principles of Nuclear Magnetism. Oxford University Press, Oxford, England (1961)

74. Allen L., Eberly J.H.: Optical Resonance and Two-Level Atoms. Wiley, New York (1975)

75. Folman R. et al.: Controlling cold atoms using nanofabricated surfaces: atom chips. Phys. Rev. Lett. 84, 4749 (2000)

76. Folman R. et al.: Microscopic atom optics: from wires to an atom chip. Adv. At. Mol. Opt. Phys. 48, 263–356 (2002)

77. Fortagh J., Zimmermann C.: Magnetic microtraps for ultracold atoms. Rev. Mod. Phys. 79, 235 (2007)

78. Lin Y., Teper I., Chin C., Vuletić V.: Impact of the Casimir-Polder potential and Johnson noise on Bose-Einstein condensate stability near surfaces. Phys. Rev. Lett. 92, 50404 (2004)

79. Aigner S. et al.: Long-range order in electronic transport through disordered metal films. Science 319, 1226–1229 (2008)

80. Haslinger S. et al.: Electron beam driven alkali metal atom source for loading a magneto-optical trap in a cryogenic environment. Appl. Phys. B 102, 819 (2011)

81. Haslinger, S.: Cold atoms in a cryogenic environment. PhD thesis, TU-Wien (2011)

82. Lukin M.D. et al.: Dipole blockade and quantum information processing in mesoscopic atomic ensembles. Phys. Rev. Lett. 87(3), 037901 (2001)

83. Nielsen M., Chuang I.: Quantum Computation and Quantum Information. Cambridge University Press, Cambridge (2000)

84. Lambropoulos P., Petrosyan D.: Fundamentals of Quantum Optics and Quantum Information. Springer, Berlin (2006)

85. Bensky G. et al.: Universal dynamical decoupling from slow noise with minimal control. Europhys. lett. 89(1), 10011 (2010)

86. Zhang P., Wang Y.D., Sun C.P.: Cooling mechanism for a nanomechanical resonator by periodic coupling to a cooper pair box. Phys. Rev. Lett. 95(9), 097204 (2005)

87. Erez N., Gordon G., Nest M., Kurizki G.: Thermodynamic control by frequent quantum measurements. Nature 452, 724–727 (2008)

88. Gordon G. et al.: Cooling down quantum bits on ultrashort time scales. New J.

Phys. 11(12), 123025 (2009)

89. Álvarez Gonzalo, A., Bhaktavatsala Rao A., Bhaktavatsala Rao D.D., Frydman L., Kurizki G.: Zeno and anti-Zeno polarization control of spin ensembles by induced dephasing. Phys. Rev. Lett. 105(16), 160401 (2010)

8
CHAPTER

QUANTUM CONTROL WITH SPECTRAL CONSTRAINTS

Łukasz Pawela and Zbigniew Puchała

Keywords: Quantum information, Quantum computation, Control in mathematical physics

INTRODUCTION

One of the fundamental issues of the quantum information science is the ability to manipulate the dynamics of a given complex quantum system. Since the beginning of quantum mechanics, controlling a quantum system has been an implicit goal of quantum physics, chemistry, and implementations of quantum information processing.

If a given quantum system is controllable, i.e., it is possible to drive it into a previously fixed state, it is desirable to develop a control strategy to accomplish the required control task. In the case of finite dimensional quantum systems, the criteria for controllability can be expressed in terms of Lie-algebraic concepts [1–3]. These concepts provide a mathematical tool, in the case of closed quantum systems, i.e., systems without external influences.

It is an important question whether the system is controllable when the control is performed only on a subsystem. This kind of approach is called a *local-controllability* and can be considered only in the case when the subsystems of a given system interact. As examples may serve coupled spin chains or spin networks [2, 4–6]. Local-control has a practical importance in proposed quantum computer architectures, as its implementation is simpler and the effect of decoherence is reduced by decreased number of control actuators [7, 8].

A widely used method for manipulating a quantum system is a coherent control strategy, where the manipulation of the quantum states is achieved by applying semi-

Citation: Łukasz Pawela and Zbigniew Puchała, Quantum Control with Spectral Constraints, DOI: 10.1007/s11128-013-0644-3

classical potentials in a fashion that preserves quantum coherence. In the case when a system is controllable, it is a point of interest what actions must be performed to control a system most efficiently, bearing in mind limitations imposed by practical restrictions. Various constraints concerning control fields can be imposed in the realistic implementations of quantum control systems. One of the most important is the restriction on the frequency spectrum of acceptable control parameters. Such restrictions come into play, for example, in an experimental setup that utilizes an external magnetic field [9]. In the case of such systems, due to various limitations, the application of piecewise-constant controls is not accurate. The real realization of controls is somehow smoothed by some filter induced by an experimental limitations. Thus, it is reasonable to seek control parameters in the domain imposed by the experimental restrictions.

In article [10], there has been discussed how the low-pass filtering, i.e., eliminating high-frequency components in a Fourier spectra, on a numerically obtained optimal control pulses affects a quality of performed control. This approach makes a contact with experimental realizations, since it implements the limitations or real quantum control systems.

In this paper, we present a general method of obtaining a piecewise-constant controls, which is robust with respect to low-pass filtering. The above means that elimination of high frequencies in a Fourier spectra reduces the fidelity only by a small amount. We utilize this approach to obtain numerically control pulses on a Heisenberg spin chain [10–12]; however, it can be applied to a quantum system with more general interactions.

This paper is organized as follows. In Sect. 2, we provide a general description of a quantum mechanical control system. In Sect. 3, we provide the description of the simulation setup used to test our model. Section 4 contains results obtained from numerical simulations and their discussion. In Sect. 5, we provide a summary of the presented work and give some concluding remarks.

OUR MODEL

To demonstrate a method of obtaining piecewise-constant controls, which are robust with respect to low-pass filtering, we will consider an isotropic Heisenberg spin- 1/2 chain of a finite length N. The control will be performed on the first spin only. The total Hamiltonian of the aforementioned quantum control system is given by

$$H(t) = H_0 + H_c(t), \tag{1}$$

Where

$$H_0 = J \sum_{i=1}^{N-1} S_x^i S_x^{i+1} + S_y^i S_y^{i+1} + S_z^i S_z^{i+1}, \tag{2}$$

is a drift part given by the Heisenberg Hamiltonian. The control is performed only on a first spin and is Zeeman-like, i.e.,

$$H_c(t) = h_x(t)S_x^1 + h_y(t)S_y^1. \tag{3}$$

In the above, S_k^i denotes k^{th} Pauli matrix acting on the spin i. Time-dependent control parameters h_x (t) and h_y (t) are chosen to be piecewise constant. We will refer to the values of the control pulses in the ith time interval as $h_{x,i}$ and $h_{y,i}$. When we reference a general control pulse without specifying the direction, we will write hl and the corresponding control Hamiltonian will be denoted H_l. Furthermore, as opposed to [10], we do not restrict the control fields to be alternating with x and y, i.e., they can be applied simultaneously (see e.g. [13] for similar approach). For notational convenience, we set $\hbar = 1$, and after this rescaling frequencies and control field amplitudes can be expressed in units of the coupling strength J , and on the other hand, all times can be expressed in units of 1/J [10]. The system described above is operator controllable, as it was shown in [5] and follows from a controllability condition using a graph infection property introduced in the same article. The controllability of the described system can be also deduced from a more general condition utilizing the notion of hypergraphs [6]. Since the interest here is focused on operator control, a quality of a control will be measured with the use of gate fidelity,

$$F = \frac{1}{2^N}|\mathrm{Tr}(U_T^\dagger U(h))|, \tag{4}$$

where U_T is the target quantum operation and U(h)is an operation achieved by control parameters h. We choose gate fidelity as it neglects global phases. To obtain piecewise-constant controls, which are robust with respect to low-pass filtering we will minimize the power in the high frequency part of a controls Fourier spectrum. We will do so by minimizing the following functional

$$G = (1 - \mu)P - \mu F, \tag{5}$$

where F is the gate fidelity described above, μ is a weight assigned to fidelity and P is a contribution of high frequencies in the total power of the control parameters. The above can be stated as, if y is a vector of Fourier coefficients for a control parameters h of length n, i.e.

$$y = Qh, \quad \text{where} \quad Q = n^{-1/2}\{e^{2\pi ikl/n}\}_{k,l=0}^{n-1}, \tag{6}$$

then P is given by

$$P = \frac{\sum_{i=\frac{n}{2}-\Delta}^{\frac{n}{2}+\Delta}|y_i|^2}{|y|^2}, \tag{7}$$

for some Δ related to the cutoff frequency of low-pass filter.

SIMULATION SETUP

To demonstrate the beneficialness of our approach, we study three- and four-qubit spin chains. The control field is applied to the first qubit only. Our target gates are:

$$\mathrm{NOT}_N = \mathbb{1}^{\otimes N-1} \otimes \sigma_x, \tag{8}$$

the negation of the last qubit of the chain, and

$$\mathrm{SWAP}_N = \mathbb{1}^{\otimes N-2} \otimes \mathrm{SWAP}, \tag{9}$$

swapping the states between the last two qubits. For each of these cases, we find two sets of control parameters. One with frequency constraints and one without. Next, we calculate an appropriate filter and using these filtered values of control parameters, we calculate the fidelity of the quantum operation. In each case, 120 independent sets of control parameters were found. We provide an explicit example in which we set the duration of the control pulse to $\Delta t = 0.2$ and the total number of pulses in each direction to n = 128 for the three-qubit chain and n = 512 in the four-qubit case, although the presented method may be applied for arbitrary values of Δt and n. The weight of fidelity in Eq. (5) is set to μ = 1 in the unconstrained case and to μ = 0.05 in the constrained case. Although the weight of the fidelity is small, the optimization still yields high fidelity values while maintaining low contribution of high frequencies in the power spectrum. We set the cutoff frequency in Eq. (7) to $\Delta = \frac{n}{4}$.

The applied filter is a frequency filter with the cutoff frequency equal to the frequency discriminated by the functional (5). As an example, we consider this filter to be an ideal low-pass filter, which was previously studied, e.g., in [11]. Obviously, one can use other filters. A more general discussion of spectral filtering is presented in [14]. The ideal low-pass filter leaves the frequencies only in the interval $[-\omega_0, \omega_0]$: $f(\omega) = \Theta(\omega+\omega_0) - \Theta(\omega-\omega_0)$, where Θ is the Heaviside step function. We obtain the following expressions for the filtered control parameters [11]:

$$\hat{h}_k(t) = \frac{1}{\pi} \sum_{i=1}^{n} h_{k,i}[a_{i+1}(t) - a_i(t)], \tag{10}$$

where k ∈ {x, y} and t denotes time. In other words, $\hat{h}_k(t)$ is a filtered version of the control fields h_k at the time moment t.

$$a_n(t) = \mathrm{Si}\left[\omega_0(n\Delta t - t)\right], \tag{11}$$

$$\mathrm{Si}(x) = \int_0^x (\sin t/t)\mathrm{d}t. \tag{12}$$

The calculation of the gradient of the fidelity function can be found in the work by Machnes et al. [15]. Here, we only show the final result of the calculation

$$\frac{\partial F(U_k)}{\partial h_l} = \frac{1}{N}\Re\mathrm{Tr}\left\{\mathrm{e}^{-\mathrm{i}\phi}U_T^{\dagger}U_n \ldots U_{k+1}\frac{\partial U_K}{\partial h_l}U_{k-1}U_1\right\}, \tag{13}$$

Where

$$\mathrm{e}^{-\mathrm{i}\phi} = \frac{\overline{\mathrm{Tr}(U_T^{\dagger}U(h))}}{F},$$

and $\frac{\partial U_K}{\partial h_l}$ may be computed using the following formula [15, Equation (24)]

$$\langle\lambda_k|\frac{\partial U}{\partial h_l}|\lambda_m\rangle = \begin{cases} -\mathrm{i}\Delta t\langle\lambda_k|H_l|\lambda_m\rangle\mathrm{e}^{-\mathrm{i}\Delta t\lambda_k} & \text{if } \lambda_k = \lambda_m \\ \langle\lambda_k|H_l|\lambda_m\rangle\frac{\mathrm{e}^{-\mathrm{i}\Delta t\lambda_k}-\mathrm{e}^{-\mathrm{i}\Delta t\lambda_m}}{(\lambda_k-\lambda_m)} & \text{if } \lambda_k \neq \lambda_m \end{cases}, \tag{14}$$

where U is the unitary gate implemented by the control pulses

$$U = \exp(-\mathrm{i}\Delta t H) = \exp\left(-\mathrm{i}\Delta t\left(H_0 + \sum_l h_l H_l\right)\right). \tag{15}$$

In this case, H_l is the control Hamiltonian corresponding to the control pulse h_l. $|\lambda\rangle$ and λ are the eigenvectors and eigenvalues of the total Hamiltonian of the system. We conduct our simulations using the Broyden–Fletcher–Goldfarb Shanno (BFGS) method [16]. This method is commonly used in quantum control theory for optimization of control pulses [10,11,17]. We first choose an initial guess for the control field vectors. The algorithm then generates iteratively new control field vectors such that at each iteration point, the fidelity is increased. The algorithm terminates after a desired accuracy is reached. This procedure ensures convergence to a local maximum, but does not guarantee the convergence to a globally optimal sequence. Hence, we perform a number of simulations, using different initial conditions. To calculate the gradient of P, one should note an elementary fact concerning differentiation of vector valued functions. For a real vector h, we define y = Ah for some fixed matrix A. Straightforward calculations give us

$$\frac{\partial |y_k|^2}{\partial h_l} = 2\Re(\overline{A}_{kl} y_k). \tag{16}$$

In the case when matrix is a quantum Fourier transform gate A = Q defined in Eq. (6), we obtain that

$$\frac{\partial |y_k|^2}{\partial h_l} = 2\Re(\overline{Q}_{kl} y_k) = \frac{2}{\sqrt{n}}\Re(\mathrm{e}^{-2\pi ikl/n} y_k). \tag{17}$$

This calculation is used to find the gradient of the contribution of high frequencies in the total power given by Eq. (7).

Results Figure 1a shows a plot of the control parameters for the target gate $U_T = NOT_3$ before and after applying the frequency filter. These parameters were found using with the value of the weight $\mu = 1$, resulting in no penalty for high-frequency terms. Clearly, the signal after filtering differs from the original values. This is reflected in the values of the fidelity of the operation. Before filtering, the fidelity is $F > 1-10^{-12}$; however, after filtering, the value drops to $F = 0.85$. Next, in Fig. 1b, we show the plot of the control parameters for the target gate $U_T = NOT_3$, before and after applying the frequency filter. Only this time, the controls were found using the value of the weight $\mu = 0.05$ resulting in a penalty for highfrequency terms.

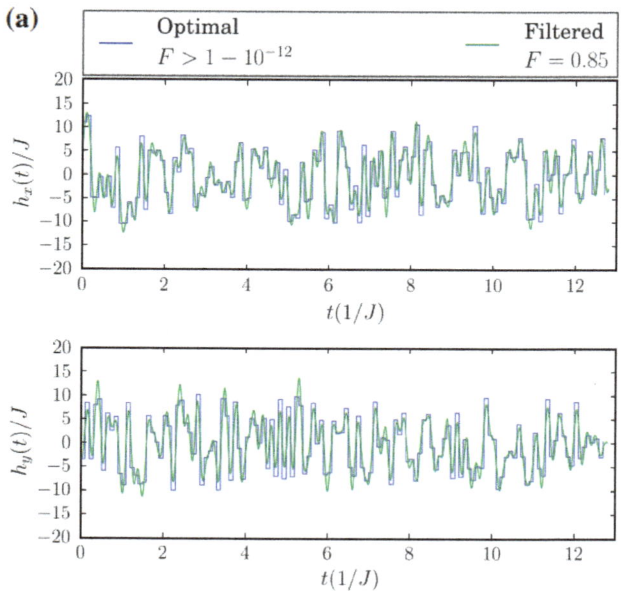

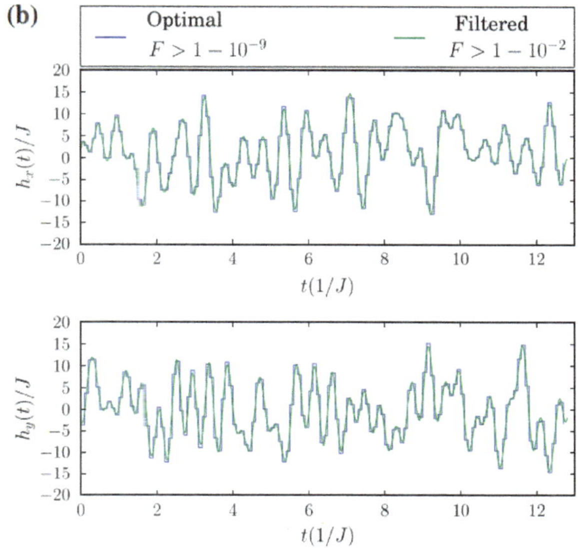

Figure 1: x and y components of the control field for target gate $U_T = NOT_3$. a Unconstrained case. b Constrained case

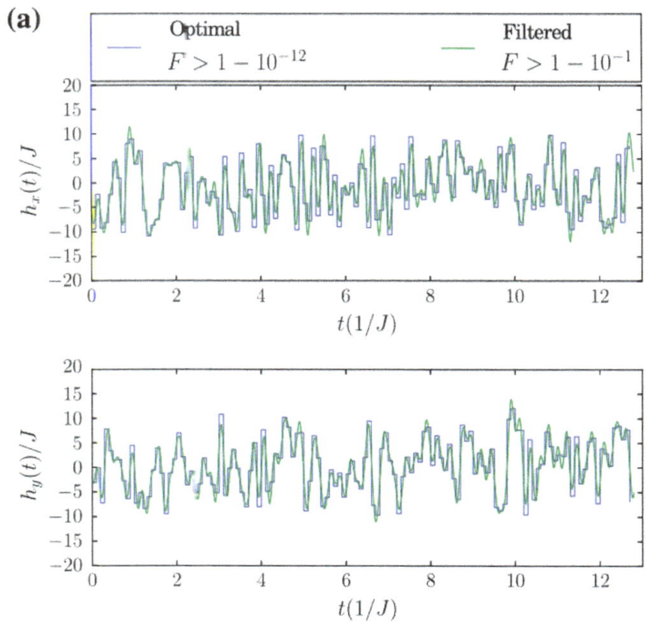

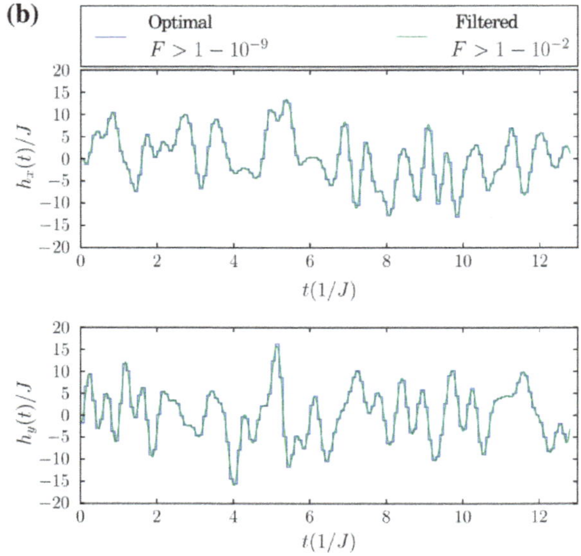

Figure 2 x and y components of the control field for target gate UT = SWAP3. a Unconstrained case. b Constrained case.

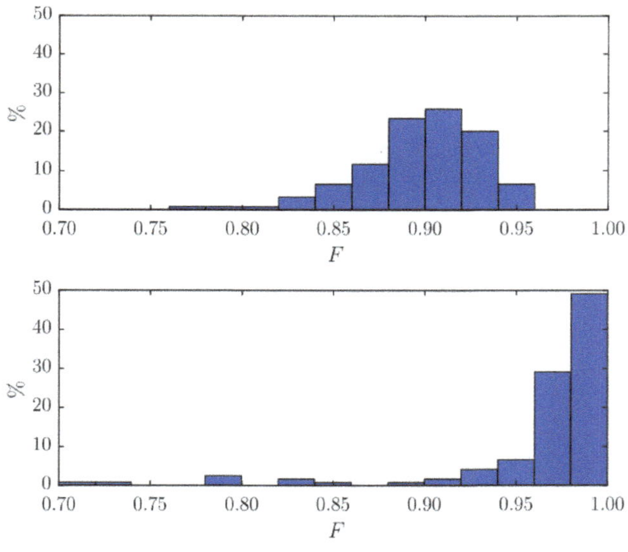

Figure 3: A histogram of the fidelities after filtering. Top without considering spectral constraints, bottom with spectral constraints considered. Target gate $U_T = NOT_3$.

A short glance reveals that the filtered parameters are almost the same as the original ones. This is reflected by the fidelity of the operation. Before filtering, the fidelity is $F > 1 - 10^{-9}$, and after filtering, it drops only to $F > 1 - 10^{-2}$, which is still a satisfactory value. Hence, these sets of control parameters are well-suited for use in computations. Figure 2a, b shows analogical results for the target gate UT = $SWAP_3$. The qualitative results in this case are the same as for the NOT_3 target gate discussed earlier. Fidelity values before and after filtering are the same order as for the NOT_3 gate. Again, we reach a conclusion that these sets of control parameters are well suited for use in computations. Results obtained for the four-qubit chain (not shown here) qualitatively resemble the results for the three-qubit chain. In this case, a penalty for high-frequency terms of the control parameters also leads to higher fidelities after filtering. Figure 3 depicts the histograms of fidelities of filtered control parameters for the NOT3 gate for different initial vectors of control fields. These vectors are drawn at random; coordinates are stochastically independent and have a uniform distribution on an interval [−10, 10]. The top plot shows the histograms for control parameters found without spectral constraints and the bottom one shows what happens when one takes spectral constraints into consideration. Clearly, a typical set of control parameters has a higher fidelity of operation when one considers spectral constraints in the optimization phase. Strictly speaking, around 80% of the control parameter sets have a fidelity greater than 0.96, whereas in the unconstrained case, there are no control parameter sets that have such high fidelities. These facts lead to a conclusion that optimization with spectral constraints will lead to high fidelity of experimental realizations of the NOT_3 gate. Analogical results for the SWAP3 are shown in Fig. 4. The top plot shows the histograms for control parameters found without spectral constraints, and the bottom one shows what happens when one takes spectral constraints into consideration. In this case, around 75% of all constrained control parameter sets have a fidelity higher than 0.96 after filtering. Also, there are no unconstrained control parameter sets with fidelities in this range. Hence, spectral constraints imposed during the optimization step have led to control parameters far less sensitive to experimental equipment limitations.

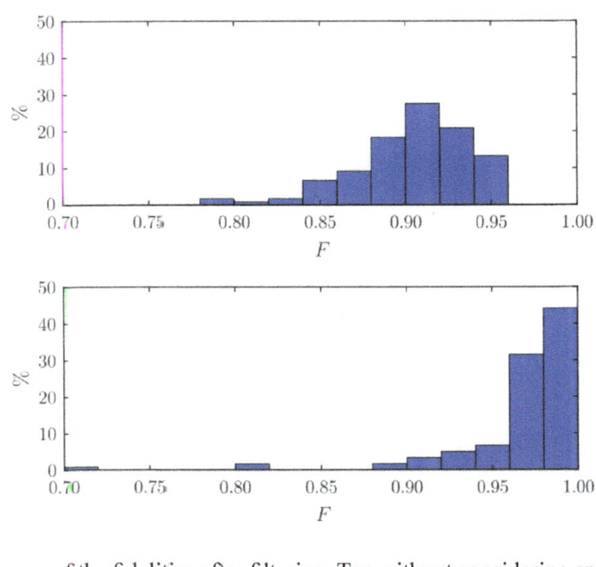

Figure 4 A histogram of the fidelities after filtering. Top without considering spectral constraints, bottom with spectral constraints considered. Target gate $U_T = SWAP_3$.

CONCLUSIONS

We investigated the impact of spectral constraints imposed on the control parameters of a quantum operation on the fidelity of the quantum operation which they implement. In order to compare our approach with the unconstrained case, we apply an ideal low-pass filter to the control parameters. We have shown that imposing spectral constraints on the control parameters leads to higher average fidelity of the quantum operation after appropriate filtering, than in the unconstrained case. These results are independent of the type of quantum operation and the number of qubits in the system under consideration. Furthermore, the requirement for smooth control parameters does not result in the increase in time necessary to conduct a quantum operation. Comparing with other research in the field [11], our times are on the same order. Further work on this subject might take into account more subtle parameters of the experimental setup than the frequency cutoff of signal sources. For instance, one could wish to find control parameters that are far from transient characteristics of the experimental setup. This may lead to an enhancement of fidelities of operations achieved experimentally by eliminating unwanted signal roughness.

ACKNOWLEDGMENTS

Work by Ł. Pawela was partially supported by the Polish National Science Centre under the Grant Number N N514 513340. Z. Puchała was partially supported by the Polish Ministry of Science and Higher Education under the project number IP2011 044271. The development of numerical methods was supported by the Grant N N514 513340. Analytical studies were supported by the project IP2011 044271. Numerical simulations presented in this work were performed on the "Leming" and "Swistak" ′ computing systems of The Institute of Theoretical and Applied Informatics, Polish Academy of Sciences.

REFERENCES

1. Albertini, F., D'Alessandro, D.: The Lie algebra structure and controllability of spin systems. Linear Algebra Its Appl. 350, 213 (2002). ISSN 0024-3795, http://www.sciencedirect.com/science/article/ pii/S0024379502002902
2. d'Alessandro, D.: Introduction to Quantum Control and Dynamics. Chapman & Hall, London (2008)
3. Elliott, D.: Bilinear Control Systems: Matrices in Action. Springer, Berlin (2009). ISBN 1402096127
4. Burgarth, D., Giovannetti, V.: Full control by locally induced relaxation. Phys. Rev. Lett. 99, 100501 (2007). doi:10.1103/PhysRevLett.99.100501
5. Burgarth, D., Bose, S., Bruder, C., Giovannetti, V.: Local controllability of quantum networks. Phys. Rev. A 79, 60305 (2009). ISSN 1094-1622, doi:10.1103/PhysRevA.79.060305
6. Puchała, Z.: Local controllability of quantum systems. Quantum Inf. Process. (2012). doi:10.1007/ s11128-012-0391-x

7. Montangero, S., Calarco, T., Fazio, R.: Robust optimal quantum gates for Josephson charge qubits. Phys. Rev. Lett. 99, 170501 (2007). doi:10.1103/PhysRevLett.99.170501
8. Fisher, R., Helmer, F., Glaser, S.J., Marquardt, F., Schulte-Herbrüggen, T.: Optimal control of circuit quantum electrodynamics in one and two dimensions. Phys. Rev. B 81, 085328 (2010). doi:10.1103/ PhysRevB.81.085328
9. Chaudhury, S., Merkel, S., Herr, T., Silberfarb, A., Deutsch, I., Jessen, P.: Quantum control of the hyperfine spin of a Cs atom ensemble. Phys. Rev. Lett. 99, 163002 (2007). doi:10.1103/PhysRevLett. 99.163002
10. Heule, R., Bruder, C., Burgarth, D., Stojanovi´c, V. M.: Local quantum control of Heisenberg spin chains. Phys. Rev. A 82, 052333 (2010). doi:10.1103/PhysRevA.82.052333
11. Heule, R., Bruder, C., Burgarth, D., Stojanovi´c, V.M.: Controlling qubit arrays with anisotropic XXZ Heisenberg interaction by acting on a single qubit. Eur. Phys. J. D 63, 41 (2011), ISSN 1434-6060, 1434-6079, http://www.springerlink.com/content/b4rr1wg3400442lv/
12. Miszczak, J., Gawron, P., Puchała, Z.: Qubit flip game on a Heisenberg spin chain. Quantum Inf. Process. (2012). doi:10.1007/s11128-011-0322-2
13. Khaneja, N., Reiss, T., Kehlet, C., Schulte-Herbrüggen, T., Glaser, S.J.: Optimal control of coupled spin dynamics: design of NMR pulse sequences by gradient ascent algorithms. J. Magn. Reson. 172, 296 (2005), ISSN 1090-7807, http://www.sciencedirect.com/science/article/pii/S1090780704003696
14. Werschnik, J., Gross, E.: Quantum optimal control theory. J. Phys. B: Atom. Molecul. Opt. Phys. 40, R175 (2007)
15. Machnes, S., Sander, U., Glaser, S.J., de Fouquières, P., Gruslys, A., Schirmer, S., Schulte-Herbrüggen, T.: Comparing, optimizing, and benchmarking quantum-control algorithms in a unifying programming framework. Phys. Rev. A 84, 022305 (2011). doi:10.1103/PhysRevA.84.022305
16. Press, W., Flannery, B., Teukolsky, S., Vetterling, W.: Numerical Recipes in FORTRAN 77: Volume 1 of Fortran Numerical Recipes: The Art of Scientific Computing, vol. 1. Cambridge University Press, Cambridge (1992)
17. Burgarth, D. Maruyama, K., Murphy, M., Montangero, S., Calarco, T., Nori, F., Plenio, M.B.: arXiv 0905.3373 (2009), Phys. Rev. A 81, 040303(R) (2010), http://arxiv.org/abs/0905.3373

9
CHAPTER

QUANTUM COMPUTATION ON GATE-DEFINED SEMICONDUCTOR QUANTUM DOTS

LI HaiOu, Yao Bing, Tu Tao and Guo GuoPing

Keywords: Quantum dot, quantum computing

The emergence of the computer has changed the human lifestyle greatly, and the development of science is increasing the demand for more powerful computers. The processing power of a classical computer depends on the number of transistors on the microprocessor. Thus, to improve the speed of the computer, one should place as many transistors as possible on each microprocessor. However, this approach is limited by the physical constraints on transistors. Specifically, quantum effects pose difficult obstacles for further decreasing the size of transistors. This suggests the idea that it may be possible to exploit these quantum phenomena for computing purposes, by building a processor whose function relies on coherent quantum behaviors. Here comes the quantum computer. Quantum computation is the study of information processing tasks that can be accomplished using quantum mechanical systems. There are five basic requirements for realizing quantum computation in physical devices [1]: (1) Qubit. A two-level quantum system is called a quantum bit, or qubit. The two eigenstates |0> and |1> act as the basic units of the qubit, just like the bits 0 and 1 in classical computation. For practicality, the quantum computer should consist of several hundred qubits, so we need a scalable physical system. (2) Initialization. It is important that the qubits are initialized to a known value before starting the computation. (3) Manipulation. Quantum computation is based on the transformation of the quantum state. Thus the system should be able to perform any unitary transformation (a universal set of quantum gate operations). One can obtain the desired states or results through a sequence of gate operations on the initial states. It has been proved that the

Citation: Li HaiOu, Yao Bing, Tu Tao and Guo GuoPing, Quantum Computation on Gate-defined Semiconductor, Quantum Dots, DOI: 10.1007/s11434-012-5091-5.

combination of full single qubit control with the two qubit CNOT gate can result in all possible quantum algorithms. (4) Readout. The system should be able to measure the final states after the unitary transformation of the initial states. (5) Coherence. The quantum computing system should have a long decoherence time, i.e., many gate operations and measurements can be performed before decoherence occurs. There are several systems with the potential to realize quantum computation. Each system has its own advantages and difficulties. Qubits can be encoded in the states of photons, atoms, nuclear or artificial atoms such as superconductor junctions and semiconductor quantum dots. Some systems can be manipulated conveniently or have long quantum coherence times, while others seem more scalable. We focus on quantum computation in gate controlled semiconductor quantum dots, which is regarded as one of the most promising candidates for quantum computation. This system is more attractive than the others in that it offers the advantages of large scale integration and compatibility with current computer technology. In what follows, we will give a brief introduction to this kind of quantum dot. We will then demonstrate the impressive achievements in the rapidly developing field of quantum computation based on gate controlled quantum dots. Some outlooks will be presented in the last section.

GATE CONTROLLED SEMICONDUCTOR QUANTUM DOTS

Quantum dots are also called "artificial atoms" [2], with a typical size ranging from several nanometers to several microns. The movement of electrons is restricted due to size limitations on quantum dots, which results in discrete energy levels. Common quantum dots can be divided into three groups: colloidal quantum dots, self-assembled quantum dots and gate controlled quantum dots [3]. In this paper, we mainly discuss the last of these. A gate controlled quantum dot is a two dimensional electron gas with some electrodes attached, and is based on the relatively mature semiconductor micro-nanofabrication method. A lateral gate defined quantum dot is depicted in Figure 1. (i) Single quantum dot. The production of quantum dots requires preparing high quality chips from GaAs/AlGaAs heterostructure by Molecule Beam Epitaxy (MBE), manufacturing the top gates by E-beam lithography (EBL) and evaporation (EV), and bringing the source and drain electrodes into contact with the electron gas (Ohmic contact) by annealing. The electrons of the conduction band gather at the interface between GaAs and AlGaAs, and can only move laterally. If we apply a negative voltage to the gate electrodes, the two dimensional electron gas directly under the gate electrode will be pushed away and form a confined region for the electrons, as shown in Figure 1. Gate controlled quantum dots are considered to be one of the most plausible systems for quantum computation since it is easy to control their shape, size and interaction strength precisely [4]. It is also obvious that this kind of quantum dot is scalable. (ii) Double quantum dot. A double quantum dot is also called an "artificial molecule", and provides a good platform for research on the interaction between artificial atoms [5].

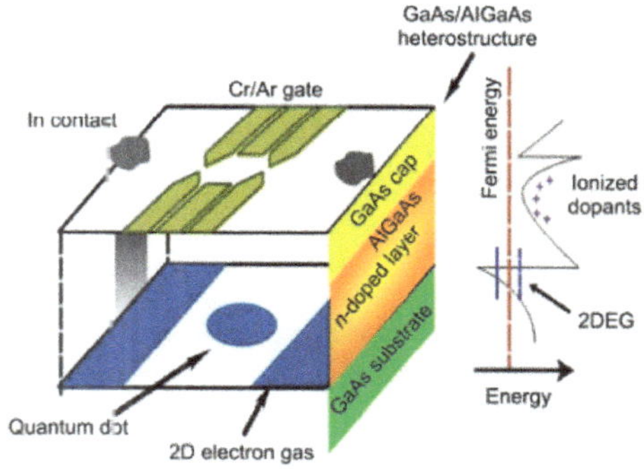

Figure 1 (Color online) Schematic diagram of GaAs/AlGaAs lateral quantum dot.

As shown in Figure 2, there are two kinds of double quantum dot which have been prepared in our group. We would like to emphasize that this is the first effort involving gate controlled quantum dots in China [6].

SPIN BASED QUANTUM COMPUTATION IN GATE CONTROLLED QUANTUM DOTS

In 1998, Loss from Bessel University and Divincenzo from IBM raised the idea of spin based quantum computation in semiconductor quantum dots [7]. Because of weak interactions with the environment, the spin state is seldom disturbed, resulting in a relatively long decoherence time even in the solid phase. The semiconductor quantum dot system is convenient for large scale integration, and is compatible with the present technology. Because of these features, quantum computation based on electron spin in the semiconductor quantum dot is the direction with the most potential in the future.

(i) Electron spin qubit. Loss and Divincenzo proposed encoding the qubit based on a single electron spin, i.e. the qubits |0> and |1> are represented by the states of spin up |↑> and spin down |↓>.

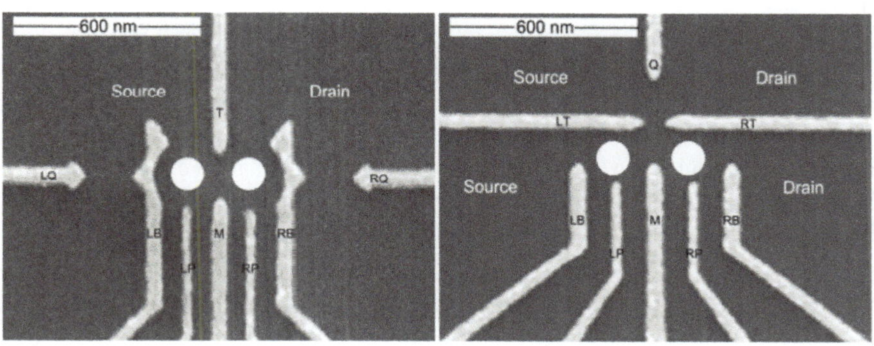

Figure 2: Scanning electron micrograph of GaAs/AlGaAs gate controlled double quantum dot in series (by USTC group).

When there is only one electron in a quantum dot, the system is similar to a hydrogen atom. When an external magnetic field is applied, all orbits can split into a two-level structure because of the existence of electron spin. The state of spin up is parallel to the direction of the magnetic field and is the ground state, while the excited state is spin down and is anti-parallel to the magnetic field. The energy difference between these states is Zeeman energy.

(ii) Initialization of electron spin qubit. It is now possible to initialize electron spin states. In 2003, Kouwenhoven and Vandersypen's team at Delft University placed a quantum dot in a dilution refrigerator with an environment temperature of 15 mK and applied a strong magnetic field of 10 T to the system [8]. This allowed the Zeeman energy of the quantum dot to be 200 μeV, which is much larger than the thermal energy of 25 μeV and smaller than interval between the orbital energy of 1.1 meV and the charging energy of 2.5 meV. Thus, the electron can have two states of spin up |↑> and spin down |↓>, which can be regarded as the qubits |0> and |1>. This is the first time that electron spin qubits have been prepared and characterized experimentally.

(iii) Gate operations of electron spin qubits. In 2006, Kouwenhoven and Vandersypen's team applied a microwave pulse with a frequency that could resonate with the Zeeman energy level of the quantum dot system to realize a single qubit operation [9]. They demonstrated that this oscillating magnetic field is perpendicular to the static one, resulting in rotation of the electron spin in the time domain, i.e. |↑> changes to |↓>. This method is also called the electron spin resonance (ESR) technique and the result is also called Rabi oscillation. In 2005, the Marcus and Yacoby groups at Harvard University applied electrical pulse techniques to realize the coupling of two electron spins in double quantum dots [10]. The two electron spins in the close quantum dots are coupled by a tunneling barrier, which can be expressed by the Hamiltonian of the Heisenberg exchange interaction. The interaction strength depends on the overlap between the wave functions of the two electrons, and can be controlled by the gate electrodes. Thus the spins of the two electrons can be swapped or even entangled by adjusting the time the pulse is applied to the gate electrodes. For example, they demonstrated the SWAP operation |↑↓> to |↓↑> for two adjacent spins in 180 ps. By combining the rotation operation of a single electron spin and the interaction operation of two electron spins, one can realize any quantum gate operation of electron spins.

(iv) Measurement of electron spin qubit. One of the critical problems with qubits is how to measure spin states. In 2004, Kouwenhoven and Vandersypen's team tuned the Fermi level of the source and drain between the spin up and spin down states [11]. The electron that is spin down |↓> will then tunnel out of the quantum dot while the electron that is spin up |↑> will stay in the quantum dot. This difference can be distinguished by the quantum point contact (QPC) electrometer beside the quantum dot.

(v) Long coherence time of electron spin qubit. Decoherence time of the system should be much longer than the operation time of the quantum gate operations. Generally, only if 104 operations can be completed during the decoherence time of the system can quantum computation be realized. In the present system, spin relaxation and dephasing both result in the loss of quantum information. The spin relaxation time T_1 and dephasing time T_2 are strongly related to the material. Spin relaxation comes mainly from spin-orbit coupling, while spin dephasing comes from the hyperfine interaction between the spin and nuclear spin of order 106 around the quantum dots. It has been observed that T_1 and T_2 for GaAs/AlGaAs quantum dots are about 1 ms [11] and 10 ns [10], respectively. In particular, the dephasing coming from nuclear spin can be reduced by the dynamical decoupling method [12] or dynamical nuclear polarization method [13], and as a result the spin dephasing time T_2 can be up to 1 μs, or 105 longer than that of the spin SWAP gate operation (about 180 ps). In summary, the basic requirements for quantum computation based on electron spin qubit are fulfilled in both conceptual and practical terms, as illustrated in Figure 3 [14]. These are exciting times for electron spin quantum computing, as it continues to drive a rapidly expanding field that has a promising future [15].

SOME OUTLOOK OF QUANTUM COMPUTATION IN QUANTUM DOTS

Based on the significant progress achieved in this research area, we now discuss what may need to be done to investigate the fundamental quantum mechanical processes in solids and provide pathways towards applications in quantum information processing.

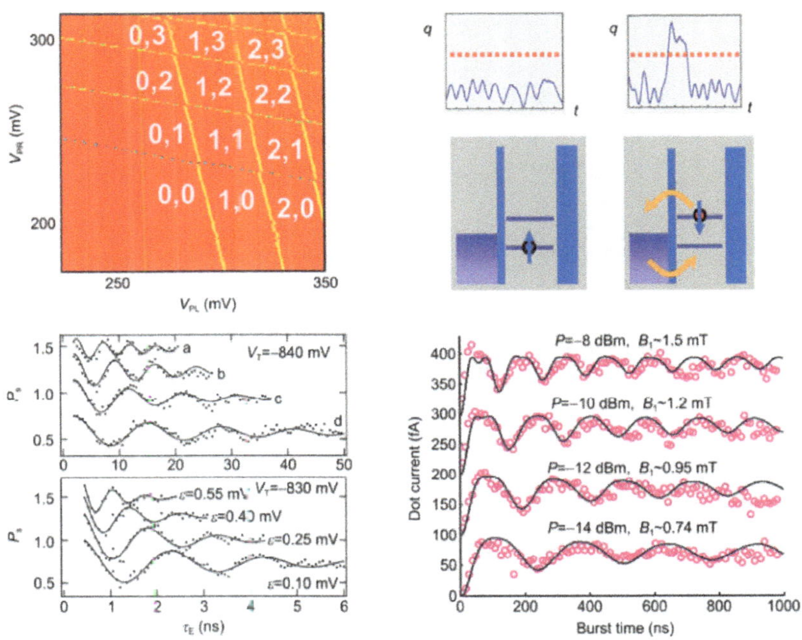

Figure 3: (Color online) Milestones of electron spin qubits in GaAs/AlGaAs quantum dots.

(i) New qubits. Many quantum coding methods, such as quantum error correcting codes, decoherence-free subspaces and dynamic decoupling, have been invented to avoid disturbing quantum information. In order to avoid decoherence caused by nuclear spin, in 2005, the Marcus, Yacoby and Lukin groups at Harvard University introduced an alternative coding proposal using spin qubits in two electron singlet and triplet states [16]. In 2009, the Yacoby group demonstrated that a universal single qubit operation could be realized by nuclear spin polarization [17]. Due to the spin blockade effects, the singlet state will transfer while the triplet state remains unchanged and one can realize the measurement of a single qubit with the nearby QPC detector. Furthermore, this coding method protects the qubits from the low frequency noise and reduces the effect of the hyperfine interaction.

(ii) New architectures. Quantum dots can be coupled directly with each other by capacitive interaction. Recently [18], the Marcus group has tried to realize inter-control between the double quantum dots by capacitive coupling as shown in Figure 4. Our group has also proposed achieving quantum information processing in a scalable architecture consisting of a switchable Ising-like interaction between any two adjacent double quantum dot qubits [19,20].

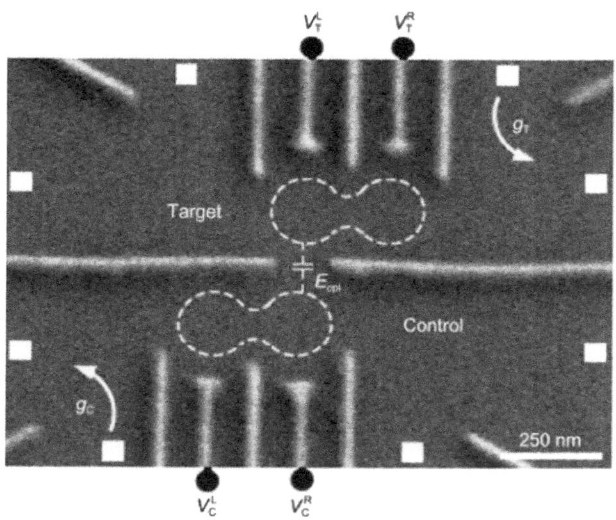

Figure 4: A pair of double quantum dots device.

However, this coupling approach is not easily scalable beyond three or four qubits because of the limitations of the architecture design and the measurement crosstalk. Furthermore, the coupling strength between distant qubits is very weak, while performing gate operations between an arbitrary pair of distant qubits is highly desirable in a quantum computer. An efficient way to achieve this goal is to couple the qubits to a quantum databus, which distributes quantum information among the qubits. As shown in Figure 5, we proposed placing the double quantum dots on the side of a superconducting transmission line resonator (TLR) [21,22]. The TLR functions as a useful quantum databus, and in this way, we can couple many double quantum dot spin qubits and be able to prepare the desired entangled states. Our proposal is an attractive architecture for quantum information processing on a chip. (iii) New materials. Most experiments have focused on quantum dots made from III-V semiconductors, such as GaAs/AlGaAs. However, the coherence of electron spins in these materials is limited by hyperfine interactions with nuclear spins. Recently, research efforts have produced quantum dots using new materials such as Si/SiO_2 [23], Si/ SiGe [24], Ge/Si core/shell nanowires [25], carbon nanotubes, and graphene [26,27], which seem ideally suited to overcome the above limitation as the most abundant nuclei in these new materials have spin zero. For example, Jiang's team at UCLA were the first to find the spin relaxation time T1 in Si/SiO_2 quantum dots, with T1 equal to approximately 1 ms [23].

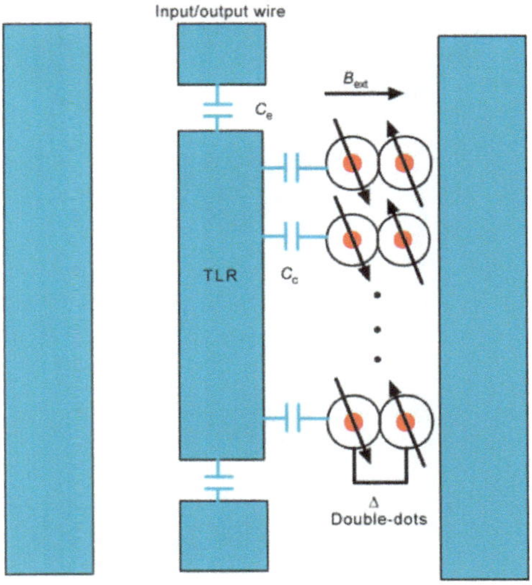

Figure 5: (Color online) Schematic diagram of a TLR and several double quantum dots coupled in circuit (by USTC group).

Our group has also made progress towards quantum computation on graphene based quantum dot devices. From the theoretical aspect, we proposed forming a one-dimensional graphene quantum dot chain from graphene nanoribbon, so that universal quantum computation can be implemented through always-on interaction between the dots [28]. On the experimental side, we have designed and fabricated the first twin-dot structure in which the larger dot serves as a single electron transistor (SET) to read out the charge state of the nearby gate controlled small dot [29], as shown in Figure 6. Owing to the high SET sensitivity, we can probe Coulomb charging as well as the excited states spectra of graphene quantum dots, even though the current through the graphene quantum dot is too small to be measured by conventional transport means. In addition to graphene double quantum dots in series, we reported the first realization of gate controlled parallel coupled graphene double quantum dot in both single layer and bilayer graphene [30], shown in Figure 7.

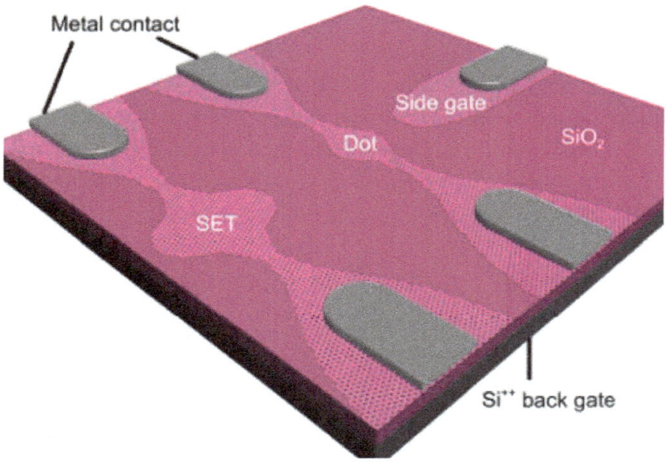

Figure 6: (Color online) Schematic of a representative device with a graphene quantum dot and a nearby SET as a charge sensor (by USTC group).

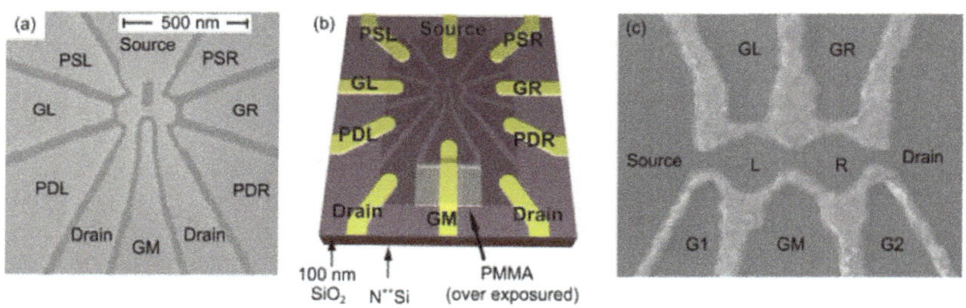

Figure 7: (Color online) Scanning electron micrograph of graphene double quantum dot (a) and in parallel (b), in series (c) (by USTC group).

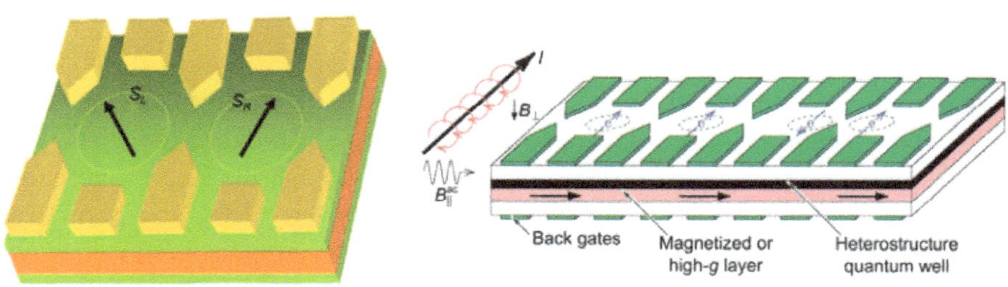

Figure 8: (Color online) There is a long way to go to realize large scale quantum computation from two electron spin qubits.

The graphene double quantum dot and integrated charge sensor serve as an essential building block for forming a solid-state qubit free of nuclear spin. In conclusion, what was considered almost impossible a decade ago, in both conceptual and practical terms, is now a reality. Electron spin qubits can be isolated, initialized, coherently manipulated and read out using electrical techniques. All the requirements for quantum computation based on electron spins in gate controlled quantum dots can be realized in principle, but there is still a long way to go. Much more work is needed before a real quantum computer is produced [31], shown in Figure 8.

REFERENCES

1. Nielsen M A, Chuang I L. Quantum Computation and Quantum Information. New York: Cambridge University Press, 2000
2. Kouwenhoven L, Marcus C. Quantum dots. Phys World, 1998, 6: 35–40
3. Kastner M A. The single-electron transistor. Rev Mod Phys, 1992, 64: 849–858
4. DiVincenzo D P. Double quantum dot as a quantum bit. Science, 2005, 309: 2173–2174
5. Van der Wiel W G, DeFranceschi S, Elzerman J M, et al. Electron transport through double quantum dots. Rev Mod Phys, 2003, 75: 1–22
6. Cao G, Li H O, Tu T, et al. Electron states in parallel double quantum dots with a tunable inter-dot coupling. Chin Phys Lett, 2009, 26: 097302
7. Loss D, DiVincenzo D P. Quantum computation with quantum dots. Phys Rev A, 1998, 57: 120–126
8. Hanson R, Witkamp B, Vandersypen L M K. Zeeman energy and spin relaxation in a one-electron quantum dot. Phys Rev Lett, 2003, 91: 196802
9. Koppens F H L, Buizert C, Tielrooij K J, et al. Driven coherent oscillations of a single electron spin in a quantum dot. Nature, 2006, 442: 766–771
10. Petta J R, Johnson A C, Taylor J M, et al. Coherent manipulation of coupled electron spins in semiconductor quantum dots. Science, 2005, 309: 2180–2184
11. Elzerman J M, Hanson R, Willems van Beveren L H, et al. Single-shot read-out of an individual electron spin in a quantum dot. Nature, 2004, 430: 431–435
12. Bluhm H, Folleti S, Neder I, et al. Dephasing time of GaAs electronspin qubits coupled to a nuclear bath exceeding 200 μs. Nat Phys, 2011, 7: 109–113
13. Reilly D J, Taylor J M, Petta J R, et al. Suppressing spin qubit dephasing by nuclear state preparation. Science, 2008, 321: 817–820
14. Hanson R, Kouwenhoven L P, Petta J R, et al. Spins in few-electron quantum dots. Rev Mod Phys, 2007, 79: 1217–1265
15. Hanson R, Awschalom D D. Coherent manipulation of single spins in semiconductors. Nature, 2008, 453: 1043–1049
16. Taylor J M, Engel H A, Dur W, et al. Fault-tolerant architecture for quantum computation using electrically controlled smiconductor spins. Nat Phys, 2005, 1: 177–183

17. Folleti S, Bluhm H, Mahalu D, et al. Universal quantum control of two-electron spin quantum bits using dynamic nuclear polarization. Nat Phys, 2009, 5: 903–908
18. Van Weperen I, Armstrong B D, Laird E A, et al. Charge-states conditional operation of a spin qubit. Phys Rev Lett, 2011, 107: 030506
19. Guo G P, Zhang H, Tu T, et al. One-step preparation of cluster states in quantum-dot molecules. Phys Rev A, 2007, 75: 050301
20. Zhang H, Guo G P, Tu T, et al. Quantum computation and Bell-state measurements with double dot molecules. Phys Rev A, 2007, 76: 012335
21. Lin Z R, Zhu F Y, Tu T, et al. Generation of quantum-dot cluster states with superconducting transmission line resonator. Phys Rev Lett, 2008, 101: 230501
22. Guo G P, Zhang H, Hu Y, et al. Dispersive coupling between the superconducting transmission line resonator and the double quantum dots. Phys Rev A, 2008, 78: 020302
23. Xiao M, House M G, Jiang H W. Measurement of the spin relaxation time of single electron in a silicon metal oxide semiconductor based quantum dot. Phys Rev Lett, 2010, 104: 096801
24. Shaji N, Simmons C B, Thalakulam M, et al. Spin blockade and lifetime-enhanced transport in a few-electron Si/SiGe double quantum dot. Nat Phys, 2008, 4: 540–544
25. Hu Y J, Churchill O H, Reilly D J, et al. A Ge/Si heterostructure nanowire-based double quantum dot with integrated charge sensor. Nat Nanotech, 2007, 2: 622–625
26. Ponomarenko L A, Schedin F, Katsnelson M I, et al. Chaotic Dirac billiard in graphene quantum dots. Science, 2008, 320: 356–358
27. Wang L J, Cao G, Tu T, et al. Ground states and excited states in a tunable graphene quantum dot. Chin Phys Lett, 2011, 28: 067301
28. Guo G P, Lin Z R, Li X P, et al. Quantum computation with graphene nanoribbon. New J Phys, 2009, 11: 123005
29. Wang L J, Cao G, Tu T, et al. A graphene quantum dot with a single electron transistor as an integrated charge sensor. Appl Phys Lett, 2010, 97: 262113
30. Wang L J, Guo G P, Wei D, et al. Gates controlled parallel-coupled double quantum dot on both single layer and bilayer graphene. Appl Phys Lett, 2011, 99: 112117
31. Ladd T D, Jelezko F, Laflamme R, et al. Quantum computers. Nature, 2010, 464: 45–53

10

CHAPTER

QUANTUM STORAGE OF OPTICAL SIGNALS AND COHERENT MANIPULATION OF QUANTUM STATES BASED ON ELECTROMAGNETICALLY INDUCED TRANSPARENCY

Wang Hai, Xie ChangDe and Peng KunChi

Keywords: Electromagnetically induced transparency (EIT), atomic ensembles, quantum storage, enhanced cross-phase modulation, coherent manipulation of the spin wave

Photons travel fast and are insensitive to the surrounding environment, thus they are often used as carriers of information. However, photons are difficult to trap and store, and the interactions between them are quite weak, which prevent their use in some applications in quantum information processing (QIP). In recent years, electromagnetically induced transparency (EIT) and its related effects have been studied in depth both theoretically and experimentally [1,2]. These studies have shown that EIT effects in three-level or multi-level atomic systems can be applied to significantly reduce the group velocity of light, to trap and store photons using an atomic ensemble and to greatly enhance the optical cross-Kerr nonlinearity. The typical EIT effects can be demonstrated using a three-level L-type atomic system, as shown in Figure 1, in which a strong coupling laser beam couples to the atomic transition from the $|s\rangle$ state to the $|e\rangle$ state and a weak probe beam couples to the atomic transition from the $|g\rangle$ state to the $|e\rangle$ state. Under these conditions, the coherent superposition states $|s\rangle$ and $|g\rangle$ are produced and a transparency window for the probe beam appears at a two-photon resonance (Figure 2). This effect is called electromagnetically induced transparency (EIT) or dark resonance [2]. Based on the EIT dynamic process, the quantum states of light have been stored successfully in atomic ensembles [3,4]. Recent studies show that EIT and its related

---

Citation: Wang Hai, Xie ChangDe and Peng KunChi, Quantum Storage of Optical Signals and Coherent Manipulation of Quantum States Based on Electromagnetically Induced Transparency, doi: 10.1007/s11434-012-5090-6.

effects are important tools in QIP [5–9]. Also, EIT effects can be used to enhance the nonlinear optical interaction between photons [1].

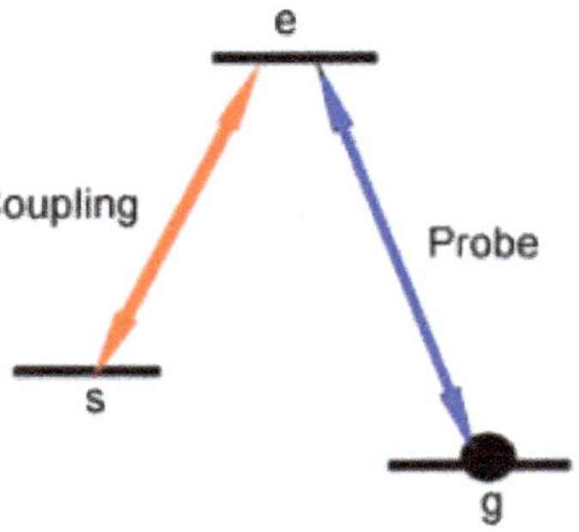

Figure 1 (Color online) Three-level L-type atomic system.

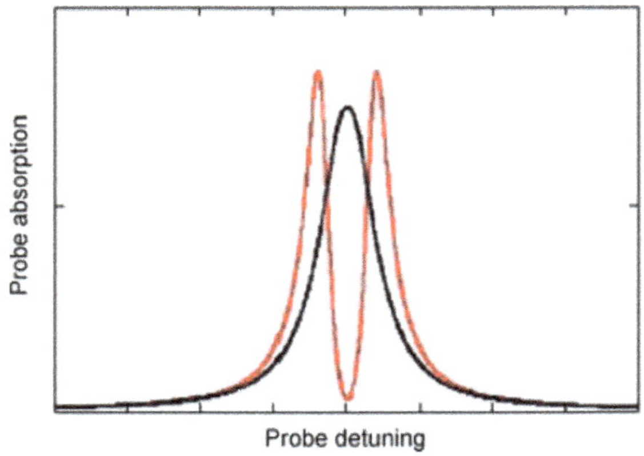

Figure 2: (Color online) Transmission spectra corresponding to the cases without EIT (black) and with EIT (red).

Because of the reduced linear absorption at resonance, EIT effects can be used in resonant nonlinear optical processes, in which the nonlinear optical coefficients of the atomic media are much larger than those under the normal off-resonant conditions [1]. In this paper, we briefly review the experimental studies of enhanced cross-phase modulation based on double EIT [10], the quantum interference of stored dual-channel spinwave excitations [11], and the coherent manipulation of the spin wave vector for the polarization of photons [12] in a four-level tripod atomic system, which were conducted by our group over the last few years.

ENHANCED CROSS-PHASE MODULATION BASED ON DOUBLE EIT IN A FOUR-LEVEL TRIPOD ATOMIC SYSTEM

Cross-Kerr nonlinearity, also known as cross-phase modulation (XPM), has been paid increasing attention because potential applications have been found in demonstrating quantum phase gates [13–15], generating quantum entanglement of single photons [16], constructing a neardeterministic controlled-not gate [17], performing nondestructive Bell-state detection [18], and implementing quantum computation in a new protocol [19]. However, the absence of sufficient XPM in conventional media is an obstacle to its application in QIP. EIT technology provides a promising avenue to solve this problem [1]. The large cross-Kerr nonlinearity in a four-level N-type EIT system has been theoretically predicted [20] and experimentally demonstrated in cold atoms [21]. However, the large cross-Kerr nonlinearity between the probe and trigger (signal) pulses may not always occur in the N-type EIT system because of their group velocity mismatch [22–24]. Further experimental research on this effect is highly anticipated. With this motivation, we carried out an experimental study of double EIT and the enhanced cross-phase modulation between the two fields in a four-level tripod EIT system of the D1 line of ^{87}Rb atoms [10].

The four-level tripod atomic system is shown in Figure 3. The probe field E_P of frequency ω_P is left-circularly polarized (s^-), and coupling occurs at the transitions from levels $|a_{i+1}>$ to $|e_i>$ (i=1, 2). The coupling field E_C of frequency w_C is also s^--polarized, to drive the transitions from levels $|b_{i+2}>$ to $|e_i>$ (i=1, 2, 3). The trigger field E_T of frequency ω_T is right-circularly-polarized (s^+) and couples to the transitions of levels $|b_i>$ to $|e_i>$ (i=1, 2, 3). In this case, the system is coherently prepared into two four-level tripod-type systems [25]. One system is formed by the levels $|a_2>$-$|b_3>$-$|b_1>$-$|e_1>$ (S1), and the other system is formed by the levels $|a_3>$-$|b_4>$-$|b_2>$-$|e_2>$ (S2). The s^- probe transition and s^+ trigger transition share a common excited state, $|e_1>$ ($|e_2>$), which induces coupling between the probe and trigger fields in such a coherently prepared tripod system, and the cross-Kerr phase modulation between the two fields is then enhanced.

The double EIT windows for the probe and trigger fields are shown in Figure 4. These windows are obtained by scanning the probe frequency across ω_{ae} (Figure 4(a)) and scanning the trigger frequency around ω_{be} (Figure 4(b)). The measured XPM phase shifts and the transmissions at the different coupling detuning (D_C) values under the conditions of weak probe and trigger fields ($P_p \approx P_T \approx 14$ µW) are shown in Figure 5. The curves A and B in Figure 5 plot the measured XPM coefficients $n_P^{(2)}$ and $n_T^{(2)}$ obtained at Δ_P -D_C »–0.5 MHz as a function of D_C. The curves C and D in the inset of Figure 5 plot the simultaneously measured transmission values, also at Δ_P -D_C »–0.5 MHz. For increasing D_C , the absolute values of $n_P^{(2)}$ and $n_T^{(2)}$ at Δ_P -D_C »–0.5 MHz decrease, and the accompanying transmission values of the probe and trigger beams increase, which allows the choice of the appropriate coupling frequency detuning to achieve a substantial XPM phase shift with smaller absorption. Figure 5 shows that (2) T n is much larger than $n_P^{(2)}$. This asymmetry of $n_T^{(2)}$ and $n_P^{(2)}$ does not agree with the predictions in [14,15], in which $n_T^{(2)}$ and $n_P^{(2)}$ are symmetrical and can be exchanged each other.

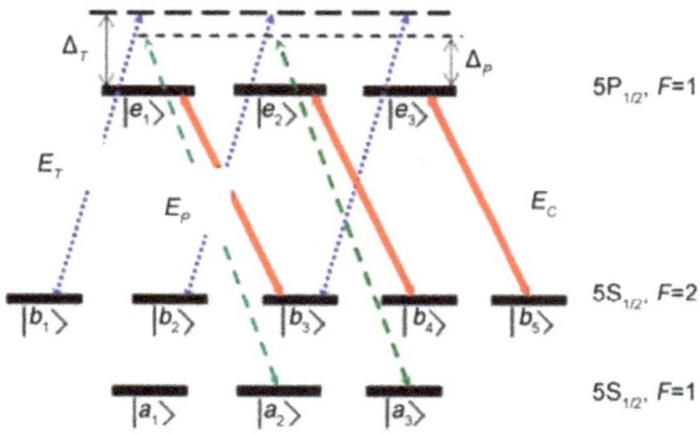

Figure 3: (Color online) Relevant atomic energy level diagram of the D1 line in the ^{87}Rb atom.

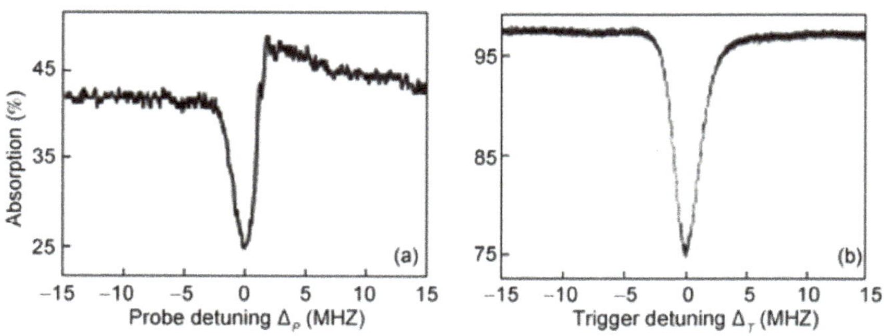

Figure 4: The measured simultaneous EIT windows for probe (a) and trigger (b) beams under the action of a coupling beam.

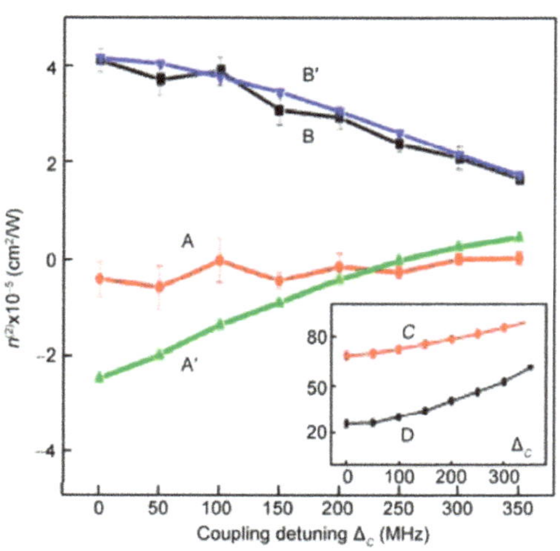

Figure 5 (Color online) The probe (A, A') and trigger (B, B') XPM coefficients as functions of coupling detuning.

One possible reason for this disagreement is that the large difference between the probe Rabi frequencies ($\Omega_{P1} = \Omega_{P2} = 1.15$MHz and the trigger Rabi frequencies (Ω_{T1}=2.83MHz Ω_{T2} = 2 MHz) results in a difference between the populations of the two ground states $|a_2\rangle$ and $|b_1\rangle$ ($|a_3\rangle$ and $|b_2\rangle$). The populations calculated using eq. (3) in [25] are $\rho_{a2,a2} \approx 0.38$, $\rho_{a3,a3} \approx 0.3$, $\rho_{b1,b1} \approx 0.12$ and $\rho_{b2,b2} \approx 0.2$ for $\Omega_{P1} = \Omega_{P2} = 1.15$ MHz, $\Omega_{T1} = 2.83$ MHZ and $\Omega_{T2} = 2$ MHz. The curves A' and B' are the calculated fitting curves for $n_P^{(2)}$ and $n_T^{(2)}$ at Δ_P -D_C =–0.5 MHz using eq. (1) in [10], and are in reasonable agreement with the experimental results. The enhanced XPM in the tripod system presented here is based on simultaneous EIT of the probe and trigger fields, and thus both the probe and trigger absorptions are small. For example, the trigger XPM coefficient can reach 2×10^{-5} cm^2 /W with a transmission of ~60% (see Figure 4). Although the conditional phase shift, which is defined as $\Phi_P^N + \Phi_T^N$ [13] in a quantum phase gate (QPG), still cannot reach ~p under our experimental conditions, if the laser linewidths of the probe and trigger beams are suppressed down to 5 kHz, a conditional phase shift ($\Phi_P^N + \Phi_T^N$) based on the XPM between the two single-photon pulses will reach 1.2 rad for W_C = 30 MHz, which may allow us to perform the QPG operation.

QUANTUM INTERFERENCE OF STORED DUAL-CHANNEL SPIN-WAVE EXCITATIONS IN A SINGLE TRIPOD SYSTEM

Using the EIT dynamic process, photons can be transferred into a single-channel collective spin-wave excitation (SWE) or atomic coherence (Zeeman coherence) in a typical EIT three-level L-type atomic system [2,9]. In storage experiments for classical and quantum optical signals, a prepared three-level L-type atomic system is used as the storage medium and a single-channel spin coherence is involved. The retrieved signal thus exhibits monotonic decay with a storage time on a millisecond time scale in a cold atomic cloud [9], and on a time scale of seconds in an atomic lattice [26] or Bose-Einstein condensate (BEC) [27]. In most atomic memory experiments, the atomic system includes multiple L-type subsystems and thus the dark-state polaritons include many different Zeeman coherences. Because different Zeeman coherences have different Larmor precession periods, the dark-state polaritons experience a "collapse and revival" phenomenon [28] because of the interference between the multiple SWEs. Schemes for dual-channel storage, i.e. storing an optical signal simultaneously into two adjacent SWEs in the four-level inverted-Y [29] or tripod [30] atomic configurations, have been proposed, in which the manipulation of the storage and release of the optical signals can be described well by the generalized dark-state polariton (GDSP) dynamics [29]. Recently, we experimentally demonstrated a dualchannel memory where the retrieval efficiency varied with the relative phase between the two reading light beams in a single four-level tripod system [11]. The scheme of the dualchannel memory in a single four-level tripod system is shown in Figure 6. The weak signal beam (P) is simultaneously stored in the dual-channel coherent SWEs ($\tilde{\sigma}_{ca}$ and $\tilde{\sigma}_{ba}$) by switching off the two circularly-polarized writing beams and can then be read at a later time by switching on the two circularly-polarized beams (R^+, R^-) with a certain relative phase $\delta_R = \phi_R^+ - \phi_R^-$. Because quantum interference is induced by the light-matter interaction between the two stored collective SWEs, the total readout signal shows a maximum or minimum value depending on the relative phase δ_R between the two reading light beams. By varying δ_R, the total readout signal shows a sinusoidal interference pattern. A simple theoretical description of the optical storage/ retrieval processes in the tripod system is given as follows. The generalized dark-state polaritons, including a quantum field $\hat{\varepsilon}_p(z,t)$ and a total collective atomic spin wave $\hat{S}(z,t)$, can be written as

$$\hat{\Psi}(z,t) = \cos\theta(t)\cdot\hat{\varepsilon}_p(z,t) - \sin\theta(t)\cdot\sqrt{N}\hat{S}(z,t), \tag{1}$$

Where

$$\hat{S}(z,t) = \frac{|\Omega_C^+(t)|}{\sqrt{|\Omega_C^+(t)|^2 + |\Omega_C^-(t)|^2}} e^{-i\phi_C^+}\tilde{\sigma}_{ca}(z,t) + \frac{|\Omega_C^-(t)|}{\sqrt{|\Omega_C^+(t)|^2 + |\Omega_C^-(t)|^2}} e^{-i\phi_C^-}\tilde{\sigma}_{ba}(z,t). \tag{2}$$

The mixing angle is given by tgq (t)= $\frac{g\sqrt{N}}{\sqrt{|\Omega^+(t)|^2 + |\Omega^-(t)|^2}}$; $|\Omega^+(t)|$ and $|\Omega^-(t)|$ (where C denotes W or R) are the amplitudes of the Rabi frequencies for the right- and left-circularly-polarized writing (W^+ and W^-) or reading (R^+ and R^-) beams, respectively; ϕ_C^+ and ϕ_C^- are the phases of the writing (W^+ and W^-) or reading (R^+ and R^-) fields, respectively. $\tilde{\sigma}_{ca}$ ($\tilde{\sigma}_{ba}$) is the spin coherence between the states $\langle c|$ and $\langle a|$ ($\langle b|$ and $\langle a|$). In the storage process, the writing beam (including the two components $\Omega_W^+(t)$ and $\Omega_W^-(t)$) is adiabatically switched off over the time interval [t_0, t_1] as the probe pulse enter into the cold atomic ensemble, and so the probe field $\hat{\varepsilon}_P^{in}(z,t)$ s mapped onto the total SWE $\hat{S}_W(z,t)$ and stored in the cold atomic ensemble. Under the condition where $|\Omega_W^+(t)| = |\Omega_W^-(t)|$, this total SWE $\hat{S}_W(z,t)$ can be written from eq. (2), as

$$\hat{S}_W(z,t_1) = \frac{1}{\sqrt{2}}\tilde{\sigma}_{ca}(z,t_1)e^{-i\phi_W^+} + \frac{1}{\sqrt{2}}\tilde{\sigma}_{ba}(z,t_1)e^{-i\phi_W^-} \tag{3}$$

Where $\tilde{\sigma}_{ca}(z,t_1)$ and $\tilde{\sigma}_{ba}(z,t_1)$ are two collective SWEs and are defined as $\tilde{\sigma}_{\beta\alpha}(z,t) = \frac{1}{N_z}\sum_{j=1}^{N_z}\tilde{\sigma}_{\beta\alpha}^j(t) = F(z,t)\hat{\sigma}_{\beta\alpha}e^{i\omega_{\alpha\beta}(z-ct)}$ $\times\hat{\sigma}_{\beta\alpha} = |\beta\rangle\langle\alpha|$ is a spin flip operator (α=a, β=b or c) and F(z,t) is the normalized distribution function for the SWE, which is related to $\hat{\varepsilon}_P^{in}(z,t)$ According to the dark-state polariton theory [2] in the case of $\theta \approx p/2$ (see eq. (1)), the signal field $\hat{\varepsilon}_P^{in}(z,t)$ is mapped into the collective SWE $\hat{S}_w(z,t)$, and the two calculated SWEs are

$$\tilde{\sigma}_{ba}(z,t_1) \propto \frac{1}{\sqrt{2}} \varepsilon_P^{in}(z - z_{01},\ t_0) \hat{\sigma}_{ba} e^{i\phi_W^-}, \tag{4a}$$

and

$$\tilde{\sigma}_{ca}(z,t_1) \propto \frac{1}{\sqrt{2}} \varepsilon_P^{in}(z - z_{01},\ t_0) \hat{\sigma}_{ca} e^{i\phi_W^+}, \tag{4b}$$

Where $\varepsilon_P^{in}(z - z_{01},\ t_0)$ is the slowly-varying amplitude of the input probe field at $z - z_{01}$ (where $z_{01} = \int_{t_0}^{t_1} dt' v_g(t')$) and time $t=t_0$. In a finite magnetic field, the atomic coherences (SWEs) $\tilde{\sigma}_{ba}(z,t_1)$ and $\tilde{\sigma}_{ca}(z,t_1)$ experience a Larmor precession, and after a storage time interval $t = t_2 - t_1$, they become

$$\tilde{\sigma}_{ba}(z,t_2) \propto \frac{1}{\sqrt{2}} \varepsilon_P^{in}(z - z_{01},\ 0) \hat{\sigma}_{ba} e^{i\phi_W^- + i2\Omega_L\tau} \tag{5a}$$

and

$$\tilde{\sigma}_{ca}(z,t_2) \propto \frac{1}{\sqrt{2}} \varepsilon_P^{in}(z - z_{01},\ 0) \hat{\sigma}_{ca} e^{i\phi_W^+}. \tag{5b}$$

$\Omega_L = g_F \mu_B B / \hbar$ is the Larmor procession frequency, and $2W_L$ is the frequency shift between the Zeemar sublevels $|c\rangle$ ($|F = 2, m = -1\rangle$) and $|b\rangle$ ($|F = 2, m = +1\rangle$) in the magnetic field B. At time t2, if the two reading beams, $|\Omega_R^+(t)| e^{-i\phi_R^+}$ and $|\Omega_R^-(t)| e^{-i\phi_R^-}$ (with $|\Omega_R^+(t)| = |\Omega_R^-(t)|$), are used to read the two-channel SWEs stored in the tripod atomic medium, then the collective SWE $\hat{S}(z,t)$ becomes

$$\begin{aligned} \tilde{S}_R(z,t_2) &= \frac{1}{\sqrt{2}} \tilde{\sigma}_{ca}(z,t_2) e^{-i\phi_R^+} + \frac{1}{\sqrt{2}} \tilde{\sigma}_{ba}(z,t_2) e^{-i\phi_R^-} \\ &= \frac{1}{2} \varepsilon_p^{in}(z - z_{01}, t_0) \left(\hat{\sigma}_{ca} e^{i\phi_+^+ - i\phi_R^+} + \hat{\sigma}_{ba} e^{i\phi_w^- - i\phi_R^- + i2\Omega_L\tau} \right), \end{aligned} \tag{6}$$

which will be turned into the released total optical signal. The total readout signal can then be written as

$$\varepsilon_P^{out}(z,t) \propto \left\langle \hat{S}_R(z - z_{02}, t_2) \right\rangle$$
$$\propto \frac{1}{2}\varepsilon_P^{in}(z - z_0, t_0)(e^{-i\delta_R + i\delta_W - i2\Omega_L\tau} + 1), \tag{7}$$

Where $z_{02} = \int_{t_2}^{t} v_g(t')dt'$, $z_0 = z_{01} + z_{02}$, $\delta_R = \phi_R^+ - \phi_R^-$ and $\delta_W = \phi_W^+ - \phi_W^-$

The individual signals $\varepsilon_{P^+}^{out}$ and $\varepsilon_{P^-}^{out}$ retrieved separately from the two memory channels ($\tilde{\sigma}_{ca}$ and $\tilde{\sigma}_{ba}$) can be easily calculated using eqs. (6) and (7) under conditions of $\Omega_R^-(t) = 0$ or $\Omega_R^+(t) = 0$ with $|\varepsilon_{P^+}^{out}(z,t)| = |\varepsilon_{P^-}^{out}(z,t)| = \varepsilon_{\text{single}}^{out}(z,t) \propto \frac{1}{\sqrt{2}}\varepsilon_P^{in}(z - z_0, t_0)$.
The total retrieved signal intensity can be expressed as

$$|\varepsilon_P^{out}|^2 \propto 2|\varepsilon_{\text{single}}^{out}|^2 \cos^2\left(\frac{\delta_R}{2} - \frac{\delta_W}{2} + \Omega_L\tau\right), \tag{8}$$

Where $\varepsilon_{\text{single}}^{out}$ denotes the amplitude of the readout signal retrieved from a single channel (only with $\Omega_R^+(t)$ or $\Omega_R^-(t)$). From eq. (8), we can easily see that when $\Delta = \delta_R - \delta_W + 2\Omega_L\tau = 0$ or π, $|\varepsilon_P^{out}|^2 \approx 2|\varepsilon_{\text{single}}^{out}|^2$ or $|\varepsilon_P^{out}|^2 = 0$, i.e. the total optical readout signal shows the constructive or destructive interference. The experiment was carried out using a standard magneto-optical trap with 87Rb atoms. The atoms are prepared in an ideal tripod system by selectively pumping the atomic population into the $5S^{1/2}$, F=1, m=+1 state [31]. The experimentally prepared tripod system can be described well using the simple energy diagram shown in Figure 6(a), and the configurations of the writing, reading, and signal beams are shown in Figure 6(b), with all of the beams split from a single high-power diode laser. The relative phase difference is introduced by separating the horizontal and vertical polarization components of the reading beam, and is controlled using the electro-optic modulator (EOM). Figure 7(a) shows the stored and retrieved signals from a single (left) channel (blocking the right writing/reading beams), which corresponds to the typical L-type system. Following the standard procedure for optical storage and retrieval, the retrieved signal is read after a 380 ns time delay (right peak). The right channel shows the same behavior when used singly. When both the left and right writing beams are turned on and have the same power ($|\Omega_W^+(t)| = |\Omega_W^-(t)|$), as in the single-channel case, the optical signal is stored simultaneously in two coherent SWEs ($\tilde{\sigma}_{ca}$ and $\tilde{\sigma}_{ba}$).

If a single reading beam (either R^+ or R^-) is used, the readout signal is approximately the same as in the single-channel case. Figure 7(b) and (c) present the readout signals from the individual memory channels $\tilde{\sigma}_{ba}$ and $\tilde{\sigma}_{ca}$ (with only one reading beam on), respectively. When both reading beams (R^+ and R^-) are on simultaneously and the total phase is tuned to satisfy D = 0 by adjusting the relative phase between the two reading beams (δ_R), a maximum readout signal is obtained (Figure 7(d)).

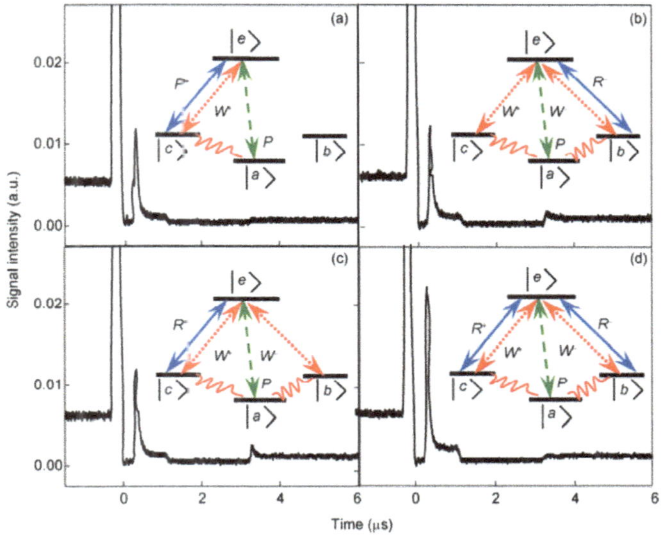

Figure 7: (Color online) Measured results for different storage-release cases. (a) The writing beam is right circularly polarized, and the reading beam is also right circularly polarized. (b)–(d) The writing beam is linearly polarized (consisting of left and right circularly polarized), and in (b) the reading beam is left circularly polarized, in (c) the reading beam is right circularly polarized, and in (d) the right and left circularly polarized reading beams are both present, and the total phase difference δf is approximately 0.

The retrieved signal intensity is twice that of the single channel case for the same storage time of 380 ns, as indicated in the calculations. These results clearly show that the quantum interference between the two stored collective SWEs was mediated by the coherent reading light beams. When the total phase difference Δ is adjusted to be p by tuning δ_R, no optical signal is retrieved from the atomic medium, even though both of the reading beams are turned on at 380 ns. After a delay time t of 3.4 μs, the atomic medium is read again with other (writing) laser beams, as shown in Figure 8(a). For comparison, Figure 8(b) presents the readout results when the stored SWEs were not perturbed before the system was read at 3.4 μs storage time. These results indicate that although the stored SWEs have been read with two reading beams at 380 ns for Δ= p, they have not been converted into optical signals because of destructive quantum interference between them and they have remained in the atomic medium, until they are read again with other laser beams at a later time. Figure 9 (square points) shows the total retrieved signal as a function of δ_R at t = 380 ns. For a fixed $\delta_W = \pi / 2$ and Larmor precession (e.g. $2W_L t$=constant), δ_R is proportional to D. The retrieval efficiency shows a sinusoidal interference pattern as δ_R is varied (see eq. (8)). At the same time, this relationship can be viewed as the population projected into the given state, $\hat{\sigma}_{+}^{\delta_R}|a\rangle$, and the interference pattern shows that the population in the $\hat{\sigma}_{+}^{\delta_R}|a\rangle$ state depends on the relative phase δ_R between the two basis states. After the stored collective SWEs are first read by a pair of reading beams (at 380 ns), the atomic medium is read again by a horizontally-polarized (writing) laser beam at 3.4 μs.

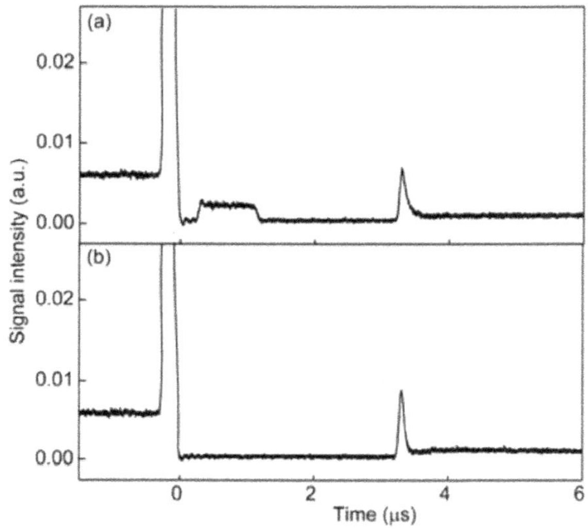

Figure 8: Measured retrieved results for the second released signals.

The signals retrieved at 3.4 μs are also shown in Figure 9 (circular points) as a function of δ_R (for the first pair of reading beams). It is clear that if all of the stored SWEs in $\hat{S}(z,t)$ are mapped completely into the signal field at the earlier time (380 ns), the second reading by the second beam will not be able to retrieve any information. However, if the first reading at t=380 ns did not retrieve information from the SWEs (because Δ=0), the second reading at 3.4 μs will have the maximum readout signal. In the tripod atomic system with a small magnetic field (~300 mG), the Zeeman sublevels $|b\rangle$ and $|c\rangle$ are split, which causes the Larmor precession for the stored collective SWEs. The total readout signal intensity as a function of storage time shows sinusoidal behavior, as shown in Figure 10 (circular points). However, such oscillatory behavior in the readout intensity can severely limit the ability to retrieve arbitrarily polarized states, and the ability to probe longlived entanglements between photons and SWEs at any desired time. Using the current phase-sensitive storage/ retrieval scheme, we demonstrate that by adjusting the relative phase difference between the two reading beams, this "collapse and revival" behavior in the readout signal intensity caused by the Larmor precession can be completely compensated (i.e. δ_R can compensate $2W_Lt$ in a total phase difference D), as shown in Figure 10 (square points). In this case, the maximum readout signal can be accessed at any storage time, and controlled by adjusting the relative phase difference between the two reading beams.

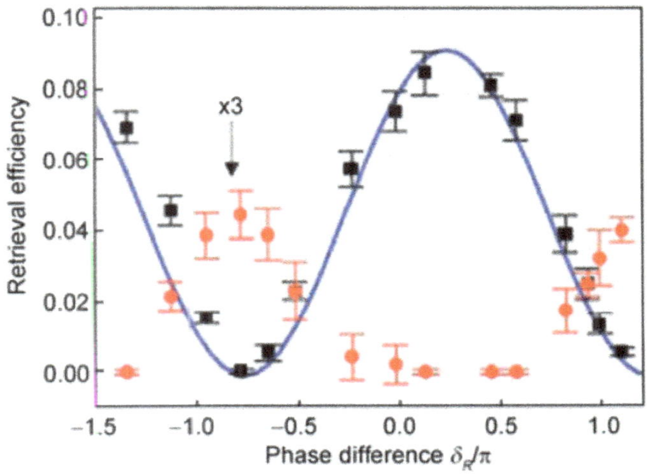

Figure 9: (Color online) Retrieval efficiency as a function of the relative phase δ_R. The blue square points are the total retrieved signal intensities at 380 ns storage time for different values of Δ. The solid blue line is the best fit to the data. The red circular points are the retrieved signal intensities at 3.4 μs storage time, after the system was first read at 380 ns with the pair of reading beams.

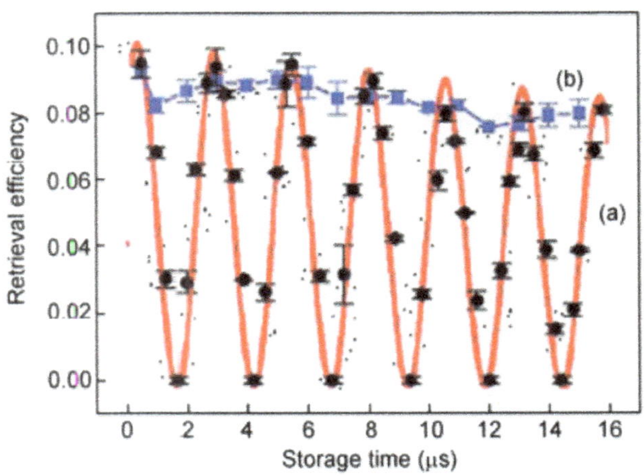

Figure 10 (Color online) Retrieval efficiency and compensation of Larmor precession. Black circular points: retrieval efficiency as a function of storage time. Curve (a) (red curve): fitting to the function $R_a = A\cos^2(\Omega_L t - \varphi/2)e^{-t/t_0}$, where the fitting parameters are A=0.1, $\Omega_L = 2\pi \times 0.21$ MHz, $\varphi = 0.3\pi$ (initial phase), and the lifetime $t_0 \approx 90$ μs. Curve (b) (with blue squares): the results with compensated Larmor precession by adjusting the relative phase δ_R.

The above two-channel storage scheme and coherent phase-sensitive readout technique have significant advantages in extending single-channel memory systems for quantum information processing, especially for qubit memory, and have overcome some important obstacles in using atomic ensembles as optical memory devices. This two-channel storage/retrieval scheme provides a unique way to measure the relative phase of the qubit memory and probe the population in a given coherent superposition state. The accessed Rabi oscillation (with a lifetime of ~90 μs) between two superposition states of the SWEs has potential for application to performing qubit gate operations. Also, by compensating the Larmor precession by tuning the relative phase between the two reading beams, the collapse and revival behavior of the total readout signal is eliminated and the total readout signal is maintained at its maximum for more than 100 μs. This phase compensation method can also be applied to experiments for the quantum memory for photon polarization states and entanglement of photons and atomic ensembles. In these experiments, if we could first measure the evolution of the relative phase of the qubit memory caused by the Larmor precession, and alter the relative phase between the two retrieved orthogonally-polarized photons using an EOM according to the measured phase data, then the changes in the photon polarization states because of the Larmor precession could be compensated, and therefore the memory time would be significantly increased.

COHERENT MANIPULATION OF SPIN WAVE VECTOR FOR POLARIZATION OF PHOTONS IN AN ATOMIC ENSEMBLE

The coherent manipulation of the quantum states in memory elements plays an important role in QIP [2,32,33]. Atomic ensembles with the ability for collectively enhanced coupling to a particular light mode [2] can serve as good quantum memory elements and have attracted considerable attention in recent years. Recent experimental studies of the collective qubit memory used to achieve atom-photon entanglement have made great progress. In these studies [34–36], two orthogonal [34] or two spatially distinct [35] spin waves, or two atomic ensembles sharing a spin-wave excitation [36], are used to encode the long-time atomic qubit. The lifetime of the qubit memory for an optical lattice spin wave has reached 3 ms [34]. QIP schemes based on atomic ensembles such as entanglement swapping [32], the multipartite entanglement of atomic ensembles [33,37], and the controlled-NOT gate [37] have been proposed. In these proposed schemes, the memory qubits may be encoded in two orthogonal spin waves, and single-bit operations are required. These single qubit gate operations are the R(q, j) and $R_z(\phi_z)$ rotations [38], which can be used to build an arbitrary rotation on the Bloch sphere. The single qubit operations have been realized in a single ion [38], in trapped neutral atoms [39,40], or in a quantum dot system [41]. In 2011, we demonstrated an experiment on the two unitary rotations R(q,j) and $R_z(\phi_z)$ on the basis of their formation by two orthogonal spin waves in an atomic ensemble [12]. The relevant ^{87}Rb levels involved in the storage, Raman manipulation and readout processes are shown in Figure 11(a)–(c), respectively, where $|\uparrow\rangle = |5^2S_{1/2}, F=1, M_F=+1\rangle$, $|\downarrow\rangle = |5^2S_{1/2}, F=1, M_F=-1\rangle$, $|s\rangle = |5^2S_{1/2}, F=2, M_F=+1\rangle$ and $|e\rangle = |5^2P_{1/2}, F'=1, M_F=0\rangle$. The spin wave associated with the coherence between the state $|s\rangle$ and the superposition state $|\Phi\rangle = \alpha|\uparrow\rangle + e^{-i\Delta}\beta|\downarrow\rangle$ (where $|\alpha|^2 + |\beta|^2 = 1$) is written as the vector is the spin wave component associated with the coherence between the state $|s\rangle$ and the state $|\uparrow\rangle$ ($|\downarrow\rangle$).

For an initial condition of $\Phi(0) = |\uparrow\rangle$, the spin wave is expressed as $\hat{S}(z,0) = \langle|\tilde{\rho}_{s\uparrow}(z,0)|\rangle\begin{pmatrix}1\\0\end{pmatrix}$.

The coherent population can be transferred between the $\hat{S}_{s\uparrow}(z,t)$ and $\hat{S}_{s\downarrow}(z,t)$ channels via a two-photon Raman transition. The Raman laser beam consists of σ^+- and σ^-- polarized light fields E_{R+} and E_{R-}, which couple to the $|\uparrow\rangle \leftrightarrow |e\rangle$ and $|\downarrow\rangle \leftrightarrow |e\rangle$ transitions. The unitary Raman rotation R(q, f)can be written as

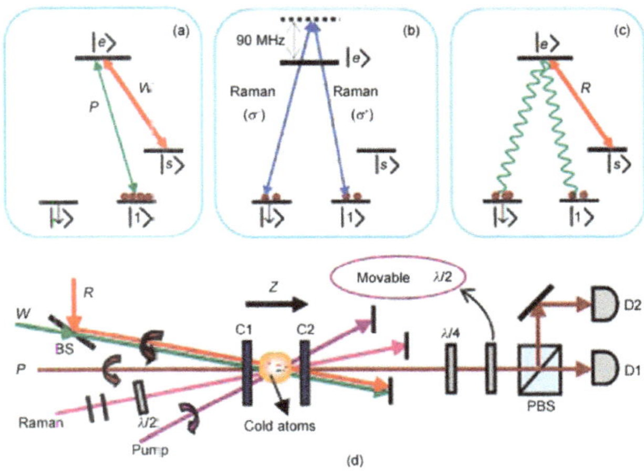

Figure 11: (Color online) The experimental scheme for rotation of the spin-wave vector. (a)–(c) Atomic level structures of the ^{87}Rb atom for optical storage, Raman manipulation, and retrieval. (d) Experimental setup, where R, W, and P denote reading, writing, and probe laser beams, respectively; BS, beam splitter; C1 and C2, coils; PBS, polarization beam splitter; D1 and D2, photodetectors.

$$R(\theta,\phi) = \begin{pmatrix} \cos\frac{\theta}{2} & -ie^{i\phi}\sin\frac{\theta}{2} \\ -ie^{-i\phi}\sin\frac{\theta}{2} & \cos\frac{\theta}{2} \end{pmatrix}.$$

where q = Ω_Rt and f=f_+- f_- is the phase difference between the Raman fields E_{R+} and E_{R-}.

In a magnetic field, the two spin-wave components $\hat{S}_{s\uparrow}(z,t)$ and $\hat{S}_{s\downarrow}(z,t)$ experience the Larmor precession and get the phase shifts ψ_1 and ψ_2 respectively, which introduce the relative phase $\varphi_z = \psi_1 - \psi_2$ after an evolution time interval t. The evolution of the spin wave is described by $\hat{S}(z,t) = R_z(\varphi_z)\hat{S}(z,0) = \langle|\tilde{\rho}_{s\uparrow}(z,0)|\rangle\begin{pmatrix}\alpha\\ e^{i\varphi_z}\beta\end{pmatrix}$, where $R_z(\varphi_z) = \begin{pmatrix}1 & 0\\ 0 & e^{i\varphi_z}\end{pmatrix}$

is the matrix rotation, and the initial spin wave is assumed to be $\hat{S}(z,0) = \langle|\tilde{\rho}_{s\uparrow}(z,0)|\rangle \begin{pmatrix} \alpha \\ \beta \end{pmatrix}$. .It is obvious that the spin wave acquires a relative phase of D = φ_z because of the Larmor precession.

We used a cold 87Rb atomic cloud to serve as the atomic ensemble. The measured optical depth of the cold atoms is about 1.5 and the trap temperature can reach ~200 μK (see [12] for the experimental details).

To perform the spin wave R(q,j) rotation induced by a Raman laser pulse, we first observe the population transfer efficiency from $\hat{S}_{s\uparrow}(z,t)$ to $\hat{S}_{s\downarrow}(z,t)$ components as a function of the Rabi frequency of the Raman laser beam. Figure 12 (the square (circle) points) presents the normalized retrieval efficiency N_+ (N_-) as a function of the Raman-Rabi frequency Ω_R , and the solid curves in Figure 12 are the best fits of the functions $(1 \pm \cos 2\pi B\Omega_R\tau_R)/2$, with the parameter B=0.82 from the experimental data. These curve fits present sinusoidal Rabi oscillations which are consistent with the theoretical predictions of $N_\pm$ = $(1 \pm \cos 2\pi\Omega_R\tau_R)/2$. At $\theta = 2\pi B\Omega_R\tau_R = \pi$ (the upper axis in Figure 12), the population transfer efficiency reaches ~97%. By applying a variable field of the magnetic pulse B(t) to the atomic ensemble, we implement a spin-wave phase operation, i.e. $R_z(\varphi_z) = \begin{pmatrix} 1 & 0 \\ 0 & e^{i\varphi_z} \end{pmatrix}$ rotation.

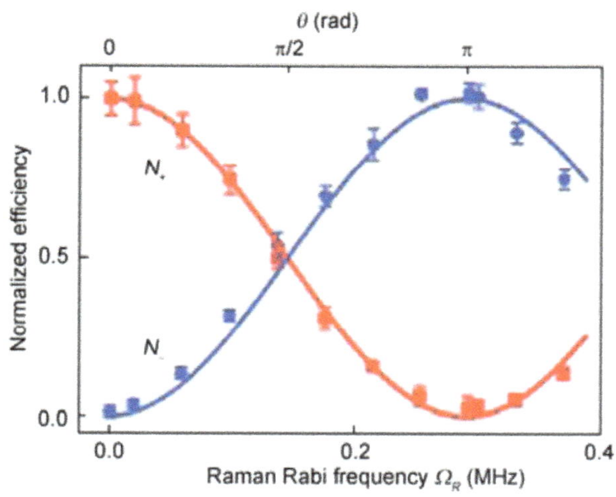

Figure 12: (Color online) The coherent population transfer efficiency between the two spin-wave components as a function of the Rabi frequency of the Raman laser beam.

We apply a current pulse I(t) to a pair of coils, C1 and C2, to generate the pulsed magnetic field B(t). The generated pulsed magnetic field B(t) is proportional to the current I(t). By changing the area $\int I(t)\mathrm{d}t$ of the current pulse, we can control the changes in $\Delta\varphi_z$ and then implement the $R_z(\Delta\varphi_z)$ rotation. By observing the interference fringe between the readout fields $\varepsilon_+^{out}(t)$ and $\varepsilon_-^{out}(t)$, we verify the implementation of the $R_z(\Delta\phi_z)$ rotation. In the experiment, we first apply a p/2 linearly-polarized Raman laser pulse (with an orientation angle of φ_R=8°) and then apply a magnetic pulse with a duration of τ~7 μs to rotate the initial spin wave $\tilde{S}(z,0)$. After these operations, the spin wave evolves into $\tilde{S}(z,t) = \frac{\langle|\tilde{\rho}_{s\uparrow}(z,0)|\rangle}{\sqrt{2}}\begin{pmatrix} 1 \\ e^{i\Delta_0 + i\Delta\varphi_z} \end{pmatrix}$, where $\Delta_0 = -\pi/2 - \phi + \varphi_{z0}$ is kept constant. At t=17 μs. we turned on the reading beam to read the spin wave. Figure 13 plots the relative retrieved efficiencies N±45° (square and circular dots) as a function of the relative magnetic pulse area A [12]. The solid lines are the curve fits of the sinusoidal functions (ab A ± cos 2π) / 2 to the measured data $N_{\pm 45°}$, with parameters $\Delta_0 = 0$, a=0.93 and contrast b=0.93. The magnetic pulse width is ~7 μs in the experiment presented here, and we believe that it can be further decreased by improving the coils and the current pulse source. To estimate the p/2-pulse fidelity, we performed a Ramsey experiment by applying two linearly-polarized p/2 Raman laser pulses with a variable time interval τ_R. At t=17 μs, we retrieve the spin wave $\hat{S}(z,t)$ into the readout signal fields out $\varepsilon_\pm^{out}$. Figure 14 plots the measured relative photon numbers $N_\pm$ as a function of the interval τ_R. The solid lines are the curve fits of the sinusoid function $(1 \pm b_\pm \cos 2\pi\tau_R / T_L)/2$ to the data $N_\pm$, with a Larmor period of $T_L = 2.82 \times 10^{-6}$ s ,and contrasts b_+=1 for N_+, as well as b_-=0.85 for N_-. According to [13], we calculate the p/2 fidelity using the lower value of contrast (b_-), which is

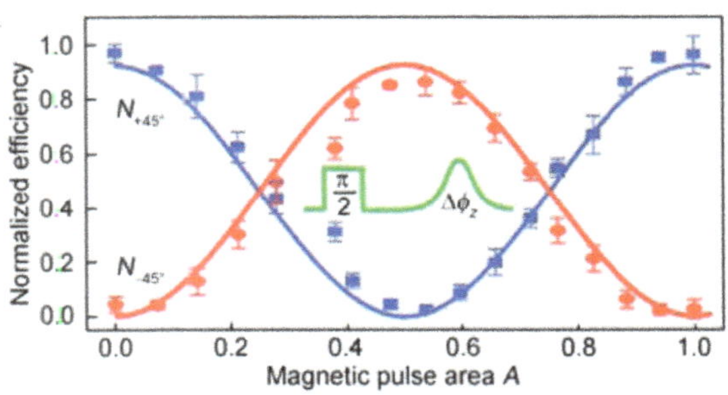

Figure 13: (Color online) The normalized retrieval efficiencies as a function of the magnetic pulse area A.

The above experiments present the R(q, j) and $R_z(\Delta\phi_z)$ rotations of the spin-wave vector on the Bloch sphere when Raman laser and magnetic field pulses are applied. For the case where the initially stored spin wave has an atomic collective excitation (i.e. $N\int\langle\tilde{\rho}_{s\uparrow}(z,0)\tilde{\rho}_{s\uparrow}^{+}(z,0)\rangle dz = 1$), the two spin-wave components sharing a single excitation become the two basis states of a qubit. The manipulation demonstrated in the work presented here may thus be used to implement a single qubit gate operation, which has potential applications in QIP based on an atomic ensemble. In conclusion, we have briefly summarized the experimental studies conducted recently by our group on enhanced cross-phase modulation based on double EIT, the quantum interference of stored dual-channel spin-wave excitations and the coherent manipulation of the spin wave vector for polarization of photons in a single tripod atomic system. The focus of this research is on the development of a long lifetime quantum memory for optical information in atomic ensembles.

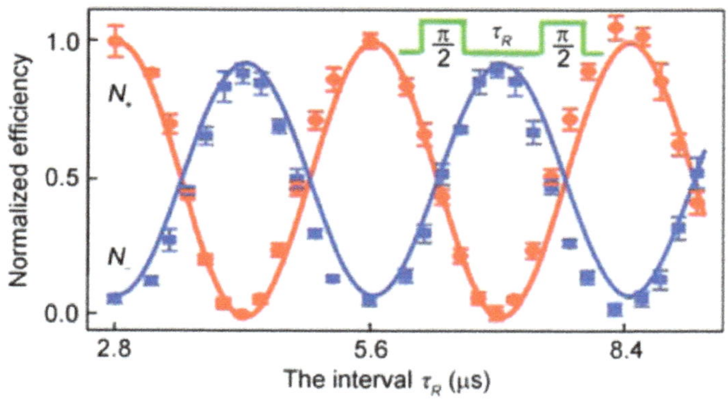

Figure 14: (Color online) Ramsey fringes for a pair of p/2 pulses with a variable separation time τ_R. Red circular points (blue square points): the normalized retrieval efficiency N_+ (N_-). The solid lines are sinusoidal fits to the data.

REFERENCES

1. Lukin M D, Imamoglu A. Controlling photons using electromagnetically induced transparency. Nature, 2001, 413: 273–276
2. Lukin M D. Trapping and manipulating photon states in atomic ensembles. Rev Mod Phys, 2003, 75: 457–472
3. Liu C, Dutton Z, Behroozi C H, et al. Observation of coherent optical information storage in an atomic medium using halted light pulses. Nature, 2001, 409: 490–493
4. Phillips D F, Fleischhauer A, Mair A, et al. Storage of light in atomic vapor. Phys Rev Lett, 2001, 86: 783–786
5. Van der Wal C H, Eisaman M D, André A, et al. Atomic memory for correlated photon states. Science, 2003, 301: 196–200
6. Matsukevich D N, Chanelière T, Bhattacharya M, et al. Entanglement of a photon and a collective atomic excitation. Phys Rev Lett, 2005, 95: 040405
7. De Riedmatten H, Laurat J, Chou C W, et al. Direct measurement of decoherence for entanglement between a photon and stored atomic excitation. Phys Rev Lett, 2006, 97: 113603
8. Zhao R, Dudin Y O, Jenkins S D, et al. Long-lived quantum memory. Nat Phys, 2009, 5: 100–104

9. Zhao B, Chen Y A, Bao X H, et al. A millisecond quantum memory for scalable quantum networks. Nat Phys, 2009, 5: 95–99
10. Li S J, Yang X D, Cao X M, et al. Enhanced cross-phase modulation based on a double electromagnetically induced transparency in a four-level tripod atomic system. Phys Rev Lett, 2008, 101: 073602
11. Wang H, Li S J, Xu Z X, et al. Quantum interference of stored dual-channel spin-wave excitations in a single tripod system. Phys Rev A, 2011, 83: 043815
12. Li S J, Xu Z X, Zheng H Y, et al. Coherent manipulation of spin-wave vector for polarization of photons in an atomic ensemble. Phys Rev A, 2011, 84: 043430
13. Turchette Q A, Hood C J, Lange W, et al. Measurement of conditional phase shifts for quantum logic. Phys Rev Lett, 1995, 75: 4710–4713
14. Rebić S, Vitali D, Ottaviani C, et al. Polarization phase gate with a tripod atomic system. Phys Rev A, 2004, 70: 032317
15. Joshi A, Xiao M. Phase gate with a four-level inverted-Y system. Phys Rev A, 2005, 72: 062319
16. Lukin M D, Imamoğlu A. Nonlinear optics and quantum entanglement of ultraslow single photons. Phys Rev Lett, 2000, 84: 1419–1422
17. Nemoto K, Munro W J. Nearly deterministic linear optical controlled-NOT gate. Phys Rev Lett, 2004, 93: 250502
18. Barrett S D, Kok P, Nemoto K, et al. Symmetry analyzer for nondestructive Bell-state detection using weak nonlinearities. Phys Rev A, 2005, 71: 060302 (R)
19. Munro W J, Nemoto K, Spiller T P. Weak nonlinearities: A new route to optical quantum computation. New J Phys, 2005, 7: 137
20. Schmidt H, Imamogdlu A. Giant Kerr nonlinearities obtained electromagnetically induced transparency. Opt Lett, 1996, 21: 1936–1938
21. Kang H, Zhu Y F. Observation of large Kerr nonlinearity at low light intensities. Phys Rev Lett, 2003, 91: 093601
22. Harris S E, Hau L V. Nonlinear optics at low light levels. Phys Rev Lett, 1999, 82: 4611–4614
23. Petrosyan D, Kurizki G. Symmetric photon-photon coupling by atoms with Zeeman-split sublevels. Phys Rev A, 2002, 65: 033833
24. Wang Z B, Marzlin K P, Sanders B C. Large cross-phase modulation between slow copropagating weak pulses in 87Rb. Phys Rev Lett, 2006, 97: 063901
25. Li S J, Yang X D, Cao X M, et al. Two electromagnetically induced transparency windows and an enhanced electromagnetically induced transparency signal in a four-level tripod atomic system. J Phys B: At Mol Opt Phys, 2007, 40: 3211–3219
26. Schnorrberger U, Thompson J D, Trotzky S, et al. Electromagnetically induced transparency and light storage in an atomic Mott insulator. Phys Rev Lett, 2009, 103: 033003
27. Zhang R, Garner S R, Hau L V. Creation of long-term coherent optical memory via controlled nonlinear interactions in Bose-Einstein condensates. Phys Rev

Lett, 2009, 103: 233602

28. Matsukevich D N, Chanelière T, Bhattacharya M, et al. Entanglement of a photon and a collective atomic excitation. Phys Rev Lett, 2005, 95: 040405
29. Joshi A, Xiao M. Generalized dark-state polaritons for photon memory in multilevel atomic media. Phys Rev A, 2005, 71: 041801 (R)
30. Raczyński A, Zaremba J, Zielińska-Kaniasty S. Beam splitting and Hong-Ou-Mandel interference for stored light. Phys Rev A, 2007, 75: 013810
31. Wang B, Han Y X, Xiao J T, et al. Preparation and determination of spin-polarized states in multi-Zeeman-sublevel atoms. Phys Rev A, 2007, 75: 051801 (R)
32. Duan L M, Lukin M D, Cirac J I, et al. Long-distance quantum communication with atomic ensembles and linear optics. Nature, 2001, 414: 413–418
33. Duan L M. Entangling many atomic ensembles through laser manipulation. Phys Rev Lett, 2002, 88: 170402
34. Dudin Y O, Jenkins S D, Zhao R, et al. Entanglement of a photon and an optical lattice spin wave. Phys Rev Lett, 2009, 103: 020505
35. Chen S, Chen Y A, Zhao B, et al. Demonstration of a stable atom-photon entanglement source for quantum repeaters. Phys Rev Lett, 2007, 99: 180505
36. Tanji H, Ghosh S, Simon J, et al. Heralded single-magnon quantum memory for photon polarization states. Phys Rev Lett, 2009, 103: 043601
37. Dong P, Xue Z Y, Yang M, et al. Generation of cluster states. Phys Rev A, 2006, 73: 033818
38. Hanneke D, Home J P, Jost J D, et al. Realization of a programmable two-qubit quantum processor. Nat Phys, 2010, 6: 13–16
39. Yavuz D D, Kulatunga P B, Urban E, et al. Fast ground state manipulation of neutral atoms in microscopic optical traps. Phys Rev Lett, 2006, 96: 063001
40. Jones M P A, Beugnon J, Gaëtan A, et al. Fast quantum state control of a single trapped neutral atom. Phys Rev A, 2007, 75: 040301
41. Press D, Ladd T D, Zhang B Y, et al. Complete quantum control of a single quantum dot spin using ultrafast optical pulses. Nature, 2008, 456: 218–221

11

CHAPTER

PERFECT COMPUTATIONAL EQUIVALENCE BETWEEN QUANTUM TURING MACHINES AND FINITELY GENERATED UNIFORM QUANTUM CIRCUIT FAMILIES

Harumichi Nishimura and Masanao Ozawa

Keywords: Quantum computation, Complexity theory, Uniform circuit families, Turing machines, Finitely generated uniform quantum circuit families

INTRODUCTION

In computational complexity theory, Turing machines and Boolean circuits are commonly used as mathematical models of computation. A Turing machine has tapes of infinite-length to treat the inputs of any length. On the other hand, a Boolean circuit can only process the inputs of a fixed length since its size is finite. To represent an algorithm carried out on inputs of any length, we need to consider a family of Boolean circuits whose n-th circuit handles the inputs of length n. In addition, to keep the computational power consistent with the Church-Turing thesis, we need the notion of uniformity of circuit families, first proposed by Borodin [6]. Roughly speaking, the uniformity of a circuit family requires that there exists a computationally simple rule for constructing all circuits in the family. It is well-known that the polynomial-time Turing machine and the polynomial-size uniform circuit family can exactly simulate each other (see, for instance, Theorem 11.5 in [22, p. 269]). We say that two classes A and B of computational models, such as the class of polynomial-time Turing machines and the class of polynomial-size uniform circuit families, are perfectly equivalent if any element of A can be exactly simulated by an element of B in a sense appropriately defined thereon, and vise versa. Thus, the class of polynomial-time Turing machines is perfectly equivalent to the class

Citation: Harumichi Nishimura and Masanao Ozawa, Perfect Computational Equivalence between Quantum Turing Machines and Finitely Generated Uniform Quantum Circuit Families, DOI 10.1007/s11128-008-0091-8.

of polynomial-size uniform circuit families. In the mid-1980s, Deutsch proposed a new parallel computing paradigm, quantum computing, which utilizes the superposition principle of quantum mechanics. As mathematical models of quantum computing, he introduced the quantum versions of Turing machines and Boolean circuits, called quantum Turing machines (QTMs) and quantum circuits [7,8]. In 1993, Bernstein and Vazirani [4] and Yao [27] reformulated Deutsch's QTM model and quantum circuit model in the form suitable for computational complexity theory. Bernstein and Vazirani introduced the complexity classes BQP, ZQP, and EQP that represent the bounded-error, zero-error, and exact quantum algorithms to set the grounds for quantum complexity theory. As the first relation between QTMs and quantum circuits, Yao showed that, for any QTM M and $T > 0$, there is a quantum circuit of size $O(T^2)$ that simulates M for T steps. This result was often mentioned to directly mean that QTMs and quantum circuit families (QCFs) are equivalent by a simple analogy with the equivalence between classical Turing machines and classical circuit families. After a few years, the notion of uniformity of QCFs was shortly mentioned in [9,24]. In particular, Shor [24] pointed out (after his private communication with R. Solovay) that uniform QCFs should satisfy a requirement, which is absent in the standard uniformity requirement, that the entries of elementary gates be polynomial-time computable. In tandem, Adleman et al. [1] showed that a polynomial-time QTM with amplitudes from the whole complex number field can compute even a non-recursive function. In the journal version [5] of [4], Bernstein and Vazirani defined the complexity classes BQP, ZQP, and EQP based on the polynomial-time QTM with amplitudes from the polynomial-time computable numbers. In our previous investigation [19], we instituted the complexity theory of uniform QCFs. For this purpose, we rigorously introduced the notion of uniformity of QCFs and based on that we defined the complexity classes, what we called BUPQC, ZUPQC, and EUPQC, which correspond to BQP, ZQP, and EQP, respectively. Using this formulation and Yao's quantum circuit construction for simulation of QTMs, we showed the following results on the computational equivalence between QTMs and uniform QCFs: (i) BUPQC = BQP [19] (QTMs and uniform QCFs are computationally equivalent in the bounded-error setting). (ii) ZQP $\subseteq$ ZUPQC and EQP $\subseteq$ EUPQC while the simulation does not work to show the converse [19,20] (the computational equivalence between the two models are open in the zero-error and exact setting). Thus, the following question still remained: In the zero-error and exact setting, what restriction for the uniform QCF guarantees the computational equivalence between the two models? Kitaev and Watrous [14] introduced the notion of uniformly generated QCFs based on Shor basis [25]. The class of uniformly generated QCFs is computationally equivalent to the class of our uniform QCFs up to bounded error simulation, and it has been used for the study of quantum interactive proof systems. However, this notion does not provide a solution to question (ii) above. In this paper, we give a complete answer to the above question. To this end, we introduce the notion of finitely generated uniform QCFs based on finite sets of elementary gates, which is a subclass of the uniform QCFs. In [19], we briefly mentioned the notion of semi-uniform QCFs, which is also based on finite sets of elementary gates. Finitely generated uniform QCFs are regarded as those semi-uniform QCFs whose amplitudes are taken from polynomial-time computable numbers. We show that the class of finitely generated uniform QCFs is perfectly equivalent to the class of polynomial-time QTMs with amplitudes from polynomial-time computable numbers. This implies that these two models can simulate each other without error, and hence not only BQP but also ZQP and EQP coincide with the corresponding classes defined through finitely generated uniform QCFs. The proof requires more minute arguments than that of the computational equivalence up to bounded error simulation. For the proof, we revisit properties of polynomial-time computable numbers, which was implicitly used in [19], and Yao's quantum circuit construction. We combine them with the exact decomposition of unitary matrices.

FINITELY GENERATED QCFS

In this section, we give a formal definition of finitely generated QCFs. For the definition of QCFs, see [19]. For convenience, we usually identify quantum gates or circuits with the unitary matrices representing them in the computational basis. For other fundamental notions of quantum computation, we refer to Gruska [11] and to Nielsen and Chuang [17]. For classical complexity theory, we refer, for example, to Papadimitriou [22]. In what follows, let N and C denote the sets of natural numbers and complex numbers, respectively. Let PC denote the set of polynomial-time computable complex numbers [12]. (Informally speaking, a complex numberr is polynomial-time computable if its real and imaginary parts can be approximated with accuracy of $1/2^n$ in time polynomial in n. See Ko [13] for its formal definition.) For two integers m and n with m < n, let [m, n]Z denote {m, m + 1,..., n}. The precise formulation of the uniformity of QCFs was introduced in our previous work [19]. Let $\mathcal{G}_u$ be the set of quantum gates such that

$$\mathcal{G}_u = \{\Lambda_1(N), R(\theta), P(\theta') \mid \theta, \theta' \in \mathbf{PC} \cap [0, 2\pi)\},$$

where $\Lambda_1(N)$ is a controlled-not gate, R(θ) is a rotation gate by angle θ, and P(θ) is a phase shift gate by angle θ. All the gates in $\mathcal{G}_u$ can be encoded by binary strings, using the codes of polynomial-time computable numbers. Here, the code of a polynomial-time computable number r is an appropriate encoding of a deterministic Turing machine (DTM) that approximates r within 2^{-n} in time polynomial in n. We can then give the code Code(C) of a quantum circuit C based on Gu. We say that a QCF C = $\{C_n\}$ based on $\mathcal{G}_u$ is polynomial-size uniform, or uniform for short, if the function $1^n \mapsto \mathrm{Code}(C_n)$ is computable by a DTM in time polynomial in n. Uniform QCFs are very suitable to represent important quantum algorithms such as the quantum Fourier transform and the amplitude amplification. However, uniform QCFs cannot be exactly simulated by any QTM [19] (even with zero-error [20]). Thus, it is questionable whether the class of languages recognized by uniform QCFs with certainty (resp. with zero-error) coincides with the class EQP (resp. ZQP) of polynomial-time QTMs. By analogy with the uniformity of classical circuits, we may imagine a "uniform" QCF based on a fixed finite set of elementary gates. For example, Kitaev and Watrous [14] introduced polynomial-time uniformly generated QCFs to define verifiers of quantum interactive proof systems: A QCF C = $\{C_x\}^1$ is polynomial-time uniformly generated if there exists a deterministic procedure that, on input x, outputs a description of C_x and runs in time polynomial in |x|. Although they used Shor's basis [25] as their set of elementary gates, in the bounded-error setting, any other universal set of elementary gates is available without changing the computational power of polynomial-time uniformly generated QCFs. However, for exact (or zero-error) algorithms, fixing a set of elementary gates may seriously reduce the computational power of "uniform" QCFs. For instance, the class of languages recognized with bounded-error by polynomial-time uniformly generated QCFs based on a universal set $\mathcal{G}_{4/5} = \{\Lambda_1(N), R(\theta), P(\theta) | \cos\theta = 4/5\}$ coincides with the class BQP [1], while the class of languages recognized with certainty by polynomial-time uniformly generated QCFs based on $\mathcal{G}_{4/5}$ coincides with the class P [18], instead of the class EQP.

We now introduced a class of QCFs, called finitely generated QCFs, an intermediate class between uniform QCFs and polynomial-time uniformly generated QCFs. A uniform QCF C is said to be finitely generated if there is a finite subset $\mathcal{G}$ of $\mathcal{G}_u$ such that C is based on G. By definition, any polynomial-time uniformly generated QCF is finitely generated, and any finitely generated uniform QCF is uniform. Finitely generated uniform QCFs have two nice properties: (i) The finitely generated uniform QCF is based on finite sets of elementary gates, different from the uniform QCF; and (ii) The definition of the finitely generated uniform QCF is independent of the choice of universal sets, different from the polynomial-time uniformly generated QCF. In [19], we provided a similar notion, called semi-uniform QCFs. A finitely generated uniform QCF can be regarded as a semi-uniform QCF such that all the components representing matrices of elementary gates are restricted to PC. An analogous concept to finitely generated uniform QCFs was also mentioned by Green et al. [10] in their study of shallow quantum circuits.

PROOF OF THE PERFECT EQUIVALENCE

In this section, we establish the perfect equivalence between the finitely generated uniform QCF and the polynomial-time QTM with amplitudes from PC. See [5,19,21,26] for the definition of QTMs. First, we recall two elementary properties on polynomial-time computable numbers given by Ko and Friedman [12]. Loosely speaking, a real function f is said to be polynomial-time computable if for any real x, the value f (x) can be approximated with accuracy of $1/2^n$ in time polynomial in n using x as an "oracle" to obtain any required bits of x (see [13] for its formal definition).

Theorem 1 (i) All roots of an analytic, polynomial-time computable function are polynomial-time computable.

(ii) PC is an algebraically closed field.

Second, we provide a standard exact decomposition of finite dimensional unitary matrices [2] with an argument on polynomial-time computable entries of matrices.[2] To this end, we use the terminology of the approximate decomposition algorithm of unitary matrices given in [5]. Although this exact decomposition is not algorithmic, it is sufficient for our purpose. Let e_j be the m-dimensional unit column vector whose j-th component is 1. We denote by $\mathrm{Near}_m(j,\theta)$ an m-dimensional unitary matrix satisfying

$$\mathrm{Near}_m(j,\theta)\mathbf{e}_k = \begin{cases} (e^{\imath\theta})\mathbf{e}_k & \text{if } k=j, \\ \mathbf{e}_k & \text{otherwise.} \end{cases}$$

We denote by $\mathrm{Near}_m(i, j, \theta)$ an m-dimensional unitary matrix satisfying

$$\mathrm{Near}_m(i,j,\theta)\mathbf{e}_k = \begin{cases} (\cos\theta)\mathbf{e}_i - (\sin\theta)\mathbf{e}_j & \text{if } k=i, \\ (\sin\theta)\mathbf{e}_i + (\cos\theta)\mathbf{e}_j & \text{if } k=j, \\ \mathbf{e}_k & \text{otherwise.} \end{cases}$$

These two types of matrices are called near-trivial [5]. We show that any unitary matrix whose components are in PC can be decomposed into near-trivial matrices whose components are also in PC.

Lemma 1 (i) Any N-dimensional unitary matrix U can be represented by the product $U_m \dots U_1$ of $m = O(N^2)$ near-trivial matrices $U_1, \dots, U_m$.

(ii) Moreover, if all the entries in U are in PC, each U_j has the entries in PC.

Proof (i) Let U be an N-dimensional unitary matrix. We can show that there is a product A of $2N - 1$ near-trivial matrices such that $AU^{(1)} = e_1$, where $U^{(1)}$ denotes the first column vector of U. Actually, we show that for any N-dimensional unit column vector v, there is a product A of $2N - 1$ near-trivial matrices such that $Av = e_1$. Let v_i denote the i-th coordinate of v. First, we use N near-trivial matrices to map v into the N-dimensional real space. Let $P_i = \mathrm{Near}_m(i, \varphi_i)$ where

$$\phi_j = \begin{cases} 2\pi - \cos^{-1}\left(\frac{\mathrm{Re}(v_j)}{|v_j|}\right) & \text{if } \mathrm{Im}(v_j) > 0, \\ \cos^{-1}\left(\frac{\mathrm{Re}(v_j)}{|v_j|}\right) & \text{if } \mathrm{Im}(v_j) < 0, \\ 0 & \text{if } v_j = 0. \end{cases}$$

Then, $P_1 \dots P_N v$ is the vector with i-th coordinate $|v_i|$. Second, we use $N - 1$ near-trivial matrices to move all of the weight of the vector into dimension 1. Let $R_i = \mathrm{Near}_m(i, i+1, \theta_i)$, where $\theta_i = \tan^{-1}\left(\sqrt{\sum_{j=i+1}^{N} |v_j|^2}/|v_i|\right)$. Then, we have $R_1 \cdots R_{N-1} P_1 \cdots P_N v = e_1$.

Now we have $AU = \mathrm{diag}(1, B)$, where B is an $(N-1)$-dimensional unitary matrix. Here, $\mathrm{diag}(A_1, \dots, A_n)$ denotes the block-diagonal square matrix composed of square matrices $A_1, \dots, A_n$ along the diagonal and 0's everywhere else. By induction, we can verify that there is the product C of $O(N^2)$ near-trivial matrices satisfying $CU = I$. Thus, U can be represented by the product of $O(N^2)$ near-trivial matrices.

(ii) In the case where all the entries in U are in PC, we should note that the above decomposition uses only four arithmetic operations, the function $x \mapsto \sqrt{x}$, the trigonometric functions and their inverse functions for each real parts and imaginary parts. Therefore, each near-trivial matrix composing U has the entries in PC by Theorem 1.

As seen in [2], any 2^n-dimensional near-trivial unitary matrix U can be decomposed into $O(n^3)$ quantum gates in $\mathcal{G}_u$ as follows: (i) Using an idea from the grey code, U can be decomposed into O(n) controlled$^{(n-1)}$ -phase shifts or rotation gates; (ii) Any controlled$^{(n-1)}$ -phase shift (or rotation gate) can be decomposed into $O(n^2)$ quantum gates in $\mathcal{G}_u$ (Corollary 7.6 in [2]). Moreover, by carefully checking the argument leading to obtain Corollary 7.6 in [2], we can see that those $O(n^3)$ quantum gates have their entries in PC provided that all the entries in U are in PC. Thus, we obtain the following exact decomposition of unitary matrices.

Proposition 1 (i) Any n-qubit gate U can be decomposed into $O(2^{2n}n^3)$ quantum gates in $\mathcal{G}_u$.

(ii) Moreover, if all the entries in U are in PC, each of the quantum gates that decomposes U has its entries in PC.

Third, we provide the exact simulation of QTMs by finitely generated uniform QCFs, combining Yao's idea for the simulation of a QTM by a quantum circuit [27] and Proposition 1. We briefly recall the definition of the simulation of QTMs by finitely generated uniform QCFs. See [19] for the detail. A quantum circuit C is said to exactly t-simulate a QTM M if the following two probability distributions D_x and D_x are equal for any string x: (i) the probability distribution D_x of the outcomes of the simultaneous measurement of the tape cells from cell −t to cell t after t steps of M for input state $|q_0, \text{tape}[x], 0\rangle$, where tape[x] represents the tape configuration such that x is written from cell 0 to cell |x| − 1; and (ii) the probability distribution D_x of the string obtained by decoding the output of C for the input of the binary string obtained by encoding x. A finitely generated uniform QCF $\{C_n\}$ is said to exactly simulate a polynomial-time QTM M if C_n t(n)-simulates M, where a polynomial t(n) is the computation time of M on input of length n.

Theorem 2 Let $M = (Q, \Sigma, \delta)$ be a polynomial-time QTM with amplitudes from PC. Then, there is a finitely generated uniform QCF $\{C_n\}$ that exactly simulates M.

Proof The basic line of this proof is in accordance with the proof in [19, Theorem 4.3], which is based on Yao's construction. For our exact simulation, we argue the decomposition of Yao's construction into elementary gates in detail. We refer to [19, Sect. 4] for the terminology of quantum circuits.

Let t(n) be the computation time of M on input of length n. First, we fix n and construct a quantum circuit $C_{\mathcal{G}}$ which t-simulates M. Henceforth, let t = t(n) for simplicity. The quantum gate determined by $C_{\mathcal{G}}$ consists of l_0 + (2t + 1)l wires (i.e., qubits), where $l_0 = \lceil \log |Q| \rceil$ and $l = 2 + \lceil \log |\Sigma| \rceil$. Its wires are indexed in order. We divide their wires into a part consisting of the first l_0 wires and 2t + 1 parts. Each of the 2t + 1 parts consists of l wires. The part consisting of the first l0 wires represents the processor configuration of M. This set of wires is called cell "P" of $C_{\mathcal{G}}$. The state of the cell P of $C_{\mathcal{G}}$ is represented by a unit vector in the Hilbert space spanned by the computational basis $\{|q\rangle\}$, where $q \in \{0, 1\}^{l_0}$. For j ∈ $[0, 2t]_Z$, the wires with indices l_0 + j_l + 1,...,l_0 + j_l + l represent the symbol in the (j − t)-th cell of M and whether the head scans this cell or not. This set of wires is called cell j − t of $C_{\mathcal{G}}$. For i ∈ [−t, t]Z, the state of the cell i of CG is represented by a unit vector in the Hilbert space spanned by the computational basis $\{|\sigma_i s_i\rangle\}$, where $\sigma_i \in \{0,1\}^{\lceil \log|\Sigma| \rceil}$ and $s_i \in \{0,1\}^2$.

The circuit $C_{\mathcal{G}}$ consists of two types of quantum gates G_1 and G_2. The quantum gate G_1 is used for simulating one step of M when its gate is connected into the cell in which the head of M exists. The quantum gate G_2 is used for resetting some wires to simulate the next step. In what follows, p, q,... denote binary strings representing elements of Σ, the symbols σ, τ, . . . denote binary strings representing elements of , and $s = \bar{0}, \bar{1}, \bar{2}$ denote 00,01,10, respectively. We denote by $|q;\sigma_1 s_1;\sigma_2 s_2;\cdots;\sigma_k s_k\rangle$ the computational basis state $|q\sigma_1 s_1 \sigma_2 s_2 \cdots \sigma_k s_k\rangle$ on the wires corresponding to all the numbers in $[1, l_0 + kl]_{\mathbb{Z}}$. Let G_1 be an $(l_0 + 3l)$-qubit gate satisfying the following conditions (i) and (ii):

(i) $G_1|w_{p,\sigma_1,\sigma,\sigma_3}\rangle = |v_{p,\sigma_1,\sigma,\sigma_3}\rangle$, where

$$|w_{p,\sigma_1,\sigma,\sigma_3}\rangle = |p;\sigma_1\bar{0};\sigma\bar{1};\sigma_3\bar{0}\rangle, \quad \text{and}$$
$$|v_{p,\sigma_1,\sigma,\sigma_3}\rangle = \sum_{q,\tau}\delta(p,\sigma,q,\tau,-1)|q;\sigma_1\bar{2};\tau\bar{0};\sigma_3\bar{0}\rangle$$
$$+\sum_{q,\tau}\delta(p,\sigma,q,\tau,0)|q;\sigma_1\bar{0};\tau\bar{2};\sigma_3\bar{0}\rangle$$
$$+\sum_{q,\tau}\delta(p,\sigma,q,\tau,1)|q;\sigma_1\bar{0};\tau\bar{0};\sigma_3\bar{2}\rangle$$

for any $(p,\sigma_1,\sigma,\sigma_3) \in Q \times \Sigma^3$; each summation $\sum_{q,\tau}$ is taken over all $(q,\tau) \in Q \times \Sigma$.

(ii) $G_1|h\rangle = |h\rangle$ for each vector $|h\rangle$ in the subspace H of $\mathbf{C}^{2^{l_0+3l}}$ spanned by three types of vectors:

(1) $|q;\sigma_1 s_1;\sigma_2 s_2;\sigma_3 s_3\rangle$, where $s_2 \neq \bar{1}$ and none of s_1, s_2, s_3 equals $\bar{2}$;
(2) $|u^1_{p,\sigma,\sigma_2,\sigma_3}\rangle = \sum_{q,\tau}\delta(p,\sigma,q,\tau,0)|q;\tau\bar{2};\sigma_2\bar{0};\sigma_3\bar{0}\rangle + \sum_{q,\tau}\delta(p,\sigma,q,\tau,1)|q;\tau\bar{0};\sigma_2\bar{2};\sigma_3\bar{0}\rangle$;
(3) $|u^2_{p,\sigma,\tau,\sigma_1,\sigma_2,\sigma_3}\rangle = \sum_{q\in Q}\delta(p,\sigma,q,\tau,1)|q;\sigma_1\bar{2};\sigma_2\bar{0};\sigma_3\bar{0}\rangle$.

Let G_2 be an $(l_0 + (2t+1)l)$-qubit gate that does nothing except for mapping all $s_i = \bar{2}$'s to $s_i = \bar{1}$'s and vice versa. Let $C_{\mathcal{G}}$ be the quantum circuit based on $\mathcal{G} = \{G_1, G_2\}$ constructed as follows. First, 2t − 1 G_1's are connected in such a way that, for j ∈ $[1, 2t-1]_Z$, the j-th G_1 is connected with cells j − t − 1, j − t and j − t + 1. The $(l_0 + (2t + 1)l)$-qubit circuit constructed from these G_1's is called C_1. Lastly, G_2 is connected with cells −t, −t +1,..., t. The $(l_0+(2t+1)l)$-qubit circuit constructed from this G2 is called C_2. Let $C_{\mathcal{G}}$ be $(C_2 \circ C_1)^t$, i.e., the t concatenations of the circuit $C_2 \circ C_1$ obtained by connecting C_2 into C_1. The quantum circuit $C_2 \circ C_1$ is illustrated in Fig. 1.

By the definitions of G_1 and G_2, it can be verified that $C_2 \circ C_1$ carries out the operation corresponding to one step of M (see [19] for the verification). Hence, $C_{\mathcal{G}}$ simulates t steps of M exactly. The (l0 + 3l)-qubit gate G1 can be exactly decomposed by a finite number of onequbit gates G_{11},..., $G_{1\alpha}$ and 1(N) from Proposition 1(i). Note that all the components of G_{11},..., $G_{1\alpha}$, 1(N) are in PC by Proposition 1(ii) since the QTM M has transition amplitudes in PC. By definition, the quantum gate G_2 can be implemented by swapping two qubits of each s_i , which can be implemented by the concatenation of three controlled-not gates. Thus, there are an (l_0 + 3l)-qubit quantum circuit $C_{u,1}$ of size O(1) and an (l_0+(2t +1)l)-bit quantum circuit $C_{u,2}$ of size 3(2t +1) based on $\mathcal{G}_u$ such that the quantum gates determined by them are G_1 and G_2, respectively.

Now, let C_a be an (l_0 + (2t + 1)l)-qubit quantum circuit obtained by decomposing each G_1 in C_1 into O(1) gates using a subset $\mathcal{G}_M = \{G_{11}, \ldots, G_{1\alpha}, \Lambda_1(N)\}$ of $\mathcal{G}_u$. Since the size of C_a is O(2t + 1),C = $(C_{u,2} \circ C_a)^t$ is a quantum circuit based on GM of size O(t2) that exactly t-simulates M. The codes of $C_{u,2}$ and Ca can be computed by a DTM in time polynomial in n from their constructions (note that the codes of polynomial-time computable numbers used in Cu,2 and Ca can be stored in the processor of the DTM since they are finite). Moreover, the code of C is also computed in time polynomial in n since C is simply the concatenation of t(n) circuits $C_{u,2} \circ C_a$. Noting that the set $\mathcal{G}_M$ of elementary gates is fixed with respect to n, we can verify that M is exactly simulated by a finitely generated uniform QCF $\{C_n\}$, where C_n is the above quantum circuit that exactly t(n)-simulates M. This completes the proof.

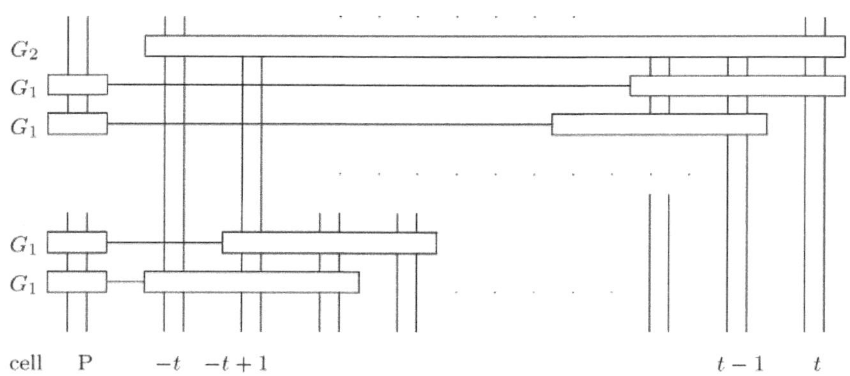

Figure 1: The quantum circuit $C_2 \circ C_1$ based on the set $\mathcal{G}$ =$\{G_1, G_2\}$ of quantum gates. This circuit simulates one step of a QTM M with amplitudes from PC.

Finally, we can exactly simulate any given finitely generated uniform QCF $\{C_n\}$ by a polynomial-time QTM M with amplitudes from PC (see Lemma 5.1 in [19]) because the set of elementary gates for C_n is finite and the quantum transition function of M can represent all the operations induced by the elementary gates. Combining this fact with Theorem 2, we obtain the perfect equivalence between the classes of polynomial-time QTMs with amplitudes from PC and finitely generated uniform QCFs.

Theorem 3 The class of polynomial-time QTMs with amplitudes from PC is perfectly equivalent to the class of finitely generated uniform QCFs.

The following corollary directly comes from Theorem 3.

Corollary 1 The class of languages recognized with certainty (resp. with zero-error and bounded-error) by finitely generated QCFs coincides with the corresponding complexity class EQP (resp. ZQP and BQP) for polynomial-time QTMs.

CONCLUDING REMARKS

We introduced a subclass of uniform QCFs, finitely generated uniform QCFs, and showed that this subclass is perfectly equivalent to the class of polynomial-time QTMs with amplitudes from polynomial-time computable numbers. Here, we shortly point out relationships among uniformity notions under other conditions.

Complexity of Coding

In the theory of circuit complexity, a number of uniformity notions (say, [3,6,23]) were proposed according to machines for constructing circuit families. We use polynomial-time DTMs as machines for computing the codes of QCFs. Although we could take polynomial-time QTMs with amplitudes from PC, instead of polynomial-time DTMs, the resulting class of uniform QCFs does not change. This is because such polynomial-time QTMs can be exactly simulated by QCFs whose coding functions are deterministically polynomial-time computable by using Yao's construction. Furthermore, we can restrict machines for computing the codes of QCFs to logspace DTMs without changing the resulting class of uniform QCFs since, in the proof of Theorem 2, the code of C is computed by an O(log n)-space DTM. This means that the finitely generated "logspace-uniform" QCF is perfectly equivalent to the finitely generated "P-uniform" (and even "EQP-uniform") QCF, like the classical case.

Parameter Of Uniform QCFS

In [14], the binary string $x \in \{0, 1\}^*$ was used as the parameter of uniform QCFs instead of the number $n \in N$. That is, a QCF $\{C_x\}$ is uniform if there is a DTM M that on input x, produces a description of C_x in time polynomial in $|x|$. However, as mentioned in [14], this variation does not change the resulting class of uniform QCFs. Actually, all the C_x's with $|x| = n$ can be simulated by only one quantum circuit C_n, which implements C_x on the target part of C n when x is provided to the controlled part of C'_n. It is easy to check that $\{C'_n\}$ is a uniform QCF in the regular meaning. S_i

Size of Elementary Gates

It is well-known that any one-qubit gates whose components are in PC and the controlled-not gate are sufficient [2] to represent any uniform QCF exactly. Therefore, uniform QCFs defined under elementary gates of larger size more than two are perfectly equivalent to our uniform QCFs.

Restriction of Transition Amplitudes

In [19], we showed the perfect equivalence between semi-uniform QCFs and polynomial-time QTMs without any restriction of amplitudes. In Theorem 2, we also have shown that both models are perfectly equivalent if their transition amplitudes are restricted to PC. However, if we restrict their transition amplitudes to other subsets of C, there is no guarantee that two models are still perfectly equivalent. This is because, in the proof of Theorem 2, we use the square root operation and the root of a function generated by trigonometric functions as well as the addition, the subtraction, the multiplication, and the division. By checking carefully the proof of Theorem 2, we can establish the perfect equivalence between two models with amplitudes from the set of algebraic numbers, provided that the size of elementary gates is allowed to be any number fixed with respect to the input length. The case of the rational number field is open since Yao's construction needs the square root operation.

In [20], the following related results on the power of finitely generated uniform QCFs have been reported by comparing it with that of uniform QCFs: (i) The quantum Fourier transform (QFT) of any order cannot be implemented with zero error by any finitely generated uniform QCF, while it can be exactly implemented by a uniform QCF by the result by Mosca and Zalka [15]. (ii) If a permutation $M_f\colon |x\rangle \mapsto |f(x)\rangle$ can be implemented with zero error by a uniform QCF, then both f and f^{-1} can be exactly computed by uniform QCFs. The first result implies that uniform QCFs cannot always be simulated with zero error by finitely generated uniform QCFs.3 The second result suggests that uniform QCFs are more useful for constructing exact quantum algorithm than finitely generated QCFs.

ACKNOWLEDGEMENTS

M. Ozawa gratefully acknowledges the financial support of the SCOPE project of the MIC, the Grant-in-Aid for Scientific Research (B) 17340021 of the JSPS, and the CREST project of the JST.

REFERENCES

1. Adleman L.M., DeMarrais J., Huang M.A.: Quantum computability. SIAM J. Comput. 26, 1524–1540 (1997)
2. Barenco A., Bennett C.H., Cleve R., DiVicenzo D.P., Margolus N., Shor P., Sleator T., Smolin J., Weinfurter H.: Elementary gates for quantum computation. Phys. Rev. A 52, 3457–3467 (1995)
3. Beame P.W., Cook S.A., Hoover H.J.: Log depth circuits for division and related problems. SIAM J. Comput. 15, 994–1003 (1986)
4. Bernstein, E., Vazirani, U.: Quantum complexity theory (preliminary abstract). In: Proceedings of the 25th ACM Symposium on Theory of Computing, pp. 11–20. ACM Press, New York (1993)
5. Bernstein E., Vazirani U.: Quantum complexity theory. SIAM J. Comput. 26, 1411–1473 (1997)

6. Borodin A.: On relating time and space to and size and depth. SIAM J. Comput. 6, 733–743 (1977)
7. Deutsch D.: Quantum theory, the Church-Turing principle and the universal quantum computer. Proc. R. Soc. Lon. Ser. A 400, 96–117 (1985)
8. Deutsch D.: Quantum computational networks. Proc. R. Soc. Lon. Ser. A 425, 73–90 (1989)
9. Ekert A., Jozsa R.: Shor's quantum algorithm for factoring numbers. Rev. Modern Phys. 68, 733–753 (1996)
10. Green F., Homer S., Moore C., Pollett C.: Counting, fanout, and the complexity of quantum ACC. Quantum Inf. Comput. 2, 35–65 (2002)
11. Gruska J.: Quantum Computing. McGraw-Hill, Berkshire (1999)
12. Ko K.-I., Friedman H.: Computational complexity of real functions. Theor. Comput. Sci. 20, 323–352 (1982)
13. Ko K.-I.: Complexity Theory of Real Functions. Birkhäuser, Basel (1991)
14. Kitaev, A., Watrous, J.: Parallelization, amplification, and exponential time simulation of quantum interactive proof systems. In: Proceedings of the 32nd ACM Symposium on Theory of Computing, pp. 608–617. ACM Press, New
15. Mosca M., Zalka C.: Exact quantum Fourier transforms and discrete logarithm algorithms. Int. J. Quantum Inf. 2, 91–100 (2004)
16. Möttönen, M., Vartiainen, J.J.: Decomposition of general quantum gates. Available via http://arxiv.org/quant-ph/0504100
17. Nielsen M.A., Chuang I.L.: Quantum Computation and Quantum Information. Cambridge University Press, Cambridge (2000)
18. Nishimura H.: Quantum computation with restricted amplitudes. Int. J. Found. Comput. Sci. 14, 853–870 (2003)
19. Nishimura H., Ozawa M.: Computational complexity of uniform quantum circuit families and quantum Turing machines. Theor. Comput. Sci. 276, 147–181 (2002)
20. Nishimura H., Ozawa M.: Uniformity of quantum circuit families for error-free algorithms. Theor. Comput. Sci. 392, 487–496 (2005)
21. Ozawa M., Nishimura H.: Local transition functions of quantum Turing machines. RAIRO Theor. Inform. Appl. 34, 379–402 (2000)
22. Papadimitriou C.H.: Computational Complexity. Addison-Wesley, Reading, MA (1994)
23. Ruzzo W.L.: On uniform circuit complexity. J. Comput. Syst. Sci. 21, 365–383 (1981)
24. Shor P.W.: Polynomial-time algorithms for prime factorization and discrete logarithms on a quantum computer. SIAM J. Comput 26, 1484–1509 (1997)
25. Shor, P.W.: Fault-tolerant quantum computation. In: Proceedings of the 37th IEEE Symposium on Foundations of Computer Science, pp. 56–65. IEEE Computer Society Press, Los Alamitos, CA (1996)

26. Yamakami, T.: A foundation of programming a multi-tape quantum Turing machine. In: Proceedings of the 24th International Symposium on Mathematical Foundations of Computer Science, Lecture Notes in Computer Science, vol. 1672, pp. 430–441 (1999)

27. Yao, A.C.-C.: Quantum circuit complexity. In: Proceedings of the 34th Annual IEEE Symposium on Foundations of Computer Science, pp. 352–361. IEEE Computer Society Press, Los Alamitos, CA (1993)

12
CHAPTER

INFORMATION TRANSFER FIDELITY IN SPIN NETWORKS AND RING-BASED QUANTUM ROUTERS

E. Jonckheere, F. C. Langbein and S. G. Schirmer

Keywords: Spin networks, Quantum ring routers, Information transfer fidelity, Simultaneous Diophantine approximation, Lenstra–Lenstra–Lovász (LLL) lattice basis reduction, Information geometry

INTRODUCTION

Efficient and controllable transport of information is crucial for information processing, both classical and quantum. While bosonic channels [1] are the most attractive option for long-distance communication, efficient on-chip interconnectivity in a quantum processor based on atomic, ionic or quantum dot-based qubits, or quantum spintronic devices [2], will require direct information transport through networks of coupled solid-state qubits. Such networks can be modeled via interacting spins and are therefore generally referred to as spin networks. Initiated by Bose's [3] seminal work, spin networks have received considerable attention in recent years (see review articles [4, 5] and references therein). Most of the work has focused on information transmission through linear chains as prototype quantum wires, starting with unmodulated chains [3] and later perfect state transfer in chains with fixed, engineered couplings [6, 7], and finally controlled state transfer in spin chains, e.g., via adiabatic passage [8], ac modulation to achieve renormalization of the couplings between adjacent qubits [9], single-node bang–bang controls [10] or global dynamic controls [11]. Perfect state transfer in more general networks has also been considered, and some interesting results

Citation: E. Jonckheere, F. C. Langbein, S. G. Schirmer, Information transfer fidelity in spin networks and ring-based quantum routers, DOI 10.1007/s11128-015-1136-4.

for complete graphs were obtained in [12]. Nonetheless, the information-theoretic properties of spin networks are not fully understood. Information encoded in excitations of a network of coupled spins propagates, even under ideal conditions when quantum coherence is maintained, in a nonclassical way determined by the Schrödinger equation. Under the best possible circumstances, this propagation of excitations determines the information transfer fidelity (ITF) between various nodes of the network.1 Perfect state transfer between two nodes can only be achieved when the information transfer fidelity between the respective nodes is unity. However, this condition is not sufficient. For example, while it is satisfied for the end nodes of a chain with uniform couplings [13], such chains are usually not considered to admit perfect state transfer except for chains of length two or three.

This raises the question of the attainability of the upper bound given by the information transfer fidelity. Attainability in general also depends on time constraints, i.e., attainable in what time, and the margins of errors we are willing to accept. In practice, some margin of error is unavoidable, and the real question of interest is therefore not whether we can achieve, e.g., perfect, i.e., unit fidelity, state transfer in time t_f, but rather whether we can achieve state transfer with a fidelity $1-\epsilon$, where ϵ is an acceptable margin of error, in a reasonable amount of time. We may be willing to accept a slightly increased margin of error for a significant reduction in the transfer time. In this work, we are interested in such fundamental questions for spin networks subject to coherent dynamics in general, and specifically simple configurations such as a circular arrangement of spins (or spin ring for short), which could serve as basic building blocks for more complex architectures.

After introducing some basic definitions and basic results in Sect. 2, the concept of asymptotic ITF, i.e., maximum information transfer fidelity attainable absent constraints on the transfer times, between nodes in a network of spins is introduced in Sect. 3. Conditions for attainability of the bounds are derived using dynamic flows on tori and the simultaneous Diophantine approximation, computationally implemented using the Lenstra-Lenstra-Lovász (LLL) algorithm. Under certain conditions, the information transfer infidelity induces a metric that captures how close two nodes in a spin network are from an information-theoretic point of view. This information transfer geometry is investigated in Sect. 4. Finally, in Sect. 5, we investigate how the information transfer geometry of a network can be changed by means of simple controls in the form of fixed biases applied to individual nodes and how this principle could be employed for dynamic routing in a spin network with ring topology without the requirement of fast-switching controls.

BASIC DEFINITIONS AND RESULTS

We consider networks of N spins arranged in some regular pattern with either XX or Heisenberg interaction [14] specified by the Hamiltonian

$$H = \sum_{i,j=1}^{N} J_{ij} \left(\sigma_i^x \sigma_j^x + \sigma_i^y \sigma_j^y + \eta \sigma_i^z \sigma_j^z\right). \tag{1}$$

We specifically focus on networks with XX coupling ($\eta = 0$) and Heisenberg coupling ($\eta = 1$), although most of the concepts and analysis in the following are not limited to these types of coupling. J_{ij} is the strength of the coupling between spin i and spin j. The factor $\sigma_i^{x,y,z}$ is the Pauli matrix along the x, y, or z direction of spin i, i.e.,

$$\sigma_i^{x,y,z} = I_{2\times 2} \otimes \cdots \otimes I_{2\times 2} \otimes \sigma^{x,y,z} \otimes I_{2\times 2} \otimes \cdots \otimes I_{2\times 2},$$

where the factor $\sigma^{x,y,z}$ occupies the ith position among the N factors and $\sigma^{x,y,z}$ is either of the single spin Pauli operators

$$\sigma^x = \begin{pmatrix} 0 & 1 \\ 1 & 0 \end{pmatrix}, \quad \sigma^y = \begin{pmatrix} 0 & -\imath \\ \imath & 0 \end{pmatrix}, \quad \sigma^z = \begin{pmatrix} 1 & 0 \\ 0 & -1 \end{pmatrix}.$$

The system Hilbert space H on which H acts is conveniently taken as $\mathbb{C}^{2^N}$. We can abstract the network of spins as a graph G = (V, E), where the vertices represent the spins and the edges indicate the presence of couplings.

A particular configuration considered in this paper is that of spin rings, i.e., spin networks defined by a circular arrangement of spins, described by a J -coupling matrix that is circulant with nearest-neighbor coupling:

$$H = \sum_{i=1}^{N-1} J_{i,i+1} \left(\sigma_i^x \sigma_{i+1}^x + \sigma_i^y \sigma_{i+1}^y + \eta \sigma_i^z \sigma_{i+1}^z\right) + J_{N,1} \left(\sigma_N^x \sigma_1^x + \sigma_N^y \sigma_1^y + \eta \sigma_N^z \sigma_1^z\right). \tag{2}$$

The term $J_{N,1}$ represents the coupling energy between the two ends, spins 1 and N, closing the ring. For networks with uniform couplings, i.e., all nonzero couplings have equal strength J (in units of Hz), we can set J = 1 by choosing time in units of J^{-1}.

Single excitation subspace

Although many of the results in the following sections are more widely applicable, we primarily concern ourselves here with the single excitation subspace of the network [5], spanned by the N single excitation quantum states {|i : i = 1,..., N}, where |i = | ↑↑ ··· ↑↓↑ ··· ↑ with ↓ in the ith position indicating that spin i carries the excitation. The natural coupling among the spins allows the excitation at i to drift toward an excitation at j with an information transfer fidelity (ITF) that can be quantified by the maximum transition probability $p_{max}(i, j)$. This concept will be precisely defined in the next section, but in this introductory exposition, we could think of "maximum" as the process of giving the transition from spin i to j the correct amount of time so that it is most likely to occur. The concepts behind these ideas are lying at the foundation of quantum mechanics as embodied in the Feynman path integral. These concepts reveal that contrary to classical least-cost-path routing that follows a single path from a source to a destination in a classical network, quantum networks follow all possible paths from the state |i to the state | j. 2.2

Eigendecomposition of the Hamiltonian

Restricted to the single excitation subspace $\tilde{\mathcal{H}} \cong \mathbb{C}^N$, the eigen decomposition of the

Hamiltonian reads $\bar{H} = \sum_k \lambda_k \Pi_k$, where λ_k for $k = 1, \ldots, \tilde{N} \leq N$ are the distinct real eigenvalues and Π_k are the projectors onto the corresponding eigenspaces.

For a spin ring of size N with uniform XX coupling between adjacent spins, J_{ij} = J for i = j ± 1, (i, j) = (1, N), (i, j) = (N, 1), and J_{ij} = 0 otherwise. In this case, the single excitation subspace Hamiltonian in the basis |i = (0, 0,..., 0, 1, 0,..., 0)^T becomes the circulant matrix

$$\bar{H}_N = \begin{pmatrix} 0 & 1 & 0 & \cdots & & 0 & 1 \\ 1 & 0 & 1 & & & 0 & 0 \\ 0 & 1 & 0 & \ddots & & 0 & 0 \\ \vdots & & \ddots & & & & \vdots \\ & & & & \ddots & & \\ 0 & 0 & 0 & & \ddots & 0 & 1 \\ 1 & 0 & 0 & \cdots & & 1 & 0 \end{pmatrix} (=: C_N), \tag{3}$$

Table 1: Eigenvalues and eigenvectors of Hamiltonian $\bar{H}$ over single excitation subspace [15] in the basis where |i = $e_{i:=}$ (0 ... 010 ... 0)^T. ρN = exp(2πι/N) and $|v_k$ j denotes the jth component of $|v^k$

	$\lambda_{k=0,\ldots,N-1}$	$\|v_k\rangle_{j=0,\ldots,N-1}$
XX coupling ($\eta = 0$)	$2\cos\left(\frac{2\pi k}{N}\right)$	$\sqrt{\frac{1}{N}}\rho_N^{k(j-1)}$
Heisenberg coupling ($\eta = 1$)	$2\cos\left(\frac{2\pi k}{N}\right) + 1$	$\sqrt{\frac{1}{N}}\rho_N^{k(j-1)}$

where the subscript N is utilized to indicate that the system has N spins. For uniform Heisenberg coupling, the Hamiltonian is the same except for the addition of a multiple of the identity, which simply shifts the eigenvalues by a constant and does not affect the eigenvector structure or differences between eigenvalues. The eigenvalues and eigenvectors of circulant matrices are well known and shown in Table 1. The N single excitation eigenvalues are conveniently parameterized by an integer k running from 0 to N − 1 or 1 to N with the cyclic condition that λ0 = λ_N .

The following lemma regarding the eigenvalues will be helpful later

Lemma 1 For a spin ring of size N with uniform XX couplings, we have:

–ForNeven,butnotdivisibleby4,thenthespectrumof$\bar{H}_N$hasmirrorsymmetryrelativeto the origin; precisely, we have $\lambda_k = \lambda_{N-k} = -\lambda_{N/2-k} = -\lambda_{N/2+k} \neq$ = 0, i.e., there are $\frac{1}{2}N - 1$ distinct pairs of double eigenvalues, and two single eigenvalues ±2, giving a total of $\tilde{N} = (N + 2)/2$ pairwise distinct eigenvalues:

$$\{-2, \lambda_k, 2 : k = 1, \ldots, \tfrac{1}{2}N - 1\}.$$

If N is divisible by 4, then the spectrum has a total of $\tilde{N} = (N + 2)/2$ pairwise distinct eigenvalues and a double eigenvalue at 0 (for k $k = \frac{1}{4}N, \frac{3}{4}N)$).

– For N odd, we have $\lambda_{N-k} = \lambda_k \neq 0$ and there are (N − 1)/2 distinct pairs of double eigenvalues and a single eigenvalue +2, giving a total of $\tilde{N} := (N+1)/2$ distinct eigenvalues:

$$\{\lambda_k, +2 : k = 1, \ldots, \tfrac{1}{2}(N-1)\}.$$

– In either case, the number of pairwise distinct eigenvalues is

$$\tilde{N} := \left\lceil \frac{N-1}{2} \right\rceil + 1 = \left\lceil \frac{N+1}{2} \right\rceil.$$

Moreover, the eigenvalues of C_N and $_{CN-}1$ are interlaced.

Proof: The listed items are trivial. The last claim is the Cauchy interlacing property [16].

For a double eigenvalue $\lambda_k = \lambda_{N-k}$, denote the projection on the corresponding eigenspace as $\Pi_k := |v_k\ v_k| + |v_{N-k}\ _{vN-k}|$, where the eigenvectors can be chosen

such that $v_{N-k} = v_k^*$. Moreover, for the single eigenvalue $\lambda_0 = +2$, define $\Pi_0 := |v_0\ v_0|$ to be its eigenprojection. If N is even, the single eigenvalue $\lambda_{N/2} = -2$ has its eigenprojection denoted as $\Pi_{N/2} := |v_{N/2}\ v_{N/2}|$. If, in addition, N is divisible by 4, denote the eigenprojection of the double eigenvalue $\lambda_{N/4} = \lambda_{3N/4} = 0$ as $\Pi_{N/4} := |v_{N/4}\ v_{N/4}| + |v_{3N/4}\ v_{3N/4}|$. With this notation, the Hamiltonian restricted to the single excitation subspace can be written as H¯

$$\bar{H} = \sum_{k=0}^{\tilde{N}-1} \lambda_k \Pi_k.$$

The above can easily be extended to the Heisenberg case by globally shifting the eigenvalues by 1.

MAXIMUM TRANSFER FIDELITY AND ATTAINABILITY

Let $|i \in \bar{\mathcal{H}}$ be a quantum state with excitation localized at spin i. The quantum mechanical probability of transition from state |i to state | j in an amount of time t is given by p

$$p_t(i, j) = |\langle i| e^{-\imath \hbar H_1 t} |j\rangle|^2,$$

where we choose energies in units of $\hbar/J$ allowing us to assume $\hbar = 1$ and omit $\hbar$ in the following. This formula is a corollary of the Feynman path integral [17,18]. To circumvent the difficulty posed by the time dependence of this probability, we proceed as in [13] and define the maximum transition probability $p_{max}(i, j)$ also referred to as information transfer fidelity (ITF):

$$p_t(i,j) = \left|\langle i|e^{-\imath H_1 t}|j\rangle\right|^2 = \left|\sum_{k=0}^{\tilde{N}-1}\langle i|\Pi_k|j\rangle e^{-\imath\lambda_k t}\right|^2$$
$$\leq \left(\sum_{k=0}^{\tilde{N}-1}|\langle i|\Pi_k|j\rangle|\right)^2 =: p_{\max}(i,j). \tag{4}$$

Clearly, $p_{max}(i, j) \leq 1$. Observe that instead of taking the sum of the absolute values of all $\langle i|\Pi_k|j\rangle$ terms, we could take the sum of the absolute values of some partial sums of such terms and derive other upper bounds. Note that the upper bound is valid for any spin network, no matter how many spins, no matter how many multiple eigenvalues, no matter the topology. Since the upper bound depends only on the eigenvectors of the Hamiltonian and since those are continuously dependent on the strengths of the couplings, the upper bound is continuous relative to the J_{ij}.

Attainability of bounds

The ITF $p_{max}(i, j)$is an upper bound on $p_t(i, j)$, which acquires its full significance if p_{max} is achievable, that is, if there exists a sequence of time samples $\{t_{i,j}(n) : n \in \mathbb{N}\}$

such that $\lim_{n\to\infty} p_{t_{ij}}(n)(i, j) = p_{max}(i, j)$. Observing that the absolute value in Eq. (4) will absorb any global phase factor, the attainability condition is that there exists $t \in [0, \infty)$ such that

$$e^{-\imath\lambda_k t} = s_k(i,j)e^{\imath\phi}, \quad \forall k = 0, \ldots, \tilde{N}-1, \tag{5}$$

where $s_k(i,j) := \mathrm{Sgn}(\langle i||\Pi_k|j\rangle) \in \{0, \pm 1\}$ is a sign factor and φ is a global phase, which is arbitrary but must be the same for all k's. Eigenspaces with $s_k = 0$ (where the (i, j) dependency is suppressed to avoid the clutter) have no overlap with the initial and/or target state and do not contribute to the sum. We shall refer to them as dark-state subspaces. They can be ignored, and we can restrict ourselves to the set $K' \subseteq \{0, 1, \ldots, \tilde{N}-1\}$of indices k for which $s_k \neq 0$. The physical interpretation of K' is the set of eigenspaces $\Pi_k\tilde{\mathcal{H}}$ that have nontrivial overlap with the initial and target states. Noting that $s_k = \pm 1$ for $k \in K'$, and exp $[-\imath\frac{\pi}{2}(s_k - 1)] = 1$ for $s_k = 1$ and exp $[-\imath\frac{\pi}{2}(s_k - 1)] = -1$ for $s_k = -1$, we can write

$$s_k = \exp\left[-\imath\pi\left(2n_k + \tfrac{1}{2}(s_k - 1)\right)\right], \quad \forall k \in K', \tag{6}$$

where $n_k \in \mathbb{Z}$ is an arbitrary integer. Inserting this into (5), taking the logarithm and dividing by $-\imath$ yield

$$\lambda_k t = 2\pi n_k + \tfrac{\pi}{2}(s_k - 1) - \phi, \quad \forall k \in K'. \tag{7}$$

This condition is not directly useful as φ can be arbitrary, but we obtain meaningful constraints if we subtract the equations in a pairwise manner, with $k \neq \ell$:

$$(\lambda_k - \lambda_\ell)t = 2\pi(n_k - n_\ell) + \tfrac{\pi}{2}(s_k - s_\ell), \quad \forall k, \ell \in K'. \tag{8}$$

We can also write the attainability constraints more explicitly:

$$\begin{aligned}
(\lambda_k - \lambda_\ell)t &= 2\pi(n_k - n_\ell), && \text{if } s_\ell = s_k,\\
(\lambda_k - \lambda_\ell)t &= 2\pi(n_k - n_\ell) + \pi, && \text{if } s_k = -s_\ell = 1,\\
(\lambda_k - \lambda_\ell)t &= 2\pi(n_k - n_\ell) - \pi, && \text{if } s_k = -s_\ell = -1.
\end{aligned}$$

These conditions are necessary and sufficient for attainability. They are physical, only involving differences of the eigenvalues, which are observable and independent of arbitrary phases. Vanishing left-hand sides in the above are not an issue, as we are only looking at the differences, which are nonzero by definition as λ_k, k ∈ K', are the distinct eigenvalues of $\bar{H}$.

Observe that all of the equations are compatible. Indeed, adding Eq. (8) for (k,) and (, m) yields (8) for (k, m). Naturally, these equations are redundant, but we obtain a set of linearly independent equations if we exclude the dark-state subspaces and restrict ourselves to a suitable subset of equations, e.g., (k_{i-1}, k_i) or (k_0, k_i) for $K' = \{k'_1, k'_2, \ldots, K'_{\bar{N}}\}$.

Example 1 (Dark States for Rings.) For ring systems with uniform XX coupling, the distinct eigenvalues are $\lambda_k = 2\cos(2\pi k/N)$. For eigenvalues of multiplicity 1, which occur for $k = 0$, and $k = \frac{1}{2}N$ if N is even, $\langle i|\Pi_0|j\rangle = (1/N) \neq 0$ and $\langle i|\Pi_{N/2}|j\rangle = (1/N)(-1)^{i-j} \neq 0$; therefore, there are no dark states associated with these eigenvalues. For eigenvalues with multiplicity 2, $\langle i|\Pi_k|j\rangle = \frac{2}{N}\cos(\frac{\pi}{2}n)$ with $n = 4k(i-j)/N$ for $k = 0, \ldots, \lceil (N-4)/2 \rceil$; therefore, there are dark states if and only if n is an odd integer. This can happen only if N is divisible by 4. The same holds of rings with uniform Heisenberg coupling as they have the same eigenspace structure and the differences between eigenvalues are the same.

Simultaneous attainability and flows on the torus

Excluding dark-state subspaces, restricting (8) to a subset S ⊆ K' × K' of linearly independent equations, and setting $\omega_{k,\ell} = (\lambda_k - \lambda_\ell)/\pi$, the attainability conditions become

$$t\omega_{k,\ell} = \tfrac{1}{2}(s_k - s_\ell) \mod 2, \quad \omega_{k,\ell} := (\lambda_k - \lambda_\ell)/\pi, \quad \forall (k, \ell) \in \mathcal{S}. \tag{9}$$

The left-hand side of the above is the solution of the flow on the torus $\dot{x} = \omega_{k\ell}$, with x(0) = 0. In this dynamic formulation, the question is whether the flow starting at x(0) = 0 passes through the point with coordinates 0 or 1, depending on whether $s_k = s_\ell$ or $s_k \neq s_\ell$, respectively. It is well known [19, Prop.1.5.1] that the flow starting at an arbitrary x(0) (which includes x(0) = 0) passes arbitrarily close to an arbitrary point on the torus if and only if the $\omega_{k,\ell}$'s are linearly independent over the rationals $\mathbb{Q}$. This property of the flow getting arbitrarily close to an arbitrary point from an arbitrary initial condition is very strong and referred to as minimality. Observe that for the flow to be minimal it suffices that starting at x(0) = 0, it gets arbitrarily close to any point. Obviously, minimality is sufficient but not necessary for attainability, as the latter only requires the flow to pass arbitrarily close to a specific point on the torus, while minimality guarantees that the flow can get arbitrarily close to any point.

Recall that Eq. (8) refers to a specific but arbitrary transfer |i→| j, as the signs depend on i, j. We could consider all Eq. (8)'s for all i ≠ j and ask the question as to whether there exists a unique t such that attainability holds for all i ≠ j. We refer to this stronger version of attainability as simultaneous attainability.

If for a given pair (i, j) there are at least three nondark eigenspaces corresponding to s_d, s_m, s_n ∈ {±1}, then there must exist a pair, say (m, n), with $s_m - s_n = 0$. In this case, setting $t = 2\tau/\omega_{mn}$ for $\tau \in \mathbb{N}$ ensures that the (m, n) Eq. (9) holds exactly and the remaining attainability equations become

$$\theta_{k\ell}\tau = \tfrac{1}{2}(s_k - s_\ell) \mod 2, \qquad \theta_{k\ell} := 2\omega_{k\ell}/\omega_{m,n}, \quad \forall (k,\ell) \in \mathcal{S}_0 := \mathcal{S} \setminus \{(m,n)\}. \tag{10}$$

The left-hand side $\theta_{k\ell}\tau$ of the preceding equation is the solution of the translation on the torus, that is, $x(\tau+1) = x(\tau) + \theta_{k\ell}$ mod 2 with initial condition x (0) = 0. By [19, Prop.1.4.1], the translation on the torus can come arbitrarily close to any point iff the elements in the set $\{1\} \cup \{\theta_{k\ell} : (k,\ell) \in \mathcal{S}_0\}$ are linearly independent over $\mathbb{Q}$. As before, the linear independence is sufficient, but not necessary for attainability. Note that we can in principle always reorder the eigenvalues so that the reference transition is (m, n) = (1, 2).

It should be noted that the attainability criteria above apply to any spin network. For specific types of networks, we can derive more explicit criteria.

Example 2: (Attainability Condition for Rings.) Given the formula for the eigenvalues for homogeneous rings, $\lambda_k = 2\cos(2\pi k/N)$, elementary trigonometry shows that

$$\omega_{k\ell} = \tfrac{1}{\pi}(\lambda_k - \lambda_\ell) = -\tfrac{4}{\pi}\sin\left(\tfrac{\pi}{N}(k+\ell)\right)\sin\left(\tfrac{\pi}{N}(k-\ell)\right). \tag{11}$$

There are $\widetilde{N} = \lfloor N/2 \rfloor + 1$ eigenspaces and $\widetilde{N} - 1$ independent transition frequencies $\omega_{k\ell}$. Choosing the subset of linearly independent equations $\mathcal{S} = \{(k, k+1) : k = 0, \ldots, \bar{N}\}$, $\bar{N} := \widetilde{N} - 2$, with the ordering of the eigenspaces as defined above, the attainability conditions can be written as

$$\tfrac{4}{\pi}\sin\left(\tfrac{\pi}{N}(2k+1)\right)\sin\left(\tfrac{\pi}{N}\right) = \tfrac{1}{2}(s_k - s_{k+1}) \mod 2, \quad \forall k = 0, \ldots, \bar{N}.$$

If $s_m = s_{m+1}$, then setting $t = 2\tau/\omega_{m,m+1}$ for $\tau \in \mathbb{N}$ ensures $\omega_{m,m+1}t = 2\tau = 0$ mod 2 and the attainability conditions become

$$\theta_k \tau = \frac{1}{2}(s_k - s_{k+1}) \mod 2, \quad \forall k = 0, \ldots, \bar{N},$$

with $\theta_k = \sin(\frac{\pi}{N}(2k+1))/\sin(\frac{\pi}{N}(2m+1))$. Notice that the signs of the projections of the initial state |i and target state | j, s_k = j|Π_k |i, depend on the choices of the latter, and it may happen that the signs s_k are alternating, $s_{k+1} = -s_k$ for all k. In this case, the problem can easily be rectified by reordering the eigenvalues, e.g., so that $s_0' = s_1'$ with the new ordering.

Example 3 (Rational Independence.) Applying the previous results to a ring of N = 5 spins, the number of pairwise distinct eigenvalues of the single excitation Hamiltonian $\bar{H}$ is $\bar{N} = 3$ and there are two linearly independent transition frequencies $\omega_{01} = -\frac{4}{\pi}\sin(\frac{1}{5}\pi)\sin(\frac{1}{5}\pi)$ and $\omega_{12} = -\frac{4}{\pi}\sin(\frac{3}{5}\pi)\sin(\pi\frac{1}{5})$. To verify linear independence of $\{\sin(\frac{1}{5}\pi), \sin(\frac{3}{5}\pi)\}$ over $\mathbb{Q}$, , we must show that the equation

$$\alpha_1 \sin\left(\frac{\pi}{5}\right) + \alpha_3 \sin\left(\frac{3\pi}{5}\right) = 0 \text{ for } \alpha_1, \alpha_3 \in \mathbb{Q},$$

has only the trivial solution $\alpha_1 = \alpha_3 = 0$ over $\mathbb{Q}$. Using sin $(\frac{\pi}{5}) = \frac{1}{4}\sqrt{10-2\sqrt{5}}$ and sin $(\frac{3\pi}{5}) = \frac{1}{4}\sqrt{10+2\sqrt{5}}$, we can rewrite the equation as $\alpha_1^2(10 - 2\sqrt{5}) = \alpha_3^2(10 + 2\sqrt{5})$. Viewing the field $\mathbb{Q}(\sqrt{5})$ as a two-dimensional vector space over $\mathbb{Q}$ with basis $1, \sqrt{5}$ gives two equations $\alpha_1^2 = \alpha_3^2$ and $\alpha_1^2 = -\alpha_3^2$, which much be simultaneously satisfied. This is possible only for $\alpha_1 = \alpha_3 = 0$. Thus, the flow on the torus is minimal, and p_{max}(i, j) is attainable for all (i, j).

Example 4: (Rational Dependence for Even Rings.) For a ring with N = 10 spins, there are $\bar{N} = 6$ distinct eigenvalues and five primary transition frequencies $\omega_{k,k+1}$ for k = 0,..., 4. Noting that sin $\left(\frac{5\pi}{10}\right) = 1$, $\sin\left(\frac{\pi}{10}\right) = \sin\left(\frac{9\pi}{10}\right) = \frac{1}{4}(-1+\sqrt{5})$ and sin $\sin\left(\frac{3\pi}{10}\right) = \sin\left(\frac{7\pi}{10}\right) = \frac{1}{4}(1+\sqrt{5})$. It is easily seen that α = (2, −2, 1) is a $\mathbb{Q}$-solution to the linear dependence equation

$$\alpha_1 \sin\left(\frac{9\pi}{10}\right) + \alpha_2 \sin\left(\frac{3\pi}{10}\right) + 4\alpha_3 \sin\left(\frac{5\pi}{10}\right) = 0.$$

Hence, the p_{max}(i, j) are not simultaneously attainable—although p_{max}(i, j) may be attainable for some (i, j).

More generally, for a ring with N even, there are $\frac{1}{2}N$ transition frequencies

$$\omega_{k,k+1} = \frac{4}{\pi}\sin\left((2k+1)\frac{\pi}{N}\right)\sin\left(\frac{\pi}{N}\right),$$

which occur in pairs $\omega_{k,k+1} = \omega_{\bar{N}-k,\bar{N}-k+1}$ with $\bar{N} = \frac{1}{2}N - 1$, precluding rational independence.

Example 5 (Rational Dependence for Odd Rings.) Similarly, we can easily verify that for a ring with N = 9 spins, the transition frequencies are not rationally independent, as we have, e.g., sin(7π/9) − sin(5π/9) + sin(π/9) = 0 and thus $\omega_{3,4} - \omega_{2,3} + \omega_{0,1} = 0$.

In general, rational independence of the transition frequencies for homogeneous rings does not hold when N is not prime.

Simultaneous Diophantine approximation

Instead of checking rational independence of $\{1\} \cup \{\theta_{k\ell} : (k, \ell) \in \mathcal{S}_0\}$ a less conservative approach is to proceed, either analytically or computationally [20,21], via the simultaneous Diophantine approximation [22–25] by finding integers pk , q such that

$$\left|\theta_{k\ell} - \frac{p_{k\ell}}{q}\right| \le \frac{c}{q^{1+\epsilon}}, \qquad \forall (k, \ell) \in \mathcal{S}_0,$$

and > 0. With τ = q, the above yields

$$|\theta_{k\ell}\tau - p_{k\ell}| \le \frac{c}{\tau^{\epsilon}}, \qquad \forall (k, \ell) \in \mathcal{S}_0.$$

In the single-dimensional case, the solution is well known to be given by the continued fraction expansion of θ. Truncating the continued fraction expansion yields convergents, i.e., rational fractions p/q, with errors bounded as |θq − p| ≤ 1/q, which is optimal among all rational approximations of denominators less than or equal to q. The major hurdle at extending this result to the multi-dimensional case is that there is an incompatibility between the unimodular property of the multi-dimensional continued fraction (MCF) solution and optimality.

Nevertheless, the celebrated Dirichlet box principle shows that there are multidimensional approximations with c = 1 and = 1/ $\bar{N}$, where in the present context $\bar{N}$ = |S0|. Moreover, there are infinitely many integer solutions q to the simultaneous Diophantine approximation; in other words, as τ is allowed to become arbitrarily large, the above error can be made arbitrarily small. The constant c = 1 can hardly be improved as for c < 1 there are "badly approximable vectors" $\theta \in \mathbb{R}^{\bar{N}}$ defined by lim inf $_{q\to\infty} q^{1/\bar{N}} d(\theta q, \mathbb{Z}^{\bar{N}}) > 0$ such that the simultaneous Diophantine approximation has only finitely many solutions [22,26, Sec. 5]. If, however, c is allowed to depend on $\bar{N}$, refined bounds (c < 1) can be derived on c($\bar{N}$) due to the existence of infinitely many solutions [23]. Specializing the approximation to $\bar{N}$ = 2, it can be shown [27] that the bound can be improved down to c = 8/13, along with = 1/2. In the one-dimensional case Hurwitz's theorem says that one can take $c = 1/\sqrt{5}$ and = 1. On a general tone, the Dirichlet approximation can only be improved slightly and at the expense of considerable extra difficulties; we will therefore work exclusively with the Dirichlet approximation in the following.

Assuming we have obtained a Dirichlet-good simultaneous Diophantine approximation, the approximate attainability conditions become

$$p_{k\ell} = \frac{1}{2}(s_k - s_\ell) \mod 2, \quad \forall (k, \ell) \in \mathcal{S}_0. \tag{12}$$

The difficulty is to find, if it exists, a simultaneous Diophantine approximation of Dirichlet accuracy that satisfies the above conditions on the numerators. The following example demonstrates that it is not, in general, possible to achieve the even/odd conditions (12) on the numerators $p_{k\ell}$ without compromising on the accuracy of the Diophantine approximation. To be more specific, arbitrary accuracy can still be achieved with Conditions (12), but a larger denominator is required to achieve the same level of accuracy in the presence of the constraints.

Example 6 (Simultaneous Diophantine Approximation with Constraints.) In Example 3, the flow on the torus for a ring with N = 5 was found to be minimal, implying that we can get arbitrarily close to an arbitrary point on the torus. By the preceding argument, this guarantees existence of simultaneous Diophantine approximations of arbitrary accuracies with prescribed even/odd numerators. Furthermore, it is readily found that

$$\theta_{12} = \frac{2\sin(3\pi/5)}{\sin(\pi/5)} = 1 + \sqrt{5} = [3; 4, 4, 4, 4, 4, ...],$$

where the final expression denotes the continued fraction expansion giving the optimal rational approximations [28, Chap. 10]. It is known that quadratic irrationality leads to continued fractions that eventually stabilize. The first convergents are

$$3, \frac{13}{4}, \frac{55}{17}, \frac{233}{72}, \frac{987}{305}, \frac{4181}{1292}, \frac{17711}{5473}, \frac{75025}{23184}, \frac{317811}{98209}, \frac{1346269}{416020}, \ldots$$

Observe that all of them have odd numerators, while the approximations we require must have even numerators since for the N = 5 ring $s_0 = s_1 = 1$. This can be rectified by using the so-called semi-convergents [29, Sec. V.4], [21]. Given two convergents ordered as $p_{n-1}/q_{n-1} < p_n/q_n$, one can easily squeeze a semi-convergent between them as follows:

$$\frac{p_{n-1}}{q_{n-1}} < \frac{p_{n-1} + p_n}{q_{n-1} + q_n} < \frac{p_n}{q_n}$$

The semi-convergent has even numerator and has the accuracy of the convergents p_{n-1}/q_{n-1} and p_n/q_n but at the cost of doubling the denominator. To prove that the semi-convergents provide approximations of arbitrary accuracy, it suffices to show that there are infinitely many n's such that $p_{n-1}/q_{n-1} < p_n/q_n$. This is a corollary of the unimodular property of continuous fractions, saying that $p_{n-1}q_n - p_nq_{n-1}$ is alternately ±1.

We propose a general iterative method to deal with the even/odd constraints. To simplify the notation, let $\theta \in \mathbb{R}^{\bar{N}}$, $p \in \mathbb{Z}^{\bar{N}}$, be column vectorizations of the $\theta_{k\ell}$s, $p_{k\ell}$s respectively, where $\bar{N} := |\mathcal{S}_0|$. We want to come up with a Dirichlet-good approximation,

$\theta \approx p/q$, where $p \in \mathbb{Z}^{\bar{N}}, q \in \mathbb{N}$, with even/odd constraints on the numerators p_i. By "Dirichlet-good," we mean that the infinity error is bounded as $\|\theta q - p\|_\infty \leq c/q^{1/\bar{N}}$, where c is a constant independent of $\bar{N}$ and q. The idea is to iteratively scale θ by (the inverse of) a diagonal matrix of positive rational numbers, $\bar{\theta} = Y(n)^{-1}\theta$, compute a Dirichlet-good approximation of $\bar{\theta}$ using, e.g., the Dirichlet box principle, or the LLL-algorithm, or Lagarias' multi-dimensional continued fractions (MCFs), and then revise the scaling to meet the even/odd constraints, with the hope that the procedure will converge. Write the Dirichlet-good approximation $\bar{\theta} \approx \bar{p}/q$ and manipulate it as follows:

$$\|\bar{\theta} q - \bar{p}\|_\infty \leq \frac{1}{q^{1/\bar{N}}},$$
$$\min_i (Y(n)_{ii}^{-1})\|\theta q - Y(n)\bar{p}\|_\infty \leq \|Y(n)^{-1}(\theta q - Y(n)\bar{p})\|_\infty \leq \frac{1}{q^{1/\bar{N}}}.$$

It follows that

$$\|\theta q - Y(n)\bar{p}(Y(n))\|_\infty \leq \frac{1}{q^{1/\bar{N}}} \max_i Y(n)_{ii}.$$

In other words, $Y(n)\bar{p}/q$ is a Dirichlet-good approximation of θ, provided $\max_i Y(n)_{ii}$ can be dominated by a bound independent of q and $\bar{N}$. Because the initial choice of Y (n) is arbitrary, it is not guaranteed that Y (n) $\bar{p}$ has the correct even/odd property. Nevertheless, we could revise Y (n) to meet those properties. If a component $\bar{p}_i$ comes out to be odd and needs to be even, we choose $Y(n+1)_{ii} = 2$. If the algorithm has converged, that is Y (n + 1) = Y (n), then the bound becomes

$$\|\theta q - Y(n+1)\bar{p}(Y(n))\|_\infty \leq 2\frac{1}{q^{1/\bar{N}}}. \tag{13}$$

Conversely, if $\bar{p}_i$ comes out to be even with 2^d in its prime number decomposition, we take $Y(n+1)_{ii} = 1/2^d$ and, at convergence, the bound on the ith component becomes

$$|\theta_i q - 2^{-d}\bar{p}_i| \leq 2^{-d}\frac{1}{q^{1/\bar{N}}}.$$

Then this procedure is repeated with the scaling $\bar{\theta} = Y(n+1)\theta$, in the hope that it converges.

Theorem 1 Given $\theta \in \mathbb{R}^{\bar{N}}$, assuming {Y (n)} converges, there exists a simultaneous Diophantine approximation θ ≈ p/q satisfying prescribed even/odd constraints on the numerators $p_i, i = 1, \ldots, \bar{N}$, with an error bound $\|\theta q - p\|_\infty \leq 2/q^{1/\bar{N}}$ that is off the usual Dirichlet bound by a factor not exceeding 2.

Example 7 (Simultaneous Diophantine Approximation with Constraints.) We consider the same situation as in Example 6, where all convergents of θ_{12} have odd numerators while attainability calls for an even numerator. We initiate the algorithm with Y (0) = 1, that is, $\bar{\theta} = \theta = 1 + \sqrt{5}$. Whatever convergent $\bar{p}/q$ we pick, it has odd numerator, and hence we take Y (1) = 2. We hence rewrite the continued fraction decomposition with $\bar{\theta} = (1/2)\theta = (1 + \sqrt{5})/2$, which gives the convergents

$$1, 2, \frac{3}{2}, \frac{5}{3}, \frac{8}{5}, \frac{13}{8}, \frac{21}{13}, \frac{34}{21}, \frac{55}{34}, \frac{89}{55}, \frac{144}{89}, \frac{233}{144}, \frac{377}{233}, \frac{610}{377}, \frac{987}{610}, \ldots$$

To secure convergence, Y (3) = Y (2) = 2, we need to pick a convergent with odd numerator, say, 377/233, and the new Diophantine approximation of θ_{12} is 2 × 377/233. This gives an error $|\theta_1 2 \times 233 - 754| = 0.0038 < 2/233 = 0.0086$, as claimed.

(Weighted) LLL-algorithm

Even though Theorem 1 guarantees that under convergence conditions, Dirichletgood simultaneous Diophantine approximations can be manipulated so as to yield numerators that have prescribed even/odd properties, we are still left with the problem of coming up with simultaneous Diophantine approximations in the first place.

One of the first computational solutions to the simultaneous Diophantine approximation was the so-called LLL-algorithm by Lenstra, Lenstra and Lovász [21,24,30]. An alternative algorithm based on geodesic multi-dimensional continued fraction expansion was proposed by Lagarias [31]. Both approaches proceed by reduction in the basis of the lattice generated by the columns of

$$B(\mathsf{s}) = \begin{pmatrix} I_{\bar{N}\times\bar{N}} & -\theta \\ 0_{1\times\bar{N}} & \mathsf{s} \end{pmatrix},$$

where s ↓ 0 is a scaling parameter. Observing that $B(s)(p, q)^T = (p - \theta q, sq)^T$, it follows that a short vector in the lattice $B(\mathsf{s})\mathbb{Z}^{\bar{N}+1}$ yields a good approximation. The numerator of this good approximation could be "fixed" by the procedure of Sect. 3.3 to satisfy the even/odd requirement. However, it is proposed to combine the two procedures into a single one—computation of a good approximation from a short lattice vector and fixing the numerator—by introducing a nonuniform diagonal scaling and work on the lattice $\Lambda(s, X)$ generated by the columns of

$$B(\mathsf{s}, X) = \begin{pmatrix} X & -X\theta \\ 0_{1\times\bar{N}} & \mathsf{s} \end{pmatrix},$$

Where X = diag $(x_1, \ldots, x_{\bar{N}})$. Note that for $\mathsf{s} = 1$ and $X = xI_{\bar{N}\times\bar{N}}$, we recover the scaling of [21]. Like the algorithm of Sect. 3.3, this procedure is not guaranteed to be successful, but if it is, it yields solutions guaranteed to be optimal relative to some criterion. The LLL-algorithm produces a basis of short Euclidean norm vectors $(b^*(\mathsf{s}, X)_1, b^*(\mathsf{s}, X)_2, \ldots, b^*(\mathsf{s}, X)_{\bar{N}+1}) =: B^*(\mathsf{s}, X)$ such that

$$\|b^*(\mathsf{s}, X)_1\| < \|b^*(\mathsf{s}, X)_j\|, \quad j = 2, \dots \bar{N} + 1.$$

The b*(s, X)1 vector is very close to the shortest one. A refined version of the LLLalgorithm captures the genuinely shortest vector of the lattice Λ(s, X) as follows: Given the reduced basis $\{b^*(\mathsf{s}, X)_i : i = 1, \dots, \bar{N} + 1\}$, it can be shown that the shortest (in the sense of the Euclidean norm) lattice vector is to be sought among all lattice vectors of the form $\sum_i \beta_i b^*(\mathsf{s}, X)_i, |\beta_i| \leq \left(2/\sqrt{3}\right)^{\bar{N}+1}$. Lagarias' theorem [24, Lemma 5] then implies that a shortest Euclidean norm vector of the lattice is a best X-weighted Diophantine approximation. Observing that

$$B(\mathsf{s}, X) \begin{pmatrix} p \\ q \end{pmatrix} = \begin{pmatrix} X(p - \theta q) \\ \mathsf{s}q \end{pmatrix},$$

and taking s ↓ 0, it becomes clear that a short vector in the lattice $B(\mathsf{s}, X)\mathbb{Z}^{\bar{N}+1}$ provides a good X-weighted Diophantine approximation:

$$q = \frac{(B^*(\mathsf{s}, X))_{\bar{N}+1,1}}{\mathsf{s}}, \quad p_i = \frac{(B^*(\mathsf{s}, X))_{i,1}}{x_i} + \theta_i q. \tag{14}$$

With the shortest vector, we construct the best approximation, that is, the approximation that minimizes

$$\|X(\theta q - p)\|_2,$$

in the same way as for the good approximation.

Before proceeding any further, we take care of a technicality: As one would expect, the simulations also suggest that q grows without bound as s decreases to zero. For the weighted LLL-algorithm, we can prove the following:

Theorem 2 For the weighted LLL-algorithm to solve

$$(\hat{p}(\mathsf{s}), \hat{q}(\mathsf{s})) = \arg \min_{(p,q)\in\mathbb{Z}^{\bar{N}+1}} \left\| B(\mathsf{s}, X) \begin{pmatrix} p \\ q \end{pmatrix} \right\|_{X\oplus 1} \tag{15}$$

where $\|\cdot\|_{X\oplus 1}$ is the Euclidean norm weighted by the direct sum of X and 1, we have $\lim_{\mathsf{s}\downarrow 0} \hat{q}(\mathsf{s}) = \infty$.

Proof Assume that there exist s_{min} and $q_{ma}x$ such that $\forall s \leq s_{min}$, we have $q \leq q_{max}$. Consider (15) for any $0 < s \leq s_{min}$. By contradicting hypothesis, $\hat{q}$ (s) $\leq q_{max}$. The above yields a Diophantine approximation of θ but not the optimal one as s $\neq$ 0. Now define

$$(\tilde{p}, \tilde{q}) = \arg \min_{(p,q)\in\mathbb{Z}^{\bar{N}+1}} \|p - \theta q\|_X$$

along with

$$\delta(\mathrm{s}) = \|\hat{p}(\mathrm{s}) - \theta\hat{q}(\mathrm{s})\|_X^2 - \|\tilde{p} - \theta\tilde{q}\|_X^2.$$

Observethatthereexistsalowerboundδminsuchthat $\delta(\mathrm{s}) \geq \delta_{\min} > 0$ as $\|\hat{p}(\mathrm{s}) - \theta\hat{q}(\mathrm{s})\|_X$ cannot reach its minimum since $\hat{q}(\mathrm{s}) \leq q_{\max}$. Now, consider the original problem (15) with $\mathrm{s} < \min\left\{\frac{\sqrt{\delta_{\min}}}{\sqrt{2}\tilde{q}}, \mathrm{s}_{\min}\right\}$. With this choice, we have

$$(\mathrm{s}\tilde{q})^2 < \frac{\delta_{\min}}{2} < \frac{\delta(\mathrm{s})}{2} < \frac{\delta(\mathrm{s})}{2} + (\mathrm{s}\hat{q}(\mathrm{s}))^2.$$

Then we have

$$\begin{aligned}
\left\|\begin{pmatrix}\tilde{p} - \theta\tilde{q}b \\ \mathrm{s}\tilde{q}\end{pmatrix}\right\|_{\mathrm{diag}(X,1)}^2 &= \|\tilde{p} - \theta\tilde{q}\|_X^2 + (\mathrm{s}\tilde{q})^2 \\
&\leq \|\hat{p}(\mathrm{s}) - \theta\hat{q}(\mathrm{s})\|_X^2 - \delta(\mathrm{s}) + (\mathrm{s}\hat{q}(\mathrm{s}))^2 + \tfrac{\delta(\mathrm{s})}{2} \\
&= \|\hat{p}(\mathrm{s}) - \theta\hat{q}(\mathrm{s})\|_X^2 + (\mathrm{s}\hat{q}(\mathrm{s}))^2 - \tfrac{\delta(\mathrm{s})}{2} \\
&= \left\|\begin{pmatrix}\hat{p}(\mathrm{s}) - \theta\hat{q}(\mathrm{s}) \\ \mathrm{s}\hat{q}(\mathrm{s})\end{pmatrix}\right\|_{X\oplus 1}^2 - \frac{\delta(\mathrm{s})}{2} \\
&< \left\|\begin{pmatrix}\hat{p}(\mathrm{s}) - \theta\hat{q}(\mathrm{s}) \\ \mathrm{s}\hat{q}(\mathrm{s})\end{pmatrix}\right\|_{X\oplus 1}^2 - \frac{\delta_{\min}}{2}.
\end{aligned}$$

The above is clearly a contradiction to the optimality of $(\hat{p}(\mathrm{s}), \hat{q}(\mathrm{s}))$

Note that the result appears trivial from Eq. (14) except that the behavior of the last component of the first vector of the reduced basis has not yet been explored in the weighted case

Comparison between the weighted LLL-algorithm, $X(\theta_{q-p})$, and the one of Sect. 3.3, $Y^{-1}(\theta q - Y\bar{p})$ indicates that a good choice of the weighting might be $X = Y^{-1}$. This is only a guiding idea, as $X = Y^{-1}$ would mean that $Y\bar{p} = p$, that is, Y × Dirichlet numerator($Y^{-1}\theta$) = Dirichlet numerator(θ), which does not hold exactly.

For practical computation of the time steps τ = q, we must find numerators pk that fulfill the odd/even constraints using the LLL-algorithm. The nonuniform variant introduced above makes it simpler to find suitable parameters X and s, but a search is still required. To automate the search, we use a standard genetic algorithm to find weight vectors X with a user-defined s that minimize the number of parity constraint violations of the $p_{k\ell}$. This works well in most cases, requiring only a few iterations (typically up to 5) for reasonably sized populations (about 200). We suggest that the standard crossover and mutation operators could be adjusted to improve the performance of the search. In particular, increasing the likelihood of changing the X values corresponding to denominators $p_{k\ell}$ that violate a constraint and increasing the likelihood of retaining X values for which the corresponding $p_{k\ell}$ do not violate the constraints may improve performance.

Example 8 (Weighted LLL-Algorithm.) Consider a ring with N = 7 spins and $p_{\max}(1, 3)$. There are four eigenspaces with projectors Πk , and three rationally independent transition frequencies $\omega_{k,k+1} = (-0.7530, -1.6920, -1.3569)/\pi$. Noting thatsk = Sgn 3|Πk |1for k = 0,..., 3 yielding $s = (s_0,s_1,s_2,s_3) = (1, -1, -1, 1)$, we choose ω12 as reference frequency and set

$$\theta = 2\omega_{12}^{-1}(\omega_{01}, \omega_{23})^T = (0.8901, 1.6039)^T \tag{16}$$

with corresponding constraints s' = (1, 1), which means that the numerators pk in the simultaneous Diophantine approximation of θ must both be odd.

Applying the classical LLL-algorithm to solve the simultaneous Diophantine approximation for θ yields rational approximations of very high accuracy, as shown in Fig. 1a. However, most of the resulting approximations pk/q do not satisfy the parity constraints. Using the weighted LLL-algorithm and varying the diagonal scaling vector X enable us to find solutions of arbitrary accuracy, as shown in Fig. 1b, all of which satisfy the parity constraints for the numerators pk .

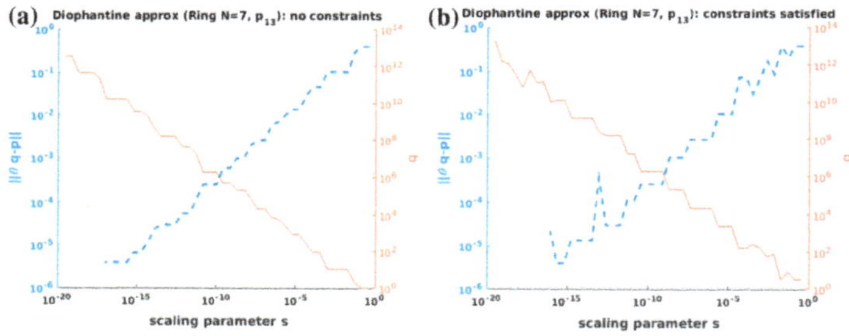

Figure 1: Behavior of LLL-algorithm applied to the simultaneous Diophantine approximation to determine attainability of $p_{max}(1, 3)$ in a N = 7 ring. The left vertical axis in both plots corresponds to the error of the approximation (thick broken line), while the right vertical axis corresponds to the transfer time (thin solid line). a Unweighted LLL-algorithm b Weighted LLL-algorithm

For the approximation p = (170921, 307989) and q = 192028, we obtain the transfer time $t_f = 2q/\omega_{12} = 7.1308 \times 10^5$ (in units of J^{-1}) and corresponding transfer fidelity

$$p_{t_f}(1,3) = \left|\sum_{k=0}^{4} e^{-\imath\lambda_k t_f}\langle 3|\Pi_k|1\rangle\right|^2 \approx 0.4122, \tag{17}$$

which is within $1 - p_{tf}(1, 3)/p_{max}(1, 3) = 2.41 \times 10^{-6}$ of the maximum transfer fidelity $p_{max}(1, 3)$.

The previous example illustrates how we can use the weighted LLL-algorithm to find optimal transfer times that yield very high transfer fidelities, and how we can control the margins of error and ensure the parity constraints are satisfied by adjusting the scaling parameter and diagonal weights in the algorithm.

Estimate of time to attain maximum probability

Our objective is to find an upper bound on the amount of time t it takes to achieve $p_t(i, j) \geq p_{max}(i, j) - prob$, i.e., $p_{max}(i, j) - p_t(i, j) \leq prob$. The approach is to translate the specification on the probability prob to a specification on the infinity norm of the simultaneous Diophantine approximation $\|\epsilon_{Da}\|_\infty$, where $\epsilon_{Da} = \theta q - p$. Proceeding from (4), recalling that $\mathrm{Sgn}\,(\langle i|\Pi_k|j\rangle) =: s_k = e^{-\imath\pi(s_k-1)/2-2\pi\imath n_k}$, where n_k is some integer, we obtain

$$\begin{aligned}
\sqrt{p_{\max}(i,j)} &= \left|\sum_{k=0}^{\tilde{N}-1}\langle i|Pi_k|j\rangle s_k\right| \\
&= \left|\sum_{k\in K'}\langle i|\Pi_k|j\rangle e^{-\imath\frac{\pi}{2}(s_k-1)-2\pi\imath n_k}\right| \\
&= \left|\sum_{k\in K'}\langle i|\Pi_k|j\rangle^{-\imath\frac{\pi}{2}(s_k-1)-2\pi\imath n_k} e^{\imath\frac{\pi}{2}(s_\ell-1)+2\pi\imath n_\ell}\right| \\
&= \left|\sum_{k\in K'}\langle i|\Pi_k|j\rangle e^{-\imath\frac{\pi}{2}(s_k-s_\ell)-2\pi\imath(n_k-n_\ell)}\right|.
\end{aligned}$$

In the second equation, the sum over k has been replaced by a sum over $k \in K'$ as states with $\langle i|\Pi_k|j\rangle = 0$ do not contribute to the sum. The third equality stems from the fact that for fixed, ℓ, $e^{\imath\frac{\pi}{2}(s_\ell-1)+2\pi\imath n_\ell}$ is a global phase factor that is absorbed by the absolute value.

Next, we introduce the attainability condition (8), which is only approximately satisfied using the simultaneous Diophantine approximation. The idea is to expose the gap between the left-hand side and the right-hand side of (8) when t is constrained to emerge from the Diophantine approximation:

$$\begin{aligned}
&\sqrt{p_{\max}(i,j)} \\
&= \left|\sum_{k\in K'}\left(\langle i|\Pi_k|j\rangle e^{-\imath(\lambda_k-\lambda_\ell)t} + \langle i|\Pi_k|j\rangle\left(e^{-\imath\frac{\pi}{2}(s_k-s_\ell)-2\pi\imath(n_k-n_\ell)} - e^{-\imath(\lambda_k-\lambda_\ell)t}\right)\right)\right| \\
&\leq \left|\sum_{k\in K'}\langle i|\Pi_k|j\rangle e^{-\imath(\lambda_k-\lambda_\ell)t}\right| + \left|\sum_{k\in K'}\langle i|\Pi_k|j\rangle\left(e^{-\imath\frac{\pi}{2}(s_k-s_\ell)-2\pi\imath(n_k-n_\ell)} - e^{-\imath(\lambda_k-\lambda_\ell)t}\right)\right| \\
&= \left|\sum_{k\in K'}\langle i|\Pi_k|j\rangle e^{-\imath\lambda_k t}\right| + \left|\sum_{k\in K'}\langle i|\Pi_k|j\rangle\left(e^{-\imath\frac{\pi}{2}(s_k-s_\ell)-2\pi\imath(n_k-n_\ell)} - e^{-\imath(\lambda_k-\lambda_\ell)t}\right)\right| \\
&\leq \sqrt{p_t(i,j)} + \sum_{k\in K'}\left|e^{-\imath\frac{\pi}{2}(s_k-s_\ell)-2\pi\imath(n_k-n_\ell)} - e^{-\imath(\lambda_k-\lambda_\ell)t}\right|.
\end{aligned}$$

It follows that $\sqrt{p_{\max}(i,j)} - \sqrt{p_t(i,j)} \leq \sum_{k\in K'}\left|e^{-\imath\frac{\pi}{2}(s_k-s_\ell)} - e^{-\imath(\lambda_k-\lambda_\ell)t}\right|$ The trivial identity

$$p_{\max} - p_t = \left(\sqrt{p_{\max}} - \sqrt{p_t}\right)\left(\sqrt{p_{\max}} + \sqrt{p_t}\right)$$

then shows that to secure $p_{max}(i, j) - p_t(i, j) \leq \epsilon_{prob}$, it suffices to require

$$\sum_{k\in K'} \left| e^{-\imath\frac{\pi}{2}(s_k - s_\ell)} - e^{-\imath(\lambda_k - \lambda_\ell)t} \right| \leq \frac{\epsilon_{\text{prob}}}{2}, \tag{18}$$

where it is observed that ∈ K' is arbitrary

The last step is to relate the left-hand side of (18) to the simultaneous Diophantine approximation error. Define

$$\epsilon_{\text{Da}}(k,\ell) := |\theta_{k\ell}q - p_{k\ell}|, \qquad \|\epsilon_{\text{Da}}\|_\infty = \max_{(k,\ell)\in\mathcal{S}} \epsilon_{\text{Da}}(k,\ell), \tag{19}$$

where S is the subset of linearly independent attainability equations chosen. By defi- nition, any constraint $c_{k\ell} := \omega_{k\ell}t - \frac{1}{2}(s_k - s_\ell) = 0$ mod 2 with $\omega_{k\ell} = (\lambda_k - \lambda_\ell)/\pi$ can be written as a linear combination of constraints with $(k',\ell') \in \mathcal{S}$, $c_{k\ell}$ = = $\sum_{(k',\ell')\in\mathcal{S}} b_{k'\ell'}c_{k'\ell'}$, with coefficients $b_{k'\ell'} \in \{0, \pm 1\}$. Furthermore, given $\omega_{mn} \in S$ with $s_m = s_n$ and setting $t = 2\tau/\omega_{mn}$ with $\tau \in \mathbb{N}$ and $\theta_{k\ell} = 2\omega_{k\ell}/\omega_{mn}$, we can write the constraints as $c_{k\ell} = \theta_{k\ell}\tau - \frac{1}{2}(s_k - s_\ell)$ for $(k,\ell) \in \mathcal{S}_0$. Given a Diophantine approximation that satisfies the parity constraints, we have $c_{k\ell} = \epsilon_{\text{Da}}(k,\ell)$ mod 2 for $(k,\ell) \in \mathcal{S}$, and $c_{k\ell} \leq \bar{N}\|\epsilon_{\text{Da}}\|_\infty$ for $(k,\ell) \notin \mathcal{S}$, where $\bar{N} = |\mathcal{S}| - 1$ is the number of independent constraints reduced by 1. Thus, we have

$$\begin{aligned}
\sum_{k\in K'} \left| e^{-\imath\frac{\pi}{2}(s_k - s_\ell)} - e^{-\imath\pi\omega_{k\ell}t} \right| &= \sum_{k\in K'} \left|1 - e^{-\imath\pi c_{k\ell}}\right| \\
&\leq |K'| \max_{k\in K'} \left|1 - e^{-\imath\pi c_{k\ell}}\right| \\
&\leq 2|K'| \left|\sin\left(\frac{\pi}{2}\bar{N}\|\epsilon_{\text{Da}}\|_\infty\right)\right|.
\end{aligned}$$

From the above string of inequalities, it follows that for the attainability accuracy prob to be reached, it is sufficient to take

$$2|K'| \left|\sin\left(\tfrac{\pi}{2}\bar{N}\|\epsilon_{\text{Da}}(q)\|_\infty\right)\right| < \frac{\epsilon_{\text{prob}}}{2}. \tag{20}$$

Although conservatively derived, the above formula is consistent with the tightly derived simulation results in Fig. 7 (right). We now summarize the situation we have reached:

Theorem 3 For homogeneous rings, the ITF specification $p_t(i, j) \geq p_{max}(i, j) - \epsilon$ prob is achieved at time $t = 2q/\omega_{mn}$ (in 1/J units) if q is chosen so that simultaneous Diophantine approximation error $\epsilon_{Da}(q) := p-\theta q$ has its infinity norm satisfying (20) and ω_{mn} is the reference transition with respect to which $\theta = (\theta_{k\ell})$ was defined in (10).

There are many simultaneous Diophantine approximation schemes. If we retain the Dirichlet-good one with even/odd constraints on the numerators, under the assumption that the algorithm of Sect. 3.3 converges, the error bound is $\|\epsilon_{Da}\|_\infty \leq 2/q^{1/\bar{N}}$, and we obtain the further sufficient condition

$$2|K'|\left|\sin\left(\frac{\pi\bar{N}}{q^{1/\bar{N}}}\right)\right| < \frac{\epsilon_{\text{prob}}}{2}$$

A minimum q that guarantees ϵ_{prob} is easily extracted from the above inequality

$$q \geq \left(\frac{\pi\bar{N}}{\sin^{-1}\left(\frac{\epsilon_{\text{prob}}}{4|K'|}\right)}\right)^{\bar{N}} \approx \left(\frac{4\pi\bar{N}|K'|}{\epsilon_{\text{prob}}}\right)^{\bar{N}}, \tag{21}$$

where the latter approximation uses $\sin(x) \approx x$ and is valid if $x = \frac{1}{4|K'|}\epsilon_{\text{prob}} \ll 1$.

As an example will soon show, contrasting the above with numerical simulations of Eq. (4) reveals that the bound $O\left(\bar{N}^{\bar{N}}\right)$ is very conservative, mainly because the continuous-time dynamics on the torus was converted to a discrete-time dynamics. The conservativeness is somewhat mitigated by the dimension reduction achieved by the elimination of dark states and symmetries that reduce the number of relevant eigenspaces. For example, for homogeneous rings $\bar{N} \approx \frac{1}{2}N$ rather than N. Further improvement of the scaling behavior could be achieved by utilizing tighter simultaneous Diophantine approximations [13, Th.2], [23], but at the expense of significantly complicating the notation. The reward for the conservativeness of this bound is that it is quite general for rings with uniform coupling and their ITF attainable by the algorithm of Sect. 3.3, as it depends on neither the eigenvalues nor the odd/even pattern. Furthermore, it becomes very general for any network subject to the mild modification of replacing $\bar{N}$ by N and |K'| by N.

Example 9 (Transfer Times—simulation results vs bounds.) For a ring with N = 5, we have $\bar{N}$ = 3 independent eigenspaces with two rationally independent transition

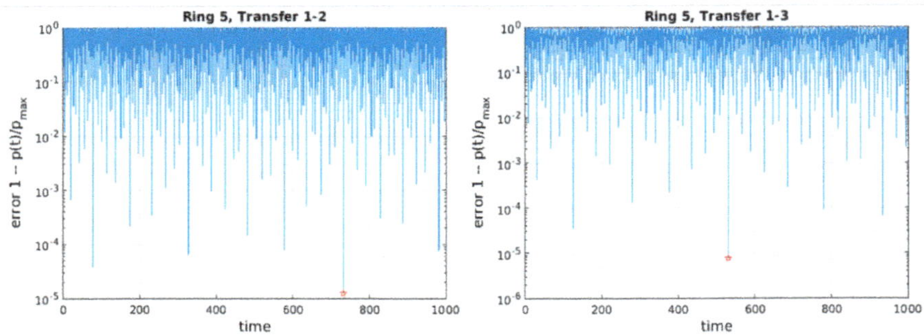

Figure 2: Simulations of transfer probabilities from $1 \rightarrow 2$ and $1 \rightarrow 3$ for a ring of size N = 5

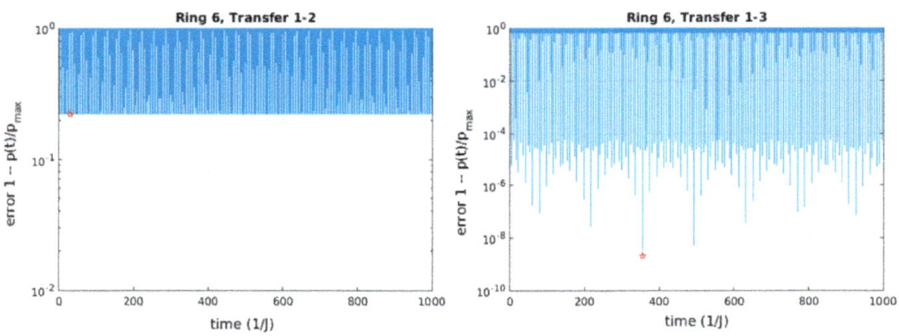

Figure 3: Simulations of transfer probabilities from $1 \rightarrow 2$ and $1 \rightarrow 3$ for a ring of size N = 6

Frequencies and there are no dark subspaces. Hence, |K'| = 3 and we have $\bar{N} = 1$ independent θ. In this case, our conservative bound implies that we can get within ϵ prob of the maximum transition probability in time $12\pi/\epsilon$prob.

In practice, simulations suggest that we can achieve very high fidelities in much shorter times. Figure 2 shows that we can achieve > 99.99% of the maximum transfer fidelity for any two nodes with distance 1 in time t = 77.28, and transfer between two nodes with distance 2 in time t = 125 (in units of 1/J). Notice that the maximum distance between any two nodes in a homogeneous ring of size N = 5 is 2, and hence any transfer can be achieved to within 0.01% of the maximum possible in time $t \leq 125$.

As observed before, for rings of size N = 6, the primary transition frequencies are not rationally independent, implying that we do not have simultaneous attainability. Indeed, Fig. 3 (left) shows that the bound $p_{max}(1, 2)$ is not attainable. Lack of simultaneous attainability does not imply that all bounds are not attainable. Indeed Fig. 3 (right) suggests near-perfect transfer between nodes of distance n = 2.

We can use simulations combined with the LLL-algorithm to estimate the minimum times required to achieve various transfers with a certain maximum error probability.

The results for rings of size N = 5 and N = 7, which satisfy the rational independence conditions for simultaneous attainability, as shown in Fig. 4, suggest a power-law scaling.

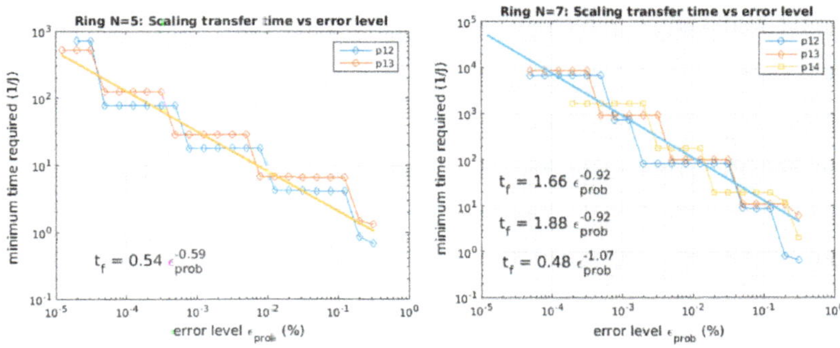

Figure 4: Transfer times estimated from simulations versus the error probability for ring of size 5 (left) and 7 (right)

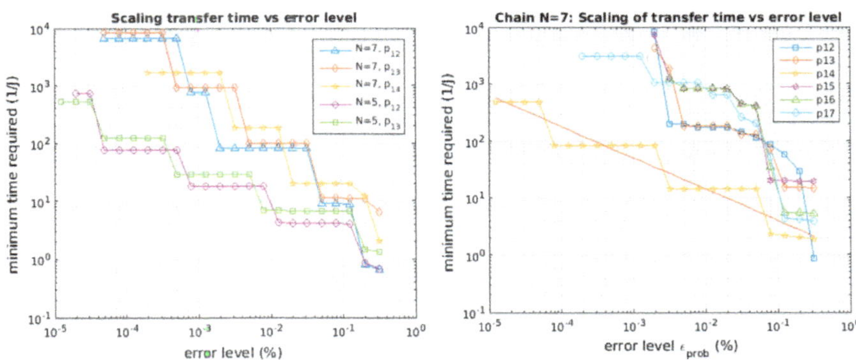

Figure 5: Comparison of scaling of transfer times with error probability for rings (left) and scaling for chains of size N = 7 (right)

Finally, comparing the scaling of the transfer times for rings of different size in Fig. 5(left) suggests that we have similar scalings for both N = 5 and N = 7 although the constant is larger for N = 7. The scaling behavior for various transfers for a chain of size N = 7 in Fig. 5(right) is similar but more complicated, and the transfer times required to get close to the upper bounds appear to be significantly longer.

Transfer time versus decoherence time

In general there is a trade-off between the error prob and the transfer time t_f required to achieve $p_{t_f}(i, j) = 1 - \epsilon_{\text{prob}}$. For actual physical realizations of quantum networks, decoherence is generally a limiting factor. In this case, the relationship between the error probability and the expected transfer time can be useful in estimating what error probabilities can be achieved based on the coherence time of the network t_{coh}.

For instance, in example 9, we showed that for a ring of size N = 5, we would achieve 99% of the maximum transfer probability between any two nodes in time $T \leq 125$ in units of the inverse coupling rate J^{-1}, and we could therefore expect to closely approximate these transfer fidelities, provided the coherence time of the system is $\gg 100J^{-1}$.

More generally, Figs. 4–5 suggest that we have the power law $t_f = c\epsilon_{\text{prob}}^{-\alpha}$ (in 1/J units) , at least for certain types of networks such as rings. In this case for the algorithm to work, it is necessary that

$$c\epsilon_{\text{prob}}^{-\alpha} \ll t_{\text{coh}}. \tag{22}$$

This means that realistically, the error probabilities ϵ_{prob} attainable are limited and we can expect

$$\epsilon_{\text{prob}} \gg (c/t_{\text{coh}})^{1/\alpha}, \tag{23}$$

and the algorithm of Sect. 3.3 could be used to construct a simultaneous Diophantine approximation compatible with this requirement.

Combining Theorem 3 and Eq. (21) also yields an upper bound on the transfer times for which the effect of decoherence should definitively be negligible

$$t_f \leq \frac{2}{\omega_{mn}} \left(\frac{\pi \bar{N}}{\sin^{-1}\left(\frac{(c/t_{\text{coh}})^{1/\alpha}}{4|K'|}\right)} \right)^{\bar{N}}, \tag{24}$$

although we would like to stress here that this bound is excessively conservative due to the approximations made. Given a concrete physical realization of a quantum network with an specific decoherence model, this information could be used to derive tighter time-dependent bounds on the transfer fidelities and realistic transfer times.

INFORMATION TRANSFER (IN-)FIDELITY METRIC AND GEOMETRY

In this section, we come back to an issue raised in Sect. 3—namely that the upper bound derived in Eq. (4) can be justified by the fact that it induces a metric on the set of vertices. Unlike the results in the previous sections, most of the results in this section apply specifically to rings, although numerical simulations suggest that similar results may hold for other homogeneous spin networks such as chains.

Definition and motivation of ITF prametric

To develop a geometric picture, we can view a spin network as a pre-metric or more precisely a prametric space[2] endowed with the prametric that quantifies the information transfer infidelity (ITI). To fix terminology, recall that given a graph $\mathcal{G} = (\mathcal{V}, \mathcal{E})$, or any set of points V for that matter, a prametric [33, p. 666], [32, p.23] is a function $d : \mathcal{V} \times \mathcal{V} \to \mathbb{R}_{\geq 0}$ such that **(i)** $d(i, j) \geq 0$ and **(ii)** $d(i, i) = 0$.

To derive a suitable prametric on the vertex set V = {|i : i = 1,..., N} from the probability p_{max} data, we inspire ourselves from a similar situation in sensor networks [34], where V is the set of sensors and a Packet Reception Rate PRR(i, j) is defined as the probability of successful transmission of the packets from sensor i to sensor j. After symmetrization of the packet reception rate, a prametric (in fact, a semi-metric [35–37]) can be defined as d(i, j) = − log PRR(i, j). Should there be a violation of the triangle inequality, say, d(i, j) > d(i, k) + d(k, j), then the distance between i and j is redefined as d(i, k) + d(k, j). The importance of the metric is that it provides a notion of network curvature, which has a dramatic impact on the traffic flow [38,39] in a paradigm that extends to quantum chains [14]. Following sensor network intuition[34], we define

$$d(i, j) = -\log p_{\max}(i, j). \tag{25}$$

Obviously, d(i, j) ≥ 0 and, as will be shown in Theorem 4, d(i,i) = 0.

We could define the time-stamped prametric by $d_t(i, j) = -\log p_t$ (|i, | j) except that in general $d_t(i,i) \neq 0$. To remedy this situation, we could define $d(i, j) = \inf_{t\geq 0} dt(i, j) = -\log \sup_{t\geq 0} p_t(i, j)$. Since, by Cauchy–Schwarz, $p_t(i,i) \leq 1$ and p_t=0(i,i) = 1, we have supt≥0 pt(i,i) = 1 and hence d(i,i) = 0. This alternate prametric definition is equivalent to the earlier one when p_{max} is attainable, but it reveals that this prametric makes the network of finite diameter ($\sup_{i,j} d(i, j) < \infty$) as $N \to \infty$ as Theorem 4 will show. This has the unfortunate consequence of preventing a genuine large-scale analysis. As Sect. 5 will show, a bias rectifies this problem (see also [14]).

Generally, this information transfer infidelity prametric is not a proper distance satisfying the triangle inequality, but for certain networks such as rings with uniform coupling this prametric will be shown to define a proper distance.

This quantum mechanical (pra)metric is quite different from the usual Euclidean distance $d_{\mathbb{E}}$ of the spins in the spintronic device. In particular, two spins that are physically close in the medium may be far quantum mechanically, and conversely. If two spins are quantum mechanically far, control is necessary to enable transmissions that are too weak or forbidden by the natural quantum mechanical couplings. This control of information can be viewed as the problem of controlling the quantum mechanical geometry of the network.

ITF distance geometry of homogeneous spin rings

It could be argued that a prametric is sufficient if we are solely interested in assessing the difficulty of communication or fidelity of information transfer between nodes in a network. However, a proper metric allows us to investigate other geometric properties such as the curvature of the network with regard to the ITF.

A prametric $d : \mathcal{V} \times \mathcal{V} \rightarrow \mathbb{R}_{\geq 0}$ is a pseudo-metric if in addition to (i) d(i, j) ≥ 0, d(i,i) = 0, it satisfies (ii) d(i, j) = d(j,i) and (iii) the triangle inequality (d(i, j) ≤ d(i, k) + d(k, j)) holds. A metric or distance is a pseudo-metric that has (iv) the separation property: d(i, j) = 0 if and only if i = j.

Theorem 4 For a quantum ring $(\mathcal{V}_N, \mathcal{E}_N)$ of N uniformly distributed spins with XX or Heisenberg couplings, d_N (i, j):= − log p_{max}(i, j) has the following properties:

1. For N odd, $(\mathcal{V}_N, d_N)$ is a metric space.

2. For N even, $(\mathcal{V}_N, d_N)$ is a pseudo-metric space that becomes metric after antipodal point identification

3. If N = p or N = 2p, where p is a prime number, then the distances on the space of equivalence classes of spins are uniform, i.e., d_N (i, j) = c_N for i $\neq$ j. Otherwise, the distances are nonuniform.

4. In all cases $\lim_{N\to\infty}$ d_N (i, j) = $= 2\log\frac{\pi}{2},\ i \neq j \mod (\frac{1}{2}N)$.

Proof To show that $(\mathcal{V}_N, d_N)$ is a pseudo-metric space, we need to verify that (i) d_N (i,i) = 0, (ii) d_N (i, j) = d_N (j,i), and (iii) the triangle inequality holds. For a metric space, we must further have (iv) d_N (i, j) $\neq$0 unless i = j.

(i) is clearly satisfied as the projectors onto the eigenspaces are a resolution of the identity, $\sum_k \Pi_k = I$, and thus for any unit vector |i, we have $\sum_{k=1}^{N} |\langle i|\Pi_k|i\rangle| = \sum_{k=1}^{N} \|\Pi_k|i\rangle\|^2$ = 1. (ii) follows from $|\langle i|\Pi_k|j\rangle| = |\langle j|\Pi_k|i\rangle|$. The proof of the remaining properties relies on the circulant matrix property of the Hamiltonian H¯ in the single excitation subspace $\bar{\mathcal{H}}_{\cdot}$, as shown in Eq. (3) and Table 1.

Observe in Table 1 the double eigenvalues λ_k = λ_{N-k}, except for k = 0 and $k = \frac{1}{2}N$ if N even. From Table 1, each of these double eigenvalues has two general complex conjugate eigenvectors. These general eigenvectors need not be orthogonal, but observing that $\langle v_k|v_\ell\rangle = \delta_{k\ell}$ and $\langle v_k|v_k^*\rangle = 0$, where v_k^* denotes the complex conjugate, it follows that

$$
\begin{aligned}
|\bar{v}_0\rangle &= |v_0\rangle = \tfrac{1}{\sqrt{N}}(1, 1, \ldots)^T, \\
|\bar{v}_k\rangle &= |v_k\rangle, \quad |v_{N-k}\rangle = |v_k^*\rangle, \quad k = 1, \ldots N' = \lfloor \tfrac{N-1}{2} \rfloor, \\
|\bar{v}_{N/2}\rangle &= |v_{N/2}\rangle = \tfrac{1}{\sqrt{N}}(1, -1, \ldots)^T, \quad \text{if } N \text{ is even,}
\end{aligned}
\tag{26}
$$

defines an orthonormal basis of $\bar{\mathcal{H}}_{\cdot}$. Furthermore, in the basis in which $\bar{\mathcal{H}}_{\cdot}$ is circulant, we have |i = e_i , where {e_i : i = 1, ..., N} is the natural basis of $\mathbb{C}^N$.

$$|\langle i|\Pi_0|j\rangle| = |\langle i|\bar{v}_0\rangle\langle\bar{v}_0|j\rangle| = \tfrac{1}{N},$$
$$|\langle i|\Pi_k|j\rangle| = |\langle i|\bar{v}_k\rangle\langle\bar{v}_k|j\rangle + \langle i|\bar{v}_{N-k}\rangle\langle\bar{v}_{N-k}|j\rangle|$$
$$= \left|\rho_N^{ki}(\rho_N^{kj})^* + (\rho_N^{ki})^*\rho_N^{kj}\right| \tfrac{1}{N} \tag{27}$$

$$= \left|\rho_N^{k(i-j)} + \rho_N^{-k(i-j)}\right| \tfrac{1}{N} = \tfrac{2}{N}\left|\cos\left(\tfrac{2\pi k(i-j)}{N}\right)\right|, \tag{28}$$

$$|\langle i|\Pi_{N/2}|j\rangle| = |\langle i|\bar{v}_{N/2}\rangle\langle\bar{v}_{N/2}|j\rangle| = \tfrac{1}{N}. \tag{29}$$

Summing over all eigenspaces $k = 0, \ldots, \lfloor N/2 \rfloor$ gives

For N = 2N' + 1, it is easy to see that $p_{max}(i, j) = 1$ if and only if i = j, hence (iv). For N = 2N' + 2, on the other hand, we also have $|\cos(\frac{2\pi k N/2}{N})| = |\cos(\pi k)| = 1,$ and thus d(i, j) = 0 for i − j = 1 2 N, i.e., the distance vanishes for antipodal points, and thus d(i, j) is at most a pseudo-metric. However, noting that we can identify antipodal points $|j\rangle$ and $|j + N' + 1\rangle$ let d be defined on the set of equivalence classes [| j] for j = 1,..., N' + 1 instead. (The antipodal identification preserves the ring structure.) At this stage, d is a semi-metric [35,36,40], that is, it satisfies all axioms of a metric except the triangle inequality. T

To prove the triangle inequality, we show that $\sqrt{p_{\max}(i,m)}\sqrt{p_{\max}(m,j)} \leq$. $\sqrt{p_{\max}(i,j)}$. The definition (4) of pmax rewritten in terms of the eigenvectors of $\tilde{H}$ using (27)–(29) gives

$$\sqrt{p_{\max}(i,m)} = \frac{1}{N}\sum_{k=0}^{N-1} \alpha_k \rho_N^{k(m-i)},$$

$$\sqrt{p_{\max}(m,j)} = \frac{1}{N}\sum_{k'=0}^{N-1} \beta_{k'} \rho_N^{k'(j-m)},$$

where $\alpha_k = s_k(i,m) = \mathrm{Sgn}\left(\rho_N^{k(m-i)} + \rho_N^{-k(m-i)}\right) \in \{\pm 1, 0\}$ is rewritten explicitly in terms of the eigenvectors rather than as in Sect. 3 and $\beta_{k'} = s_{k'}(m,j)$. Setting

$$\gamma_k = \sum_{k'=0}^{N-1} \alpha_k \beta_{k'} \rho_N^{(k'-k)(j-m)}$$

we obtain

$$\begin{aligned}
&\sqrt{p_{\max}(i,m)}\sqrt{p_{\max}(m,j)}\\
&=\frac{1}{N^2}\sum_{k,k'=0}^{N-1}\alpha_k\beta_{k'}\rho_N^{k(m-i)}\rho_N^{k'(j-m)}\\
&=\frac{1}{N^2}\sum_{k,k'=0}^{N-1}\alpha_k\beta_{k'}\rho_N^{k(j-i)+(k'-k)(j-m)}\\
&=\frac{1}{N^2}\sum_{k=0}^{N-1}\gamma_k\rho_N^{k(j-i)}=\left|\frac{1}{N^2}\sum_{k=0}^{N-1}\gamma_k\rho_N^{k(j-i)}\right|.
\end{aligned}$$

The final equality follows because the LHS and thus the RHS are known to be real and positive. Furthermore, as ρ_N is a root of unity, $|\rho_N| = 1$, and recalling $|\alpha k| = |\beta_k'| = 1, 0$,

$$\begin{aligned}
|\gamma_k| &= \left|\rho_N^{k(m-j)}\sum_{k'=0}^{N-1}\alpha_k\beta_{k'}\rho_N^{k'(j-m)}\right|\\
&\leq \left|\rho_N^{k(m-j)}\right|\cdot\sum_{k'=0}^{N-1}\left|\alpha_k\beta_{k'}\rho_N^{k'(j-m)}\right|\leq N,
\end{aligned}$$

where the last inequality allows for the presence of dark states. Again we have $\rho_N^{(N-k)(m-j)}=\rho_N^{-\bar{k}(m-\bar{j})}$, and as the LHS above is known to be real, we know that we must have $\gamma k = \gamma_{N-k}$. Hence, we can again collect exponential terms pairwise to obtain cosines, which gives for N = 2N' + 1:

$$\begin{aligned}
\left|\frac{1}{N^2}\sum_{k=0}^{N-1}\gamma_k\rho_N^{k(j-i)}\right| &= \left|\frac{\gamma_0}{N^2}+\frac{1}{N^2}\sum_{k=1}^{N'}2\gamma_k\cos\left(\tfrac{2\pi k(j-i)}{N}\right)\right|\\
&\leq \frac{|\gamma_0|}{N^2}+\frac{2}{N^2}\sum_{k=1}^{N'}|\gamma_k|\left|\cos\left(\tfrac{2\pi k(j-i)}{N}\right)\right|\\
&\leq \frac{1}{N}+\frac{2}{N}\sum_{k=1}^{N'}\left|\cos\left(\tfrac{2\pi k(j-i)}{N}\right)\right|\\
&=\sqrt{p_{\max}(i,j)}.
\end{aligned}$$

For N = 2N' + 2, we simply replace γ_0 by $\gamma_0 + \gamma N' + 1$ above to obtain

$$\left|\frac{1}{N^2}\sum_{k=0}^{N-1}\gamma_k\rho_N^{k(j-i)}\right| \leq \frac{2}{N}+\frac{2}{N}\sum_{k=1}^{N'}\left|\cos\left(\frac{2\pi k(j-i)}{N}\right)\right|$$
$$= \sqrt{p_{\max}(i,j)}.$$

This proves (iii) and hence parts (1) and (2) of the theorem.

To establish (3), we note that if N = 2N' + 1 is prime, then

$$\sum_{k=1}^{N'}\left|\cos\left(\frac{2\pi k(i-j)}{N}\right)\right| = \sum_{k=1}^{N'}\left|\cos\left(\frac{2\pi k}{N}\right)\right|.$$

If N is not p or 2p, then N and (i – j) will have factors (which can be canceled) in common for some (i – j) but not for others and hence we will obtain different distances.

To establish (4), letting N → ∞, it is easily seen that the dependency on i, j is eliminated, provided $i \neq j \mod (\frac{1}{2}N)$Hence, taking the norm of the above and then – log(·), it follows that at the infinite ring limit, the distance is uniform for i ≠ j + mod $(\frac{1}{2}N)$. Finally,

$$\lim_{N\to\infty}\sqrt{p_{\max}(i,j)} = \lim_{N\to\infty}\frac{2}{N}\sum_{k=0}^{N/2}|\cos((i-j)2\pi k/N)|$$
$$= \frac{2|i-j|}{\pi}\int_0^{\frac{\pi}{2|i-j|}}\cos(|i-j|x)dx$$
$$= \frac{2|i-j|}{\pi|i-j|}[\sin(|i-j|x)]_0^{\frac{\pi}{2|i-j|}} = \frac{2}{\pi}$$

shows that $\lim^{N\to\infty} d_N(i,j) = 2\log\frac{\pi}{2} \approx 2\times 0.4516$ for $i \neq j \mod (N/2)$.

Case 3 of Theorem 4 allows for a very specific geometrization of the quantum ring in terms of constant curvature spaces. Define the n-sphere of curvature κ as $\mathbb{S}_\kappa^n := \{x \in \mathbb{R}^{n+1} : \|x\|^2 = 1/\kappa\}$. We have the following corollary:

Corollary 1 The metric space (V_p, d_p) of p spins (p ≥ 3 prime) arranged in a homogeneous ring with uniform ITI distance $d_p(i,j) = c_p, i \neq j$, is isometrically embeddable in $\mathbb{S}_\kappa^{p-1}$ *iff*

$$\kappa \leq \left[\frac{1}{c_p}\cos^{-1}\left(-\frac{1}{p-1}\right)\right]^2. \tag{31}$$

Furthermore, it is irreducibly isometrically embeddable in $\mathbb{S}_\kappa^{p-2}$ for

$$\kappa = \left[\frac{1}{c_p} \cos^{-1} \left(-\frac{1}{p-1} \right) \right]^2. \tag{32}$$

Notes: In the above, "irreducibly embeddable" means that the embedding cannot happen into a lower-dimensional constant curvature space. By convention, $\cos^{-1}$ takes values in [π/2, π].

Proof This result is a corollary of [40, Th.63.1]. For the details, see [41, Appendix].

Note that this corollary deals with embeddability of the vertices only; however, edges can be mapped isometrically as arcs of great circles on either the sphere of curvature (31) or that of curvature (32). Also note that the symmetry of the simple p = 3 case of the circle $\mathbb{S}^1$ circumscribed to a equilateral triangle is misleading, as in very high dimension (p → ∞), Eq. (32) yields $1/\sqrt{\kappa} =: R \to \frac{c_p}{\pi/2}$, that is, all vertices are mapped to the half-sphere of radius R.

Regarding N = 2p in Case 3, we could first do the antipodal identification on the combinatorial ring $(\mathcal{V}_{2p}, \mathcal{E}_{2p})$, leading to $(\mathcal{V}_p, \mathcal{E}_p)$ring, and then embed $(\mathcal{V}_p, \mathcal{E}_p)$as in the preceding corollary.

Regarding Case 4 when N is odd, define $\epsilon := \max_{i \neq j} |d_N(i, j) - 2\log(\pi/2)|$. Then the metric space $(\mathcal{V}_N, d_N)$can be mapped isometrically on the sphere $\mathbb{S}_\kappa^{N-2}$ of radius $d_\infty / \cos^{-1} -(N - 1)-1$ up to an additive distortion not exceeding , ϵ that is, the embedding is quasi-isometric [42, 7.2.G]. The case of an even N is dealt with as before using antipodal identification. The geometry of a genuinely infinite ring (N = ∞ rather than N → ∞) is completely different and is left to future work.

The N even case can be dealt with in a different way. Rather than doing, first, a combinatorial antipodal identification $(i = j \text{ if } i - j = 0 \mod (\frac{1}{2}N))$ and, then, mapping the quotient space $\mathcal{V}_N / \sim$to the sphere, we could map the combinatorial antipodal points to geometrical antipodal points on the sphere $\mathbb{S}_\kappa^{N-2}$ with the understanding that geometrical antipodal points on the sphere are identified to yield the real projective space $\mathbb{RP}^{N-2}$. A slight generalization of (32) of Corollary 1 together with 4 of Theorem 4 yields an irreducible embedding of $(\mathcal{V}_N, d_N)$into the sphere of curvature $\kappa = \left(\left(\cos^{-1}\left(-\frac{1}{N-1}\right)\right) / \left(2\log\frac{\pi}{2}\right)\right)^2$. On the other hand, $\mathbb{RP}^{N-2}$ is usually endowed with the standard curvature 1 metric of diameter π/2. To sum up: 1

Corollary 2 For N even, there is an embedding $\mathcal{V}_N \hookrightarrow \mathbb{RP}^{N-2}$, which is quasiisometric for the scaled distance $d_N \cos^{-1}\left(-\frac{1}{N-1}\right) / \left(4\log\frac{\pi}{2}\right)$ *on* $\mathcal{V}_N$and the curvature 1 distance on $\mathbb{PR}^{N-2}$. Furthermore, for N → ∞ the distortion becomes vanishingly small.

CONTROL OF INFORMATION TRANSFER FIDELITY

To overcome intrinsic limitations on quantum state transfer or speed up transfer, one can either try to engineer spin chains or networks with nonuniform couplings [6,7], or introduce dynamic control to change the network topology [9–11].

Our analysis above shows that engineering the couplings is not strictly necessary. For an XX or Heisenberg-type chain with uniform nearest-neighbor couplings, for example, it can easily be verified that the information transfer fidelity between the end spins is unity, and attainability of the bounds means that we can achieve arbitrarily high state transfer fidelities between the end spins if we wait long enough. Engineering the couplings, however, can speed up certain state transfer tasks such as state transfer between the end spins at the expense of others.

A more flexible alternative to fixed engineered couplings is to apply control to change the network geometry and hence speed up state transfer as well as enable some transfers that either were forbidden or had poor ITF. One way this can be achieved is to apply static electromagnetic bias fields to change the energy-level splittings between the spin-up and spin-down states for different nodes in the graph, as suggested e.g., in [12]. To see how the application of such bias fields can alter the transfer fidelities and network geometry, consider a simple, concrete example of a single bias field ζ applied to node in a spin ring with uniform coupling. First, due to translation invariance, we can always relabel the nodes so that the biased node is node N. Then, assuming XX coupling, the Hamiltonian on the single excitation subspace becomes

$$\bar{H}_N^{(\zeta)} = \begin{pmatrix} 0 & 1 & \dots & 0 & 0 & 0 & \dots & 0 & 1 \\ 1 & 0 & \dots & 0 & 0 & 0 & \dots & 0 & 0 \\ \vdots & \vdots & \ddots & \vdots & \vdots & & & \vdots & \vdots \\ 0 & 0 & \dots & 0 & 1 & 0 & \dots & 0 & 0 \\ 0 & 0 & \dots & 1 & 0 & 1 & \dots & 0 & 0 \\ 0 & 0 & \dots & 0 & 1 & 0 & \dots & 0 & 0 \\ \vdots & \vdots & & \vdots & \vdots & \vdots & \ddots & \vdots & \vdots \\ 0 & 0 & \dots & 0 & 0 & 0 & \dots & 0 & 1 \\ 1 & 0 & \dots & 0 & 0 & 0 & \dots & 1 & \zeta \end{pmatrix}, \tag{33}$$

where it is observed that we have the decomposition

$$\bar{H}_N^{(\zeta)} = C_N + \zeta E_{N,N},$$

where C_N is the N × N circulant matrix defined above and $E_{N,N}$ is a N × N matrix which is zero except for a 1 at position (N, N).

Physically, applying a large bias field to the Nth node in the ring results in a large detuning that effectively eliminates this node from the ring and breaks the ring open, leaving a chain of length N − 1. Hence, in the limit $\zeta \to \infty$, we expect the transition fidelities for the first N − 1 nodes to approach those for a chain of length N − 1 while the transition fidelities between the first N −1 nodes and the final (biased) node approach 0. We now reformulate this intuitively obvious result in precise mathematical language.

Lemma 2 The eigenvalues and eigenvectors of the (N −1)×(N −1) Toeplitz matrix TN−1 made up of ones on the super diagonal and subdiagonal and zeros everywhere else are given by $\lambda_k = 2\cos\left(\frac{\pi k}{N}\right)$ *and* $|v_k\rangle_i = \sqrt{\frac{2}{N}}\sin\left(\frac{\pi k i}{N}\right)$; $k = 1, \ldots, N-1$, $i = 1, \ldots, N-1$.. Furthermore, for k even, $|v_k\rangle_1 + |v_k\rangle_{N-1} = 0$.T

Theorem 5 *Let* p_{chain}^{N-1} be the maximum transfer fidelities for a spin chain of length N −1 with uniform coupling between the nearest neighbors. *Let* $p_{\text{ring}}^{N,\zeta}$ be the maximum transfer fidelities for a ring of size N with bias ζ on the Nth node. Then

$$\lim_{\zeta\to\infty} p_{\text{ring}}^{N,\zeta}(i,j) = \begin{cases} p_{\text{chain}}^{N-1}(i,j), & if\, i,j < N; \\ 0, & i = N, j \neq N or\, i \neq N, j = N; \\ 1, & i,j = N. \end{cases} \tag{34}$$

Proof Write the characteristic polynomial of $\bar{H}_N^{\zeta}$ as $\det((\lambda I_N - C_N) - \zeta E_{N,N})$ and recall that the determinant of the sum of two matrices equals the sums of the determinants of all matrices made up with some columns of one matrix and the complementary columns of the other matrix. Applying the latter to the characteristic polynomial of $\bar{H}_N^{(\zeta)}$ yields

$$\det(\lambda I_N - \bar{H}^{(\zeta)}) = \det(\lambda I_N - C_N) - \zeta \det(\lambda I_{N-1} - T_{N-1}),$$

where T_{N-1} is the Toeplitz matrix defined in the lemma. From classical root-locus techniques, it follows that as $\zeta \to \infty$, exactly one eigenvalue $\lambda_N(\zeta)$ goes to ∞, while the remaining ones $\lambda 1(\zeta) \ldots \lambda_{N-1}(\zeta)$ converge to the roots of $\det(\lambda I_{N-1} - T_{N-1}) = 0$.

Next, we look at the eigenvectors and rewrite the eigenvector equation as

$$\left(\begin{array}{c|c} T_{N-1} & \begin{matrix} 1 \\ 0_{N-3} \\ 1 \end{matrix} \\ \hline 1\ 0_{N-3}\ 1 & \zeta \end{array}\right) \left(\begin{array}{c} |v_k(\zeta)\rangle_1 \\ \vdots \\ |v_k(\zeta)\rangle_{N-1} \\ \hline |v_k(\zeta)\rangle_N \end{array}\right) = \lambda_k(\zeta) \left(\begin{array}{c} |v_k(\zeta)\rangle_1 \\ \vdots \\ |v_k(\zeta)\rangle_{N-1} \\ \hline |v_k(\zeta)\rangle_N \end{array}\right).$$

Consider first the first $k \neq N$ equations. Since $\lim_{\zeta\to\infty} \lambda_k(\zeta)$ exists and is finite, it follows from the bottom eigenequation that $\zeta\, |v_k(\zeta)\, N$ remains bounded as $\zeta \to \infty$. Therefore, $\lim_{\zeta\to\infty} |v_k\rangle_N = 0$. Since $\lambda_k(\infty)$ is a unique eigenvalue of T_{N-1}, it follows that $\lim_{\zeta\to\infty} |v_k(\zeta)\rangle_{1:N-1}$ is the corresponding eigenvector of TN−1. It remains to 1 show that with this $|v_k\rangle_{1:N-1}$ the bottom eigenequation can be made to hold. This is easily achieved by defining

$$\lim_{\zeta\to\infty} \zeta |v_k(\zeta)\rangle_N = -\lim_{\zeta\to\infty} (|v_k(\zeta)\rangle_1 + |v_k(\zeta)\rangle_{N-1})$$

By the lemma, for k even, we have $\lim_{\zeta\to\infty} \zeta |v_k(\zeta)\rangle_N = 0$, and therefore the k < N eigenequation holds with $|v_k(\zeta)\rangle_N$ going to zero faster than 1/ζ. For k odd, $|v_k(\zeta)\rangle_N$ goes to zero as c/ζ, where $c \neq 0$ is some constant.

By the root-locus result, for ζ large enough, all eigenvalues are distinct, and we have

$$\sqrt{p_{\text{ring}}^{(\zeta,N)}(i,j)} = \sum_{k<N} |\langle i|v_k(\zeta)\rangle\langle v_k(\zeta)|j\rangle| + |\langle i|v_N(\zeta)\rangle\langle v_N(\zeta)|j\rangle|$$
$$= \sqrt{p_{\text{chain}}^{(N-1)}(i,j)} + |\langle i|v_N(\zeta)\rangle\langle v_N(\zeta)|j\rangle| \qquad (35)$$

where the second equality is understood as the $\zeta \to \infty$ limit. To complete the proof, it therefore remains to look at $|v_N(\zeta)\rangle$.

The last k = N eigenequation easily implies that $\zeta |v_N(\zeta)\rangle_{1:N-1}$ remains bounded as $\zeta \to \infty$. Therefore, $\lim_{\zeta\to\infty} |v_N(\zeta)\rangle_{1:N-1} = 0$. To normalize the eigenvector, we take $\lim_{\zeta\to\infty} |v_N(\zeta)\rangle_N = 1$. The latter together with (35) proves the theorem.

Thus, we have a systematic way to compute the asymptotic transfer probability of a ring with high bias from the transfer probability of a chain without bias.

Example 10 (Dynamic Routing.) As an illustration of how these results can be used, consider a ring of size N = 9. The maximum transfer fidelities between nodes i≠ j for this ring without bias are quite low, 0.4094 and 0.4444. However, applying a large bias to node 9 changes the maximum transfer fidelities. In particular, the maximum transfer fidelity between nodes 1 and 8, 2 and 7, 3 and 6 and 4 and 5 now approaches 1. Figure 6 shows a visual representation of the transfer fidelities for the ring without bias (left) and with bias (right). This result is consistent with Theorem 5, as using Lemma 2, it is easily verified that

$$\sqrt{p_{\text{ring}}^{(8)}(i,9-i)} = \frac{2}{9}\sum_{k=1}^{8}\left|\sin\left(\frac{\pi i k}{9}\right)\sin\left(\frac{\pi(9-i)k}{9}\right)\right| = 1. \qquad (36)$$

The example also shows that a finite bias is sufficient to enable almost perfect state transfer in practice, despite the fact that the ring only becomes a chain in the limit when an infinite bias is applied to node 9. We also used the LLL-inspired algorithm to estimate the transfer time as a function of the infidelity of the transfer. We note here that it was crucial to use the weighted LLL-algorithm to generate a range of simultaneous Diophantine approximations, which generally did not satisfy the parity constraints on the numerators, and to use the idea of combining approximations to satisfy the constraints. With this approach, we were able to find solutions satisfying all of the parity constraints on the numerators over a wide range of infidelities to estimate the transfer times required as a function of the tolerated infidelity. The results, shown in Fig. 7 (left), suggest that high fidelities are indeed attainable for modest biases, and the apparent linearity of the data in the bilogarithmic plot still suggests a polynomial scaling. However, the actual transfer times are significantly higher in this case than in previous examples. We point out here that our algorithm is not guaranteed to find the shortest possible time although Fig. 7 (right) shows that there is a good correlation between the Diophantine approximation error and the observed infidelity of the transfer, as already anticipated by Eq. (20). Furthermore, the algorithm enables us to estimate necessary transfer times far beyond the regime accessible by brute-force numerical simulations.

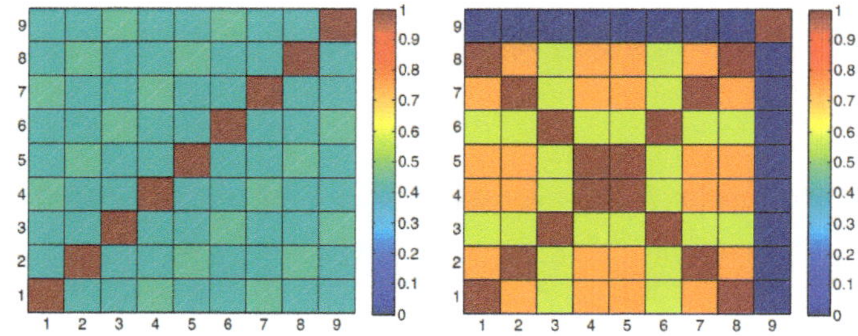

Figure 6: Visual representation of maximum transfer fidelities for a ring of size 9 without bias (left) and the same ring with a finite bias applied to node 9 (right)

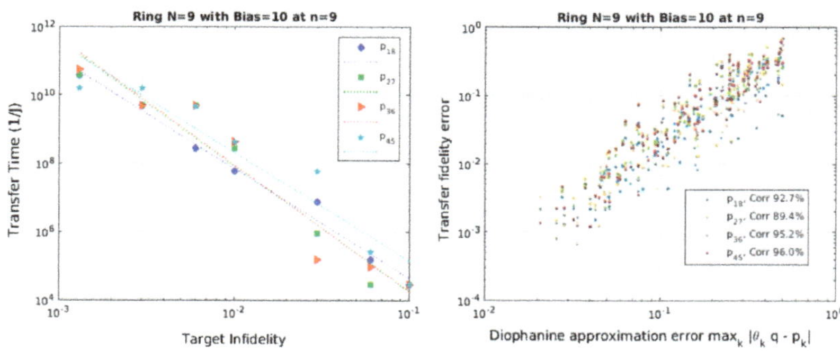

Figure 7: Scaling of transfer times for transfers between nodes (1, 8), (2, 7), (3, 6) and (4, 5) for a ring of size N = 9 with a bias of strength 10 (in units of 1/J) applied to node n = 9 (left) and correlation of fidelity error (or infidelity) and Diophantine approximation error (right)

This example shows how a dynamic routing scheme can be implemented to transfer information from any node in a ring to any other node with fidelity approaching unity by simply applying bias fields to different nodes. For transfer between nodes 1 and 8, 2 and 7, 3 and 6, or 4 and 5, it suffices to apply a large bias to node 9. If we wish to transfer information from node 1 to 4, then translation invariance of the ring allows us to shift the labels by 2, so that node 1 becomes 3 and 4 becomes 6, and applying a bias to the new node 9 will enable the transfer.

Further reflection shows that we can achieve maximum transfer fidelities approaching unity for transfer between any pair of nodes in a ring of size N, provided N is odd by simply biasing the node in the middle between the pair of spins. This is because in this case N − 2 is odd, so there must be an odd number of spins along one path around the ring and an even number between the spins around the other. By applying the bias in the middle of the path with an odd number of spins, we asymptotically reach a chain with N − 1 (even) spins. In this chain, the transfer probability between spins mirrored at the center is 1, which is specifically true for the source and target spin with an even number of spins between them in the chain.

If N is even instead, then the situation is more complicated. If there is an odd number of spins between source and target along the ring, then applying a bias at the middle creates an odd chain where source and target are connected with probability 1 as they are at mirror-symmetric positions in the ring. If there is an even number of spins between source and target, then applying a single bias cannot achieve perfect information transfer as the spins can never be at mirrored positions in the odd chain (which are the only ones in the chain perfectly connected). There are, however, multiple solutions to apply a bias at two spins that can asymptotically generate a suitable chain.

In practice, it may be possible and even preferable to simultaneously apply biases to several nodes instead of a single node to shape the overall potential landscape. This case is more difficult to treat analytically, but preliminary results [43] suggest that numerical optimization can be used in this case to optimize the applied biases to achieve significant reductions in the transfer times and the magnitude of the required bias fields, as well as to deal with practical issues such as leakage of the bias fields, i.e., the tendency of a bias applied to one node to also affect nearby nodes.

CONCLUSION

The concept of maximum transfer fidelity for information transfer between nodes in a network of interacting spins was introduced, and criteria for attainability of the bounds in terms of the transition frequencies of the network were given. Attainability was shown to be related, theoretically, to minimality of a linear flow and, computationally, to a translation on a torus. This last connection enabled us to derive upper bounds on the time required to realize transfer fidelities within ϵ_{prob} of the maximum transfer fidelity, for arbitrary $\epsilon_{\text{prob}} > 0$, via the simultaneous Diophantine approximation. Algorithms were discussed to find the required approximations.

The ultimate aim of this analysis is to understand the intrinsic limitations of information transfer in spin networks and utilize this understanding to engineer networks with favorable bounds on the information transfer fidelities and dynamic attainability properties, so that high spin transfer fidelities can be attained in short times, enabling fast transfer and minimizing the effects of noise and decoherence. An advantage of our approach of combining general ITF bounds and asymptotic attainability conditions with an algorithm to estimate the time required to achieve transfer within a set margin of error, compared to engineering the spectrum of the network Hamiltonian to admit perfect state transfer, for example, is that the latter condition is generally a too strong requirement, as in practice there are always margins of error. Therefore, it makes more sense to ask how much time is required to achieve a certain transfer fidelity for a given acceptable margin of error , and try to optimize the network topology, couplings or biases to achieve the best possible transfer times for the acceptable margins of error.

The general results were applied specifically to regular spin structures such as rings with uniform coupling. In this case, the information transfer infidelity prametric induced by maximum transfer fidelity takes on full significance as it can be shown to be a proper metric defining an information transfer infidelity geometry for the network, which is significantly different from the physical network geometry. The analysis shows that the intrinsic transfer fidelities for simple networks such as rings are often attainable asymptotically, but the times required to achieve high fidelities can be very long. The intrinsic bounds on the ITFs and transfer times can be favorably changed, however, by simple Hamiltonian engineering such as applying spatially distributed static bias fields. In particular, it was shown how such simple controls can be used to alter the information transfer fidelities and geometry of a network. It was demonstrated how this idea can be applied to enable or disable information transfer between a pair of nodes in the network. Simple bias controls are sufficient to direct information flow between nodes. By changing the biases, different transfers can be targeted, and thus a spin ring with fixed couplings can be turned into a simple quantum router for information encoded in excitations of a spin network.

Directions for future work include optimizing information transfer in spin networks via optimal control to achieve faster and more efficient dynamic routing in more complex spin networks. While this work focused on transfer of a single excitation, the concepts and analysis can also be applied to the case of encoding and simultaneous transfer of multiple excitations. This is interesting as it could increase the information transmission capacity of the network. Finally, although simulation results for similar spin systems suggest that some degree of intrinsic robustness of state transfer and the ability to mitigate the effects of noise, decoherence or fluctuations in the couplings via control [10,11,44], the sensitivity of transfer fidelities with regard to noise and deleterious effect of the environment need to be investigated for specific physical realizations of spin networks.

ACKNOWLEDGMENTS

E. A. Jonckheere was partially supported by the Army Research Office (ARO) Multi University Research Initiative (MURI) grant W911NF-11-1-0268. S. G. Schirmer and F. C. Langbein acknowledge support from the Ser Cymru National Research Network on Advanced Engineering. SGS also acknowledges funding from a Royal Society Leverhulme Trust Senior Fellowship.

REFERENCES

1. Caruso, F., Giovannetti, V., Lupo, C., Mancini, S.: Quantum channels and memory effects. Rev. Mod. Phys. 86, 1203–1259 (2014)
2. Awschalom, D.D., et al.: Quantum spintronics: engineering and manipulating atom-like spins in semiconductors. Science 339(6124), 1174–1179 (2013)
3. Bose, S.: Quantum communication through an unmodulated spin chain. Phys. Rev. Lett. 91, 07901 (2003)
4. Bose, S.: Quantum communication through spin chain dynamics: an introductory overview. Contemp. Phys. 48, 13–30 (2007)
5. Key, A.: A review of perfect, efficient, state transfer and its application as a constructive tool. Int. J. Quantum Inf. 08, 641 (2010)
6. Christandl, M., Datta, N., Ekert, A., Landahl, A.J.: Perfect state transfer in

quantum spin networks. Phys. Rev. Lett. 92, 187902 (2004)

7. Christandl, M., Datta, N., Dorlas, T., Ekert, A., Kay, A., Landahl, A.: Perfect transfer of arbitrary states in quantum spin networks. Phys. Rev. A 71, 032312 (2005)
8. Greentree, A.D., Cole, J.H., Hamilton, A.R., Hollenberg, L.C.L.: Coherent electronic transfer in quantum dot systems using adiabatic passage. Phys. Rev. B 70, 235317 (2004)
9. Zueco, D., Galve, F., Kohler, S., Hänggi, P.: Quantum router based on ac control of qubit chains. Phys. Rev. A 80, 042303 (2009)
10. Schirmer, S.G., Pemberton-Ross, P.J.: Fast high-fidelity information transmission through spin-chain quantum wires. Phys. Rev. A 80, 030301 (2009)
11. Wang, X., Bayat, A., Bose, S., Schirmer, S.G.: Global control methods for Greenberger–Horne– Zeilinger-state generation on a one-dimensional ising chain. Phys. Rev. A 82, 012330 (2010)
12. Casaccino, A., Lloyd, S., Mancini, S., Severini, S.: Quantum state transfer through a qubit network with energy shifts and fluctuations. Int. J. Quantum Inf. 7(8), 1417–1427 (2009)
13. Jonckheere, E., Schirmer, S., Langbein, F.: Geometry and curvature of spin networks. In: IEEE MultiConference on Systems and Control, pp. 786–791, Denver (2011). doi:10.1109/CCA.2011.6044395. arXiv:1102.3208v1 [quant-ph]
14. Jonckheere, E., Schirmer, S., Langbein, F.: Quantum networks: the anti-core of spin chains. Quantum Inf. Process. 13, 1607–1637 (2014)
15. Wang, X., Pemberton-Ross, P., Schirmer, S.G.: Symmetry and controllability for spin networks with a single-node control. IEEE Trans. Autom. Control 57, 1945–1956 (2012)
16. Massey, A., Miller, S.J., Sinsheimer, J.: Distribution of eigenvalues of real symmetric palindromic Toeplitz matrices and circulant matrices. J. Theor. Probab. 20, 637–662 (2007)
17. Kaku, M.: Introduction to Superstrings, ser. Graduate Texts in Contemporary Physics. Springer, New York (1998)
18. Prakash, N.: Mathematical Perspectives on Theoretical Physics: A Journey from Black Holes to Superstrings. Tata McGraw-Hill, New Delhi (2000)
19. Katok, A., Hasselblatt, B.: Introduction to the Modern Theory of Dynamical Systems. Cambridge University Press, Cambridge (1997)
20. Lagarias, J.C.: The computational complexity of simultaneous Diophantine approximation problems. In: 23rd Annual Symposium on Foundations of Computer Science, 1982. SFCS '08, pp. 32–39 (1982)
21. Kovacs, A., Tihanyi, N.: Efficient computing of n-dimensional simultaneous Diophantine approximation problems. Acta Univ. Sapientiae Inform. 5(1), 16–34 (2013)
22. Simultaneous Diophantine approximation. http://www.math.tamu.edu/~Doug.

Hensley/ SimultaneousDiophantine (2005)

23. Nowak, W.G.: On simultaneous Diophantine approximation. Rendiconti del Circolo Matematico di Palermo XXXIII, 456–460 (1984)
24. Chevallier, N.: A Survey of Best Simultaneous Diophantine Approximations. http://www.math.uha. fr/chevallier/publication/meilleures3 (2011)
25. Bosma, W., Smeets I.: Finding simultaneous Diophantine approximations with prescribed quality. In: Howe, E.W., Kedlaya, K.S., (eds.) ANTS X: Tenth Algorithmic Number Theory Symposium 2012, UC San Diego, July 9–13 2012. Preprint arXiv:1001.4455v1
26. Lagarias, J.C.: Best simultaneous Diophantine approximations II. Behavior of consecutive best approximations. Pac. J. Math. 102(1), 61–88 (1982)
27. Nowak, W.G.: A note on the simultaneous Diophantine approximation. Manuscr. Math. 36, 33–46 (1981)
28. Keng, H.L.: Introduction to Number Theory. Springer, Berlin (1982)
29. Koblitz, N.: A course in Number Theory and Cryptography, ser. Graduate Texts in Mathematics. Springer, Berlin (1988)
30. Lenstra, K., Lenstra, H.W., Lovasz, L.: Factoring with rational coefficients. Math. Ann. 261(4), 515– 534 (1982)
31. Lagarias, J.C.: Geodesic multidimensional continued fractions. Proc. Lond. Math. Soc. 3(69), 464–488
32. (1994)
33. Arkhangel'skii, A.V., Pontryagin, L.S.: General topology i; basic concepts and constructions; dimension theory. In: Gramkrelidze, R.V. (ed.) Encyclopedia of Mathematical Sciences. Springer, Berlin (1990)
34. Aldrovandi, R., Pereira, J.G.: An Introduction to Geometrical Physics. World Scientific, Singapore (1995)
35. 34. Ariaei, F., Lou, M., Jonckheere, E., Krishnamachari, B., Zuniga, M.: Curvature of indoor sensor network: clustering coefficient. EURASIP J.Wirel. Commun. Netw. 2008, 20 (2008), article ID 213185. doi:10.1155/2008/2131185
36. 35. Wilson, W.A.: On semi-metric spaces. Am. J. Math. 53(2), 361–373 (1931). http://www.jstor.org/ stable/2370790
37. Shore, S.D.: Coherent distance functions. Topol. Proceed. 6, 405–422 (1981)
38. Blumenthal, L.M.: Some embedding theorems and characterization problems of distance geometry. Bull. Am. Math. Soc. 40, 321–338 (1943)
39. Jonckheere, E., Lou, M., Bonahon, F., Baryshnikov, Y.: Euclidean versus hyperbolic congestion in idealized versus experimental networks. Internet Math. 7(1), 1–27 (2011)
40. Wang, C., Jonckheere, E., Banirazi, R.:Wireless network capacity versus Ollivier–Ricci curvature under Heat Diffusion (HD) protocol. In: American Control Conference (ACC), pp. 3536–3541, Portland, June 04–06 2014. http://eudoxus2.usc.edu

41. Blumenthal, L.M.: Theory and Applications of Distance Geometry. Oxford at the Clarendon Press, London (1953)
42. Jonckheere, E., Langbein, F.C., Schirmer, S.G.: Curvature of quantum rings. In: Proceedings of the 5th International Symposium on Communications, Control and Signal Processing (ISCCSP 2012), Rome, May 2–4 2012. doi:10.1109/ISCCSP.2012.6217863
43. Gromov, M.: Hyperbolic groups. In: Gersten, S.M. (ed.) Essays in Group Theory, ser. Mathematical Sciences, pp. 75–263. Springer, New York (1987)
44. Langbein, F., Schirmer, S., Jonckheere, E.: Time-optimal information transfer in spintronic networks. to appear. In: Proceedings of IEEE CDC (2015), Preprint arXiv:1508.00928
45. Burgarth, D.: Quantum state transfer and time-dependent disorder in quantum chains. Eur. Phys. J. Special Top. 151(1), 147–155 (2007)

13

CHAPTER

TRAJECTORY TRACKING CONTROL OF QUANTUM SYSTEMS

Cong Shuang and Liu JianXiu

Keywords: trajectory tracking, unitary transformation, free-evolutionary target system, regulation

In recent years, state preparation and steering problem have been extensively studied [1–3] due to their great important applications in quantum teleportation [4], quantum computation and quantum information [5]. Quantum trajectory tracking is of equal importance as quantum state steering. It requires controlled state to be able to follow a time-variant target state, which is more complicated than the control of state steering. On the other hand, the convergence of quantum tracking is as much important as its control law design. The probability of state in a quantum system makes a stable control not enough anymore. The asymptotic convergence of control strategy in such a case is needed to make sure that the experiments are practicable. Therefore the convergence of the control of quantum system directly determines whether a proper control law is designed. At the same time the condition of convergence can provide the methods and ways to derive an effective control law. Especially in quantum systems, the complexity of controlled system model, various characteristics of different quantum states and the probability of controlled variables lead to great difficulties in state transferring. Furthermore, non-autonomic system appears in tracking control, all of which show many difficulties in a quantum system's trajectory tracking control. In [6], with the help of some concepts in system control, we studied trajectory tracking of quantum state: introducing an error state e(t) by defining subtraction between target $\hat{\rho}_f(t)$ and system

Citation: Cong Shuang and Liu JianXiu, Trajectory tracking control of quantum systems, DOI: 10.1007/s11434-012-5194-z.

state $\hat{\rho}(t)$ We changed the tracking control goal of $\hat{\rho}(t)$ following $\hat{\rho}_f(t)$ problem of error state e(t) from the arbitrary initial error state e(0) to its stable state e(tf) which is a zero matrix and denoted by $e(t_f)=O$. To design a proper control law, we also introduced a unitary transformation: $U(t)=\exp(-itH_0)$ to remove the drift item with H_0 which makes it difficult to determine the sign of the first order derivative of Lyapunov function.

We find that the unitary transformation introduced in [6] can not only eliminate the drift item with H0 but also turn the time-dependent target state $\hat{\rho}_f(t)$ of original system into a stationary state $\hat{\rho}_{f0}$ In other words, with the help of the unitary transformation the system state $\hat{\rho}(t)$ tracking target state $\hat{\rho}_f(t)$ simultaneously becomes the steering problem of the system state toward a target initial state $\hat{\rho}_{f0}$ from which we think the procedure of introducing error state e(t) is not necessary in quantum tracking control, which motivates this paper. In this work, the unitary transformation will be applied to the original system, which changes free-evolutionary target system $\hat{\rho}_f(t)$ into a time-independent state $\hat{\rho}_{f0}$ We achieve the tracking goal by steering system initial state $\hat{\rho}_0$ to $\hat{\rho}_{f0}$ In such a case, the stable final state $\hat{\rho}_{f0}$ of steering problem is no longer a zero matrix, which is much different from the case in [6]. Therefore the design of control law and the proof of convergence are more difficult than that in the former.

DESCRIPTION OF SYSTEM MODEL

The controlled system in this paper is based on quantum Liouville equation acting on the Hilbert space:

$$i\hbar\frac{\partial}{\partial t}\hat{\rho}(t)=\left[H,\hat{\rho}(t)\right],\quad \hat{\rho}(0)=\hat{\rho}_0, \tag{1}$$

$$i\hbar\frac{\partial}{\partial t}\hat{\rho}_f(t)=\left[H_0,\hat{\rho}_f(t)\right],\quad \hat{\rho}_f(0)=\hat{\rho}_{f0}, \tag{2}$$

Where $H=H_0+\sum_{m=1}^{M}f_m(t)H_m$. H_0 and H_m represent system's internal (or free) and external (or control) Hamiltonian, respectively, and all of them will be assumed to be time-independent. $f_m(t)$ are time-dependent external control fields. We choose the Planck constant $\hbar=1$ for convenience. Eq. (1) is the controlled system and eq. (2) describes the target system which is time-dependent and free-evolutionary. For systems (1) and (2), the expected control goal is: for arbitrary initial state of system (2), state $\hat{\rho}(t)$ of (1) will follow target system (2). Since the evolution of a closed quantum systems is unitary, the spectrum of $\hat{\rho}(t)$ is timeinvariant, viz. $Tr[\hat{\rho}^n(t)]=Tr[\hat{\rho}_f^n(t)]$.

The transformation deduced by a linear unitary operator is called unitary transformation. Here we introduce unitary transformation $U(t)=\exp(-itH_0)$ in original systems (1) and (2), and the controlled state $\hat{\rho}(t)$ and target state $\hat{\rho}_f(t)$ become

$$\rho(t) = U^+(t)\hat{\rho}(t)U(t),$$
$$\rho_f(t) = U^+(t)\hat{\rho}_f(t)U(t), \tag{3}$$

Where "+" denotes conjugate, $\hat{\rho}(t)$ and $\hat{\rho}_f(t)$ are system state and target state before unitary transformation, respectively. Because of $U(0) = U^+(0) = 1$ (t=0), one has $\rho(0) = \hat{\rho}(0) = \hat{\rho}_0$, $\rho_f(0) = \hat{\rho}_f(0) = \hat{\rho}_{f0}$ after transformation. More over, target system (2) is free-evolutionary under free Hamiltonian H0. The solution of (2) is $\hat{\rho}_f(t) = U(t)\hat{\rho}_f(0)U^+(t)$ where U(t) is the evolution matrix and $U(t) = \exp(-itH_0)$. Substituting this solution into (3), the transformed target state $\rho_f(t)$ becomes

$$\rho_f(t) = U^+(t)U(t)\hat{\rho}_f(0)U^+(t)U(t) = \hat{\rho}_f(0) = \hat{\rho}_{f0}. \tag{4}$$

Therefore the original systems (1) and (2) after unitary transformation will be

$$i\hbar\frac{\partial}{\partial t}\rho(t) = \left[\sum_m f_m(t)H_m(t), \rho(t)\right], \quad \rho(0) = \hat{\rho}_0, \tag{5a}$$

$$\rho_f(t) = \hat{\rho}_{f0}. \tag{5b}$$

Eq. (5) is the new controlled and target systems from which one can see that:

Remark 1. State $\rho_f(t)$ in original target system (2) evolves according to Liouville equation, which is time-dependent and non-stationary. After unitary transformation the trajectory of target system in (5) becomes time-independent and stationary which is a constant value and equals to the initial state $\hat{\rho}_{f0}$ of target system.

Remark 2. After the transformation the control Hamiltonian H_m, which is time-independent, becomes $H_m(t)$ with time-dependent and determines the evolution of states in interaction picture. Unitary transformation in this paper has two meanings, one of which is the rotation of the basic vectors in Hilbert space. We say a unitary operator leads to the rotation of the Hilbert space. New state operators $\rho(t)$ and $\rho_f(t)$ are used to describe the transformed quantum system (5). Under unitary transformation, all physical properties are unchanged. The other meaning is that only control Hamiltonian remained in new system (5a) makes every item in the first order derivation of Lyapunov function controllable, which is convenient to design control law. By introducing unitary transformation, the original trajectory tracking problem of system state $\hat{\rho}(t)$ following free evolutionary target state $\hat{\rho}_f(t)$ becomes the steering problem of state $\rho(t)$ being regulated to $\hat{\rho}_f(0)$ in system (5).

DESIGN OF CONTROL LAW

Among many control methods, the Lyapunov-based method is simple and easy to design. The basic idea of Lyapunovbased method is to select V(x) as Lyapunov function, which must satisfy the following three conditions: (i) V(x) is continuous and its first-order partial derivatives is also continuous in its definition; (ii) V(x) is positive semi-definite, i.e. V(x)≥0; (iii) The first order time derivative of the Lyapunov function is negative semi-definite, i.e. V x &() 0 £ . If they are true then the control system is stable. In system control field one often uses the Lyapunov-based method to design the control law because this controller design method can guarantee the whole control system is stable. We will use the Lyapunov-based method next. Select the average value of an observable quantity as the Lyapunov function here

$$V = tr\left(P\rho\right) + C, \tag{6}$$

where P is the observable operator, and it is not necessarily positive definite as it not always corresponds to an actual observable quantity. P is also called a virtual mechanical quantity operator. C is a constant used to adjust the value of Lyapunov function. According to the third condition of Lyapunov function and system (5), one can obtain

$$\dot{V} = -\sum_{m} f_m\left(t\right) tr\left(iH_m(t)\left[\rho\left(t\right), P\right]\right). \tag{7}$$

For the sake of simplicity and availability we let each item in the right side of (7) of summation sign be non-positive in order to ensure V £ 0. The control law can be derived as

$$f_m(t) = k_m tr\left(iH_m\left(t\right)\left[\rho\left(t\right), P\right]\right), \quad k_m > 0, \tag{8}$$

where k_m>0 is the control gain which is used to adjust the convergence speed of the system state. Generally, control law (8) is only a stable control and cannot always guarantee the system converges to its target state. Because of the probability control in quantum system, only asymptotic stable control can guarantee the system to achieve the target state completely. Next we will study the conditions of convergence of the control law (8). Firstly, an appropriate observable P will be constructed to ensure the stable target state $\hat{\rho}_f(0)$ to be the minimum point of (6).

One can rewrite the condition (ii) and (iii) as

$$V\left(\rho_f\right) = \min_{\rho} V\left(\rho\right), \tag{9a}$$

$$\dot{V}\left(\rho_f\right) = 0. \tag{9b}$$

These conditions ensure a specific form of P. In those cases the target state is a stable point in sense of Lyapunov. According to (9b), states $\rho(t)$, which induce the control fields $f_m(t)$ to be zeros, belong to the set

$$\mathcal{R} \equiv \{\rho : tr(iH_m(t)[\rho, P]) = 0, \forall m, t\}. \tag{10}$$

Different from the method of introducing error state in [6], where stable error state is always zero matrix, the method in this paper is easy to get the condition of operator P. However no clear information can be obtained from (9).

The coherent vectors will be used here to construct observable operator P.

Rewrite (9) as

$$tr(P\rho_f) = \min_{\rho} V(\rho), \tag{11a}$$

$$\rho_f \in \mathcal{R}. \tag{11b}$$

A Hermite matrix is decomposed by means of a unitary Lie algebra. In such a way, a matrix is expressed as coherent vectors. The bases of a unitary Lie algebra are described as

$$\{X_1, \cdots X_m\}. \tag{12}$$

And the following conditions hold

$$tr(X_l) = 0, \tag{13a}$$

$$tr(X_l X_j) = \delta_{lj}. \tag{13b}$$

For , $\forall l, \exists \gamma_j, \gamma_k \neq 0$, one has

$$\left[\sum_j \gamma_j X_j, \sum_k \gamma_k X_k\right] = iX_l. \tag{13c}$$

Then an n-dimensional state density matrix can be rewritten as

$$\rho = I/n + \sum x_l X_l. \tag{14}$$

In such a way, the coherent vector R_r of a density matrix r can be expressed as

$$R_\rho = \left(x_1, x_2, \cdots x_{n^2-1}\right)^T. \tag{15}$$

The module of coherent vector indicates the purity of density matrix, which reaches its maximum $\sqrt{(n-1)/n}$ when r is pure state and zero of maximum mixed-state I/n. Because of the unitary characteristic of closed quantum system, the purity in evolution is unchanged, then at any time it is satisfied

$$\sum_l |x_l|^2 = C\text{, } C \text{ is a constant.} \tag{16}$$

We assume that system state and target state both are controllable in this paper. According to the theoretical analysis mentioned above, the target state and observable operator P are unfolded as follows:

$$\rho_f = I/n + \sum_j f_j X_j, \tag{17a}$$

$$P = c_0 I + \sum_k c_k X_k. \tag{17b}$$

Substituting (17) into (11), one has

$$tr\left(P\rho_f\right) = c_0 + \sum_k c_k f_k. \tag{18}$$

If we want to get the minimum point of (18), then

$$c_k = \lambda f_k,\ \lambda < 0, \tag{19}$$

which implies coherent vector R_P of operator P is opposite in direction from that of target state ρ_f:

$$R_P = \lambda R_{\rho_f},\ \lambda < 0. \tag{20}$$

Therefore the simplest method to construct the observe operator P is to choose

$$P = -\rho_f. \tag{21}$$

The structure of P in (21) is same as that in [6] which ensures ρ_f is the minimum point of Lyapunov function (6) and control law (8) is stable. The other procedure of convergence proof is the same of that in [7]. We do not repeat it here.

NUMERICAL SIMULATION EXPERIMENTS AND RESULTS ANALYSIS

In order to demonstrate the effectiveness and advantage of the method proposed in this paper, two examples and results analysis will be given in this section. A five-level energy quantum system used both in [8] and [9] is selected, where the free Hamiltonian is

$$H_0 = \begin{pmatrix} 1 & 0 & 0 & 0 & 0 \\ 0 & 1.2 & 0 & 0 & 0 \\ 0 & 0 & 1.3 & 0 & 0 \\ 0 & 0 & 0 & 2 & 0 \\ 0 & 0 & 0 & 0 & 2.15 \end{pmatrix}.$$

Eigenvalues of H0 are $\lambda_1=1$, $\lambda_2=1.2$, $\lambda_3=1.3$, $l_4=2$, $l_5=2.15$. The corresponding eigenstates are

$$|\lambda_1\rangle = \begin{bmatrix} 1 \\ 0 \\ 0 \\ 0 \\ 0 \end{bmatrix},\ |\lambda_2\rangle = \begin{bmatrix} 0 \\ 1 \\ 0 \\ 0 \\ 0 \end{bmatrix},\ |\lambda_3\rangle = \begin{bmatrix} 0 \\ 0 \\ 1 \\ 0 \\ 0 \end{bmatrix},\ |\lambda_4\rangle = \begin{bmatrix} 0 \\ 0 \\ 0 \\ 0 \\ 1 \end{bmatrix} \text{ and } |\lambda_5\rangle = \begin{bmatrix} 0 \\ 0 \\ 0 \\ 0 \\ 1 \end{bmatrix}.$$

It is assumed that initial state of system (1) is $\hat{\rho}_0$. The free evolutionary target system (2) has an initial value $\hat{\rho}_{f0}$.

According to the conditions mentioned above, after unitary transformation new system (5) replaces the original systems (1) and (2). The tracking problem has become state steering one with new state r( )t in system (5) from its initial state $\hat{\rho}_0$ to a stationary target state $\hat{\rho}_{f0}$.

In this system, all the transition frequencies are different viz. $\omega_{jk} \neq \omega_{pq}, (j,k) \neq (p,q)$ where $\omega_{jk} = \lambda_j - \lambda_k$, which satisfies assumption 1 of convergence conditions in [7]. According to assumption 2, control Hamiltonians must be full connected, which means ten control Hamiltonians are needed for this system. However in most cases, we do not need so many control Hamiltonians, the number of which is determined by the concrete structure of $\hat{\rho}_0$ and $\hat{\rho}_{f0}$.

Example A. The initial state of system (1) is an eigenstates

$$|\lambda_1\rangle = \begin{bmatrix} 1 \\ 0 \\ 0 \\ 0 \\ 0 \end{bmatrix} \text{ of } H_0 \text{, then } \hat{\rho}_0 = \begin{pmatrix} 1 & 0 & 0 & 0 & 0 \\ 0 & 0 & 0 & 0 & 0 \\ 0 & 0 & 0 & 0 & 0 \\ 0 & 0 & 0 & 0 & 0 \\ 0 & 0 & 0 & 0 & 0 \end{pmatrix}.$$

The probability superposition of eigenstates $|\lambda_2\rangle$ and $|\lambda_3\rangle$ is the initial state of free-evolutionary target system (2), viz.

$$|\psi_{f0}\rangle = 1/\sqrt{2}|\lambda_2\rangle + 1/\sqrt{2}|\lambda_3\rangle,$$

$$\hat{\rho}_{f0} = |\psi_{f0}\rangle\langle\psi_{f0}| = 0.5\times\begin{pmatrix} 0 & 0 & 0 & 0 & 0 \\ 0 & 1 & 1 & 0 & 0 \\ 0 & 1 & 1 & 0 & 0 \\ 0 & 0 & 0 & 0 & 0 \\ 0 & 0 & 0 & 0 & 0 \end{pmatrix}.$$

Considering the specific forms of $\hat{\rho}_0$ and $\hat{\rho}_{f0}$, the density matrixes of initial state and target state only have differences in the population of energies 1, 2 and 3 and the interference item of energies 2 and 3. So the designed control law is aimed to these different items, that is to say, interaction only existed between energies 1, 2 and 3. Three control Hamiltonians are enough to deal with the problem in this section shown as

$$H_1 = \begin{pmatrix} 0 & 1 & 0 & 0 & 0 \\ 1 & 0 & 0 & 0 & 0 \\ 0 & 0 & 0 & 0 & 0 \\ 0 & 0 & 0 & 0 & 0 \\ 0 & 0 & 0 & 0 & 0 \end{pmatrix}, H_2 = \begin{pmatrix} 0 & 0 & 1 & 0 & 0 \\ 0 & 0 & 0 & 0 & 0 \\ 1 & 0 & 0 & 0 & 0 \\ 0 & 0 & 0 & 0 & 0 \\ 0 & 0 & 0 & 0 & 0 \end{pmatrix},$$

$$H_3 = \begin{pmatrix} 0 & 0 & 0 & 0 & 0 \\ 0 & 0 & 1 & 0 & 0 \\ 0 & 1 & 0 & 0 & 0 \\ 0 & 0 & 0 & 0 & 0 \\ 0 & 0 & 0 & 0 & 0 \end{pmatrix}. \tag{22}$$

Next, we consider the diagonal elements ρ_{nn}(n=1, 2…5) of density matrix which is the population of state, and off diagonal elements ρ_{23}, which represents interference item.

Here we mainly observe the tracking of interference item ρ23. According to (22), the control fields corresponding to each Hamiltonian are f_1, f_2 and f_3, respectively. Control law can be obtained from (8) where k=0.5, f_0=0.015. To demonstrate the tracking of system (1) to system (2) we keep data of control function $f_m(t)$, m=1, 2, 3 at all times and apply them to original system (1). It is known that the population of target system (2) is unchanged under unitary evolution. Figure 1 displays the effectiveness of state steering of r(t) from its initial state $\hat{\rho}_0$ to final state $\hat{\rho}_{f0}$ of system (5), in which Figure 1(a) is the evolution of diagonal items ρ_{nn}, n=1, 2, 3 and interference item ρ_{23} of state r(t). From Figure 1(a) one can see that both of them evolves towards relevant items of target state $\hat{\rho}_{f0}$.Figure 1(b) is the tracking of interference item in which dot line is free evolutionary trajectory of $\hat{\rho}_{f23}$ of target system state $\hat{\rho}_f(t)$ and solid line is tracking curve of $\hat{\rho}_{23}$ of controlled system state $\hat{\rho}(t)$. Figure 1(c) shows the control fields $f_m(t)$. We set a performance index to describe the distance between initial state and target state: $v=\|\rho_f(t)-\rho(t)\|^2$. Here, v=1.4125× 10^{-6} ($t=30$ a.u.). Example B. We choose the same example as in [8], where the initial state of system (1) is a superposition state $|\psi_0\rangle = 1/\sqrt{3}|\lambda_1\rangle + \sqrt{2}/\sqrt{3}|\lambda_4\rangle$. The initial target state of system (2) is an eigenstate $|\lambda_3\rangle = [0,0,1,0,0]^T$. Here we choose the same parameters k=0.15 and

$$P=\begin{pmatrix} 0.8 & 0 & 0 & 0 & 0 \\ 0 & 1.1 & 0 & 0 & 0 \\ 0 & 0 & 0.4 & 0 & 0 \\ 0 & 0 & 0 & 1.2 & 0 \\ 0 & 0 & 0 & 0 & 0.6 \end{pmatrix}$$

as those in [8], but a different control Hamilton is selected of the form:

$$H_c=\begin{pmatrix} 0 & 0 & 1 & 0 & 0 \\ 0 & 0 & 0 & 0 & 0 \\ 1 & 0 & 0 & 1 & 0 \\ 0 & 0 & 1 & 0 & 0 \\ 0 & 0 & 0 & 0 & 0 \end{pmatrix},$$

where interactions exist between energies 1 and 3 and between energies 4 and 3, respectively. The simulation experimental results are shown in Figure 2, which demonstrates the effectiveness of state steering of r(t) from its initial state $\hat{\rho}_0$ to the final one $\hat{\rho}_{f0}$.Figure 2(a) shows the population evolution, Figure 2(b) is the control law and Figure 2(c) shows the interference item ρ_{14} in which dashed line indicates the item of target state and solid line indicates the corresponding item of r(t).

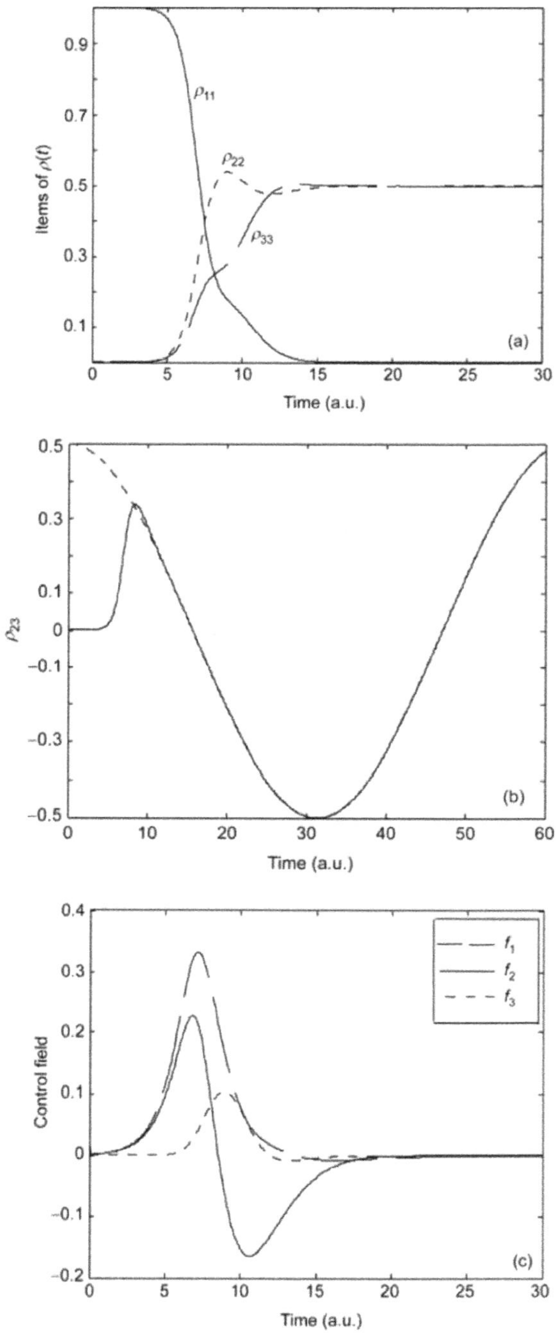

Figure 1: (a) Population of state r (t) in system (5); (b) tracking of ρ_{23} for original systems (1) and (2); (c) control fields $f_m(t)$.

To achieve the same goal one needs a control time of 400 a.u. in [8], but form Figure 2 one can see that here a different control Hamiltonian designed results in a shorter time of less than 150 a.u. In experiment B, we can divide the control Hamiltonian

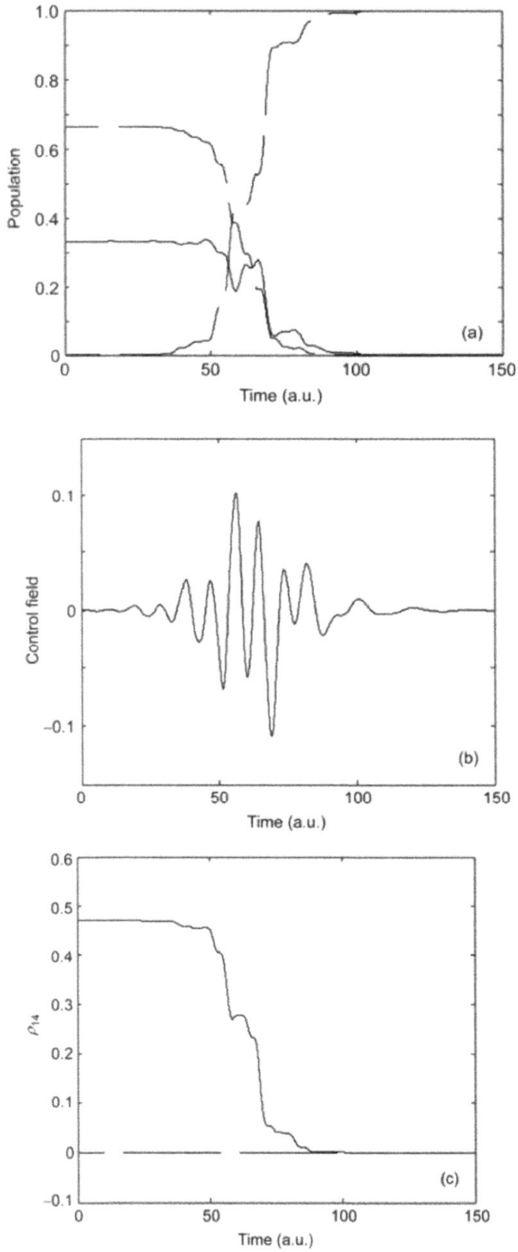

Figure 2: (a) Population of state r(t) in system (5); (b) control fields $f_1(t)$; (c) tracking of interference item ρ_{14}.

into two parts, as in experiment A. If we set

$$H_1 = \begin{pmatrix} 0 & 0 & 1 & 0 & 0 \\ 0 & 0 & 0 & 0 & 0 \\ 1 & 0 & 0 & 0 & 0 \\ 0 & 0 & 0 & 0 & 0 \\ 0 & 0 & 0 & 0 & 0 \end{pmatrix},\ H_2 = \begin{pmatrix} 0 & 0 & 0 & 0 & 0 \\ 0 & 0 & 0 & 0 & 0 \\ 0 & 0 & 0 & 1 & 0 \\ 0 & 0 & 1 & 0 & 0 \\ 0 & 0 & 0 & 0 & 0 \end{pmatrix}, \tag{23}$$

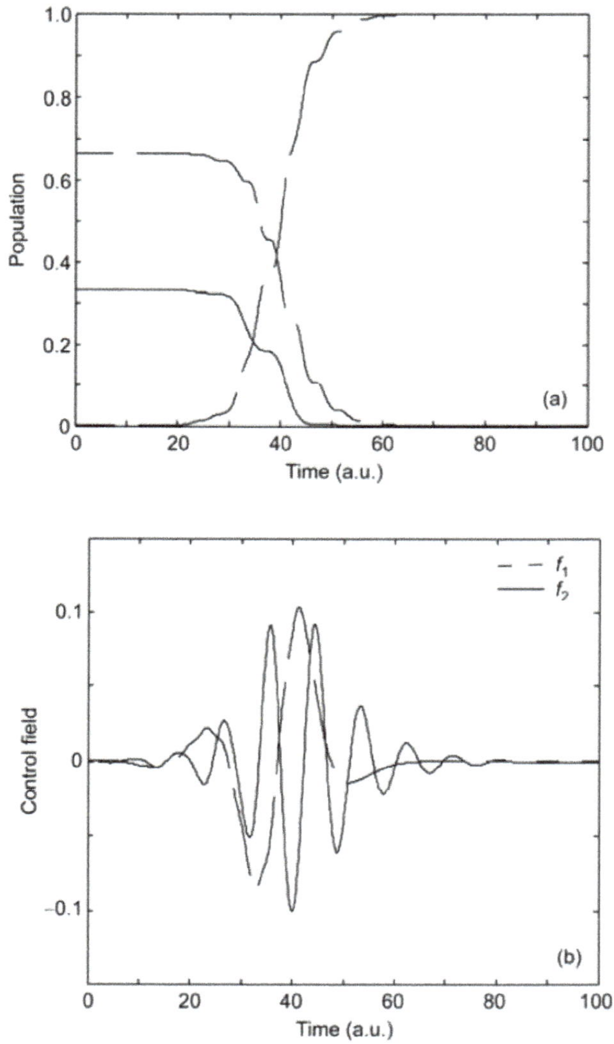

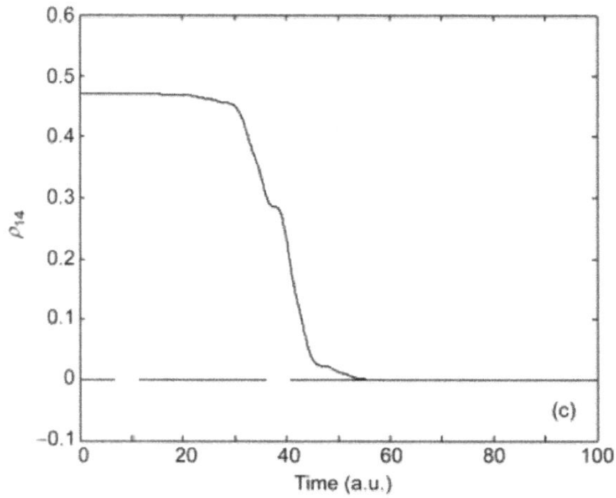

Figure 3: (a) Population of state r(t) in system (5); (b) control fields f(t); (c) tracking of interference item ρ14.

where $H_c=H_1+H_2$. Then two control fields will be needed. The results are shown in Figure 3, where the three figures correspond to the ones in Figure 2. We can see the time to complete control task is shorter than that in Figure 2, which is less than 100 a.u. In Figure 2, there is only one control Hamiltonian included all energy transforms, so control action on all energy is the same, which brings restrictions to control fields and lengthens the time to complete control task, sometimes even causes larger error.

CONCLUSION

In this paper we have developed unitary transformation to change tracking problem of quantum state into steering one, which leads to the same results as in [6]. The special unitary transformation makes tracking problem in closed quantum system easier than that in classical one. If a closed quantum system is considered, the tracking problem of a free-evolutionary quantum system, which is a very common situation, becomes the regulation of controlled system initial state to target system initial state. It is a special case of the tracking problem. The error state is not necessary. But it is not the case for a non-free-evolutionary target system, such as a controlled target system and a predefined target time-dependent function, or in open quantum system [10,11], which need to use the method in [6]. We also demonstrate that different structure of control Hamiltonians, which in fact describes different control paths such as the comparison analysis between experiment B-2 and [8], affects control time. The control Hamiltonians with same structure, such as experiment B-2 and B-3 in this paper, may have shorter control time when they are organized separately instead of as a whole.

REFERENCES

1. Cong S, Zhang Y Y. Superposition state preparation based on Lyapunov stability theorem in quantum systems. J Univ Sci Tech China, 2008, 38: 821–827

2. Resch K, Lundeen J, Steinberg A. Quantum state preparation and conditional coherence. Phys Rev Lett, 2002, 88: 113601

3. Shi T, Li Y, Song Z, et al. Quantum-state transfer via the ferromagnetic chain in a spatially modulated field. Phys Rev A, 2005, 71: 032309

4. Wang X. Quantum teleportation of entangled coherent states. Phys Rev A, 2001, 64: 022302

5. Sillanpää M A, Park J I, Simmonds R W. Coherent quantum state storage and transfer between two phase qubits via a resonant cavity. Nature, 2007, 449: 438–442

6. Liu J X, Cong S. Trajectory tracking of quantum states based on Lyapunov method. In: The 9th IEEE International Conference on Control & Automation, Santiago, Chile, 2011. 318–323

7. Lou Y S, Cong S, Yang J, et al. Path programming control strategy of quantum state transfer. IET Control Theory Appl, 2011, 5: 291–298

8. Kuang S, Cong S, Lou Y S. Population control of quantum states based on invariant subsets under a diagonal Lyapunov function. In: Joint 48th IEEE Conference on Decision and Control and 28th Chinese Conference, Shanghai, 2009. 2486–2491

9. Ramakrishna V, Salapaka M, Dahleh M, et al. Controllability of molecular systems. Phys Rev A, 1995, 51: 960–966

10. Rong X, Wang Y, Yang J H, et al. Dynamical decoupling of electron spins in phosphorus-doped silicon. Chin Sci Bull, 2011, 56: 591–597

11. Cao W C, Liu X S, Bai H B. Bang-bang control suppression of amplitude damping in a three-level atom. Sci China Ser G-Phys Mech Astron, 2008, 51: 29–37

14
CHAPTER

AVAILABLE CONTROL IN DYNAMICAL DECOUPLED QUANTUM SYSTEMS

Pan Yu, XI ZaiRong & Cui Wei

Keywords: Dynamical decoupling, PDD, UDD, Universal control

A quantum bit, commonly represented by a two-level quantum system, is constantly interacting with its surrounding environment. However, the coupling between qubit and environment is detrimental to quantum coherence, which is the crucial resource enabling quantum computation [1]. To protect qubits in quantum registers, dynamical decoupling (DD) methods were introduced to suppress decoherence by isolating the qubit from its environment [2]. During recent years, both theory and experimental progresses have been made to demonstrate the power of various DD sequences in preserving quantum states and generating coherent operations [3–9]. General DD methods employ sequence of discrete pulses to eliminate system-bath coupling. On the other hand, certain couplings need to be switched on for a finite time in order to perform quantum control operations. For example, qubit-qubit interaction is required when implementing twoqubit gates. Besides its ability to keep quantum coherent states in quantum registers, to perform high-fidelity quantum computation on DD-proteced qubits without extra resources, dynamical decoupling process should leave the controlled interaction intact. We will see the ability to accomplish this task is closely related to the selected DD sequence used in different decoupling schemes. The first type of DD sequence is provided by nuclear magnetic resonance spectroscopy, known as "periodic DD" (PDD) [2], which averages out the non-unitary evolution of a qubit by applying a predetermined sequence periodically. In a single period, the pulse locations are equidistant, and cycle time Tc

Citation: Pan Yu, Xi ZaiRong and Cui Wei, Available Control in Dynamical Decoupled Quantum Systems, doi: 10.1007/s11434-012-5138-7.

is required to be as small as possible to ensure the higher-order terms vanish where the propagator during Tc can be expanded into a standard Magnus series. Recently, a kind of optimized DD sequence, Uhrig DD (UDD) [3], makes an enormous progress over PDD by canceling qubit-bath coupling order by order with each additional pulse. In UDD, pulse locations are not equidistant but optimized and restriction on Tc → 0 is relaxed. It was first discovered for a pure dephasing spin-boson model, and later proved to be applicable to qubit relaxation as well [10]. In this paper, we first investigate the possibility of performing universal control over periodically decoupled quantum systems. Afterwards in Section 2, we analyze the controllability of Uhrig decoupled system. We manage to find out the available qubit-qubit interaction hamiltonian which can be used for quantum control and give a rough estimation of the control errors. Conclusions are put in Section 3.

CONTROL OF PERIODICALLY DECOUPLED SYSTEM

The general hamiltonian describing the qubit-bath coupling can be written as

$H = H_{SB} + H_B,$

where H_{SB} and H_B are interaction and bath hamiltonians respectively. We assume all the DD pulses are applied instantaneously, meaning that the corresponding hamiltonian can be turned on for negligible time with full strength. These control operations are called bang-bang controls [2, 11]. Obviously the realizable bang-bang controls constitute a finite subgroup of all the unitary transformations on the qubit subspace. A decoupling group $\mathcal{G}$ is chosen to be a finite subgroup of the bang-bang controls, with group elements $g_j, j = 0, \ldots, |\mathcal{G}| - 1$. To implement dynamical decoupling, cycle time T_c is divided into $|\mathcal{G}|$slices, $T_c = |\mathcal{G}|\Delta t$, and decoupling pulses $D_j = g_j g_{j-1}^\dagger$ are applied exactly at $j\Delta t$. Thus the evolution in a cycle time T_c can be expressed as

$$U(T_c) = \prod_{j=0}^{|\mathcal{G}|-1} g_j^\dagger U_0(\Delta t) g_j = e^{-iH_{\text{eff}}T_c}, \tag{1}$$

$U_0(\Delta t)$ denoting the free evolution between decoupling pulses. H_{eff} is the resulting effective hamiltonian, which can be expanded as Magnus series

$$H_{\text{eff}} = \sum_j H^{(j)}, \tag{2}$$

where $H^{(j)}$ is jth order contribution. As indicated in [2], in the limit of fast cycling time $T_c \to 0$, H_{eff} approaches the zeroth term

$$H_{\text{eff}} \to H^{(0)} = \frac{1}{|\mathcal{G}|} \sum_{g_j \in \mathcal{G}} g_j^\dagger H g_j = P_{\mathcal{G}}(H_{SB}). \tag{3}$$

This process is similar to the group averaging procedure, projecting the original interaction hamiltonian HS B into the centralizer of $\mathcal{G}$:

$$Z(\mathcal{G}) = \{O | [O, g_j] = 0 \quad \forall g_j \in \mathcal{G}\}.$$

The averaged qubit operators are made invariant under the group action. More specifically, the effective interaction hamiltonian must obey the commutation relation

$$[H_{\mathrm{eff}}, H_D] = 0, \tag{4}$$

where HD is the hamiltonian generating the decoupling group. If no knowledge is available on the interaction hamiltonian H_{SB}, we have to perform the so-called maximal averaging [2]. Here transformations g_j of the control group $\mathcal{G}$ span the whole space of bounded operators on the qubit subspace and the centralizer $Z(\mathcal{G})$ is consequently trivial, which means $H_{eff} = \lambda I \otimes H_B$. In this case, whatever "slow" controls we impose on the qubits, they will be quenched completely through dynamical decoupling. To achieve universal control, either a universal set of operations can be implemented bang-bang, or a universal set of hamiltonian can be switched very fast. In addition, these controls have to be synchronized with decoupling pulses, posing great challenge to present experimental technique. But if we have certain knowledge of H_{SB}, the situation will be quite different. We can choose the minimum decoupling group $\mathcal{G}$ such that

$$P_{\mathcal{G}}(H_{SB}) = \sum_j^N g_j H_{SB} g_j^\dagger = 0. \tag{5}$$

As long as G is not full rank, or in other words the qubit component in HS B does not span the whole qubit subspace, the centralizer $Z(\mathcal{G})$ is non-trivial, $Z(\mathcal{G})$ {I}, which corresponds to selective averaging. As a matter of fact, any hamiltonians contained in $Z(\mathcal{G})$ can be used for quantum control. By virtue of eq. (1), it is always possible to apply slowly any hamiltonian $H_c \in Z(\mathcal{G})$ in parallel with the decoupling cycle [12]. As an example to illustrate how to perform universal control under PDD, we consider two qubits which are coupled to environment by a linear interaction of the form

$$H_{SB} = \sum_{\alpha=x,y,z} \sum_{j=1,2} \sigma_j^\alpha \otimes \mathcal{B}_j^\alpha.$$

σ_α are Pauli operators. To perform selective decoupling, we choose decoupling group to be $\mathcal{G} = \{I, \otimes_j \sigma_j^x, \otimes_j \sigma_j^y, \otimes_j \sigma_j^z\}$ which contains 3 collective π-rotations. It can be easily verified that $P_{\mathcal{G}}(H_{SB}) = 0$. Noting that $\sum_j \sigma_j^x, \sum_j \sigma_j^y, \sum_j \sigma_j^z$ generates the group elements $\otimes_j \sigma_j^x, \otimes_j \sigma_j^y, \otimes_j \sigma_j^z$ respectively, two-qubit Heisenberg interaction $H_c = J\vec{\sigma}_1 \cdot \vec{\sigma}_2$ can be turned on simultaneously to implement two-qubit gates for quantum computation since $[H_c, \sum_j \sigma_j^\alpha] = 0$. It is trivial to extend these results to general Heisenberg interaction $H_c = \sum_{i<j} J_{ij} \vec{\sigma}_i \cdot \vec{\sigma}_j$ for large number of qubits. Heisenberg interaction is commonly used to swap any pair of qubits [13], and able to perform universal quantum computation on the decoupled system when combined with fast single-qubit operations.

In fact, universal quantum computation is possible using only the Heisenberg interaction without physical level single-qubit gates, provided qubits are encoded into appropriate decoherence-free subspaces [14–16]. Here we have to mention that since $P_{\mathcal{G}}$ is a zeroth-order approximation of effective hamiltonian H_{eff}, the control hamiltonian extracted from $Z_{\mathcal{G}}$ cannot be performed perfectly. Higher-order corrections will produce non-negligible control errors. A rigorous bound has been derived for these error, which sets a scalability limit on the protection of quantum computation using periodic dynamical decoupling [17].

CONTROL OF UHRIG DECOUPLED SYSTEM

A powerful DD scheme was introduced recently by Uhrig to suppress pure dephasing and general decoherence, which optimizes relative pulse locations according to

$$T_j = T\sin^2\frac{j\pi}{2(N+1)}, \quad \text{for} \quad j = 1, 2, \ldots N, \tag{6}$$

to eliminate qubit-bath coupling up to $O(T^{N+1})$. T is the qubit evolution time. However, the possibility of performing universal control while the qubits are protected by UDD has not been explored yet. We begin by considering a pure dephasing model

$$H = C + \sigma_z \otimes Z$$

where σ_z is the qubit Pauli matrix along the z-direction and C and Z are arbitrary bath operators. This hamiltonian does not include qubit flip operators σ_x or σ_y so that it leads to pure dephasing. To counter dephasing process, a sequence of π-pulses are applied at T_j obeying eq. (6), which can be expressed as the following hamiltonian:

$$H_{DD} = \sum_{j=1}^{N} \pi\delta(t - T_j)\frac{\sigma_x}{2}. \tag{7}$$

Right after each π-pulse, spin σ_z changes its sign, evolving in the opposite direction. Actually this is the UDD version of selective averaging. The qubit state-dependent propagators under Nth order UDD are

$$U_{\pm}^{N} = e^{-i[C\pm(-1)^N Z](T-T_N)} \ldots e^{-i[C\mp Z](T_2-T_1)} e^{-i[C\pm Z]T_1}. \tag{8}$$

Using these expressions, qubit dephasing is characterized by

$$U_{-}^{N\dagger} U_{+}^{N} = I + U_{-}^{N\dagger}\delta U.$$

To analyze δU, $U_{\pm}^{N}$ are transformed in the interaction picture

$$U_{\pm}^{N} = \mathrm{e}^{-iCT} \Im \mathrm{e}^{-i \int_0^T \pm F_N(t) Z_I(t) \mathrm{d}t},$$

(9)

Where $\Im$ is the time-ordering operator, the modulation function FN(t) = $(-1)^j$ for t ∈ [T_j, T_{j+1}] with $T_0 = 0$ and $T_{N+1} = T$, and

$$Z_I(t) = \mathrm{e}^{iCt} Z \mathrm{e}^{-iCt} = \sum_{p=0}^{\infty} \frac{(it)^p}{p!} [C, [C, ...[C, Z]...]].$$

Thus the difference δU is given by

$$\delta U = 2\mathrm{e}^{-iCT} \sum_{n=2k+1}^{\infty} (-i)^n \Delta_n,$$

(10)

With

$$\Delta_n = \int_0^T F_N(t_N)... \int_0^{t_2} F_N(t_1)[Z_I(t_N)...Z_I(t_1)]\mathrm{d}t_1...\mathrm{d}t_N.$$

Yang and Liu [10] proved the odd-order terms vanish $\Delta_{2k+1} = O(T^{N+1})$, which verifies that qubit pure dephasing is eliminated to Nth order.

While performing high-fidelity universal quantum control, the qubit-qubit interaction has to be turned on in parallel with the UDD sequence to implement two-qubit quantum gates. Here we explore the available qubit-qubit couplings under dynamical decoupling by considering two-qubit hamiltonian of the form

$$H = H_c + C + (\sigma_z^1 + \sigma_z^2) \otimes Z,$$

where H_c is the qubit-qubit interaction hamiltonian. We have to analyze U_+^N and U_-^N separately to see what kinds of Hc can be applied under UDD. If the available H_c are state independent, then these are the controls we are seeking to implement quantum compuation. First to suppress pure dephasing, the UDD π-pulses are given by

$$H_{DD} = \sum_{j=1}^{N} \frac{\pi}{2} \delta(t - T_j)(\sigma_x^1 + \sigma_x^2).$$

(11)

Observing eqs. (8) and (10), we find that C and Z are not necessarily bath operators. In fact for any pair of operators satisfying the following conditions after each UDD pulse

$$C \to C, Z \to -Z,$$

the conclusion $\Delta_{2k+1} = O(T^{N+1})$ also hold true, which means that pure dephasing induced by qubit-bath coupling is canceled to order N. Meanwhile, qubit-qubit couplings included in Z are eliminated as well. We again consider the Heisenberg interaction $H_c = \sigma_x^1\sigma_x^2 + \sigma_y^1\sigma_y^2 + \sigma_z^1\sigma_z^2.$ Making use of the commutation relations $[H_{DD}, \sigma_x^1\sigma_x^2 + \sigma_y^1\sigma_y^2 + \sigma_z^1\sigma_z^2] = 0$, $\sigma_x^1\sigma_x^2 + \sigma_y^1\sigma_y^2 + \sigma_z^1\sigma_z^2$ is incorporated into C. The two-qubit hamiltonian with Heisenberg coupling can be simplified as

$$H = C_0 + (\sigma_z^1 + \sigma_z^2) \otimes Z_0,$$

with

$$C_0 = C + \sigma_x^1\sigma_x^2 + \sigma_y^1\sigma_y^2 + \sigma_z^1\sigma_z^2.$$

The propagator $U_\pm^N$ becomes

$$U_\pm^N = \mathrm{e}^{-i[C_0 \pm (-1)^N Z_0](T-T_N)} \ldots \mathrm{e}^{-i[C_0 \mp Z_0](T_2-T_1)} \mathrm{e}^{-i[C_0 \pm Z_0]T_1}. \tag{12}$$

In interaction representation U_+^N is expanded into Taylor series

$$U_+^N = \mathrm{e}^{-iC_0T} \sum_{n=0}^{\infty} (-i)^n \Delta_n. \tag{13}$$

Although odd terms Δ_{2k+1} with $2k + 1 \leqslant N$ vanish as usual, now even terms Δ_{2k} play a crucial role in the controlled system evolution. Below we will show that the error rate of the control operation is eliminated to at least second order in T. Defining the relative pulse locations δ_j by $T_j = T\delta_j$, we combine the exponentials in eq. (12) term by term using the formula [18]

$$\mathrm{e}^{TA}\mathrm{e}^{TB} = \mathrm{e}^{TA+TB+\frac{T^2}{2}[A,B]+O(T^3)}. \tag{14}$$

Setting

$$A = -i[C_0 + (-1)^N Z_0](1 - \delta_N),$$
$$B = -i[C_0 + (-1)^{N-1} Z_0](\delta_N - \delta_{N-1}),$$

we have

$$e^{-i[C_0+(-1)^N]T(1-\delta_N)}e^{-i[C_0+(-1)^{N-1}]T(\delta_N-\delta_{N-1})} =$$
$$e^{-i\{C_0T(1-\delta_{N-1})-(-1)^{N-1}Z_0T[(1-\delta_N)-(\delta_N-\delta_{N-1})]\}+\frac{T^2}{2}C_1+O(T^3)},$$

With

$$\begin{aligned} C_1 &= [A, B] \\ &= 2(-1)^N(1-\delta_N)(\delta_N-\delta_{N-1})[C_0, Z_0]. \end{aligned}$$

Because $[C_0, Z_0] = [C, Z_0]$ only contains a combination of bath operators, the second order term $\frac{T^2}{2}C_1$ has no relevance with the qubit evolution. In the next we will prove that after all the exponentials are combined, the second order terms are added to zero, bounding the control error rate up to $O(T^2)$ at least. To do this, we repeat above procedure by setting

$$\begin{aligned} A_1 &= -i\{C_0T(1-\delta_{N-1})-(-1)^{N-1}Z_0T[(1-\delta_N)-(\delta_N-\delta_{N-1})]\} \\ &\quad + \frac{T}{2}C_1 + O(T^2), \\ B_1 &= -i[C_0 + (-1)^{N-2}Z_0](\delta_{N-1}-\delta_{N-2}). \end{aligned}$$

Again using eq. (14), and to second order in T, we obtain the expression of three exponentials combined

$$e^{-i\{C_0T(1-\delta_{N-1}+\delta_{N-1}-\delta_{N-2})+(-1)^{N-2}Z_0T[(1-\delta_N)-(\delta_N-\delta_{N-1})+(\delta_{N-1}-\delta_{N-2})]\}+\frac{T^2}{2}C_1+\frac{T^2}{2}C_2+O(T^3)} \tag{15}$$

With

$$C_2 = 2(-1)^{N-1}(\delta_N-\delta_{N-1})(\delta_{N-1}-\delta_{N-2})[C_0, Z_0].$$

By simply repeating this procedure N times, we find that the approximation form of U_+^N is

$$U_+^N = e^{-iC_0T\lambda_1-iZ_0T\lambda_2} + T^2[C_0, Z_0]\lambda_3 + O(T^3), \tag{16}$$

where the parameters λ_i can be written in compact form

$$\lambda_1 = \sum_{n=0}^{N}(-1)^n\delta_{N-n},$$

$$\lambda_2 = \sum_{n=0}^{N}(-1)^n(\delta_{N-n+1} - \delta_{N-n}),$$

$$\lambda_3 = \sum_{n=0}^{N}(-1)^{N-n}(\delta_{N-n+1} - \delta_{N-n}) \cdot (\delta_{N-n} - \delta_{N-n-1}).$$

It can be easily verified that $\lambda_1 = 1$, $\lambda_2 = 0$ and $\lambda_3 = 0$ for UDD pulse locations δ_j, so the propagator is reduced to

$$U_+^N = e^{-iC_0T+O(T^3)}. \tag{17}$$

We can also prove the same result for U_-^N in a similar way. Now we can say that at the end of the evolution, Heisenberg interaction $\sigma_x^1\sigma_x^2 + \sigma_y^1\sigma_y^2 + \sigma_z^1\sigma_z^2$ generates the control operation of two-qubit system which is expressed by

$$U(T) = e^{-i(\sigma_x^1\sigma_x^2+\sigma_y^1\sigma_y^2+\sigma_z^1\sigma_z^2)T+O(T^3)}, \tag{18}$$

indicating the possibility of performing universal control on Uhrig decoupled quantum systems. As in PDD, the available control operations should satisfy $[H_{DD}, H_c] = 0$. At this stage, the control error rate is bounded by $O(T^3)$. Since phase error is eliminated to Nth order by UDD, naturally we would expect to lower the bound of control error to $O(T^{N+1})$ as well. However, the odd order contributions $\sim T^{2k+1}$ cannot be eliminated under UDD sequence. To make our controls more accurate, these higher order terms have to be considered. Generally they will introduce a drift term in the qubit evolution hamiltonian in eq. (18). If the drift term can be calculated explicitly, we can make corrections to the control hamiltonian to reduce control errors even further. Currently we are working towards this goal.

The above derivations are based on pure dephasing model. We now briefly discuss the general decoherence problem, where the interaction hamiltonian is modeled as $H = \sigma_x \otimes X + \sigma_y \otimes Y + \sigma_z \otimes Z$. In spite of the great advantages of UDD in dealing with pure dephasing or relaxation, it is difficult to design UDD sequence to cancel both of them. A near-optimal method has been proposed using sequence of the form [19]

$$Z^N X_N(T_N - T_{N-1})ZX_N(T_{N-1} - T_{N-2})...ZX_N(T_1).$$

X, Z are π-pulses along axesz and x, and $X_N(T)$ denotes a Nth order UDD using X pulses in a period of time T. Z^N means if N is odd, a Z-pulse is added at the end of the sequence. This so-called quadratic DD (QDD) cancels pure dephasing in inner intervals

$T_N - T_{N-1}, T_{N-1} - T_{N-2}..., T_1$ of a common UDD, and relaxation on the secondary level by Z pulses. It is easy to see that operations that commute with X and Z can be performed in parallel with QDD. In paticular, Heisenberg interaction can be implemented.

CONCLUSION

In conclusion, we have proven that the condition for successfully performing control operations over periodically decoupled quantum systems also applies to the UDD case. Universal high-fidelity quantum computation is possible using the Heisenberg interaction while coherence is preserved by PDD or UDD sequence. Other proposed schemes which intend to implement high-fidelity quantum gates against decoherence, such as fault-tolerant quantum computation, quantum error-correction and decoherence-free subspace [20,21], all require an unrealistic amount of resources. On the contrary, control while performing dynamical decoupling is less resource-consuming, thus made a suitable candidate to implement quantum computation in some circumstances.

REFERENCES

1. Nielsen M A, Chuang I L. Quantum Computation and Quantum Information. Cambridge: Cambridge University, 2000
2. Viola L, Knill E, Lloyd S. Dynamical decoupling of open quantum systems. Phys Rev Lett, 1999, 82: 2417
3. Uhrig G S. Keeping a quantum bit alive by optimized π-pulse sequences. Phys Rev Lett, 2007, 98: 100504
4. Du J, Rong X, Zhao N, et al. Preserving electron spin coherence in solids by optimal dynamical decoupling. Nature, 2009, 461: 1265
5. De Lange G, Wang Z H, Riste D, et al. Universal dynamical decoupling of a single solid-state spin from a spin bath. Science, 2010, 330: 60
6. Peng X H, Suter D, Lidar D A. High fidelity quantum memory via dynamical decoupling: Theory and experiment. J Phys B: At Mol Opt Phys, 2011, 44: 154003
7. Rong X, Wang Y, Yang J H, et al. Dynamical decoupling of electron spins in phosphorus-doped silicon. Chin Sci Bull, 2011, 56: 591–597
8. Yang W, Wang Z Y, Liu R B. Preserving qubit coherence by dynamical decoupling. Front Phys, 2011, 6: 2
9. Cao W C, Liu X S, Bai H B, et al. Bang-bang control suppression of amplitude damping in a three-level atom. Sci China Ser G-Phys Mech Astron, 2008, 51: 29–37
10. Yang W, Liu R B. Universality of Uhrig dynamical decoupling for suppressing qubit pure dephasing and relaxation. Phys Rev Lett, 2008, 101: 180403
11. Viola L, Lloyd S. Dynamical suppression of decoherence in two-state quantum systems. Phys Rev A, 1998, 58: 2733
12. Viola L, Lloyd S, Knill E. Universal control of decoupled quantum systems. Phys Rev Lett, 1999, 83: 4888

13. Burkard G, Loss D, DiVincenzo D P. Coupled quantum dots as quantum gates. Phys Rev B, 1999, 59: 2070
14. Kempe J, Bacon D, Lidar D A, et al. Theory of decoherence-free fault-tolerant universal quantum computation. Phys Rev A, 2001, 63: 042307
15. Bacon D, Kempe J, Lidar D A, et al. Universal fault-tolerant quantum computation on decoherence-free subspaces. Phys Rev Lett, 2000, 85: 1758
16. DiVincenzo D P, Bacon D, Kempe J, et al. Universal quantum computation with the exchange interaction. Nature, 2000, 408: 339
17. Khodjasteh K, Lidar D A. Rigorous bounds on the performance of a hybrid dynamical-decoupling quantum-computing scheme. Phys Rev A, 2008, 78: 012355
18. Sagle A A, Walde R E. Introduction to Lie Groups and Lie Algebras. New York: Academic Press, 1973
19. West J R, Fong B H, Lidar D A. Near-optimal dynamical decoupling of a qubit. Phys Rev Lett, 2010, 104: 130501
20. Gaitan F. Quantum Error Correction and Fault Tolerant Quantum Computing. Boca Raton: CRC Press, 2008
21. Lidar D A, Chuang I L, Whaley K B. Decoherence-free subspaces for quantum computation. Phys Rev Lett, 1998, 81: 2594

15

CHAPTER

QUANTIFYING CHANNELS OUTPUT SIMILARITY WITH APPLICATIONS TO QUANTUM CONTROL

Łukasz Pawela and Zbigniew Puchała

Keywords: Quantum control, Fidelity, Superfidelity

INTRODUCTION

Recent applications of quantum mechanics are based on processing and transferring information encoded in quantum states. The full description of quantum information processing procedures is given in terms of quantum channels, i.e. completely positive, trace- preserving maps on the set of quantum states. In many areas of quantum information processing, one needs to quantify the difference between ideal quantum procedure and the procedure which is performed in the laboratory. This is especially true in the situation when one deals with imperfections during the realization of experiments. These imperfections can be countered, in a quantum control setup, using various techniques, such us dynamical decoupling [1–4], sliding mode control [5] and risk sensitive quantum control [6,7]. A different approach is to model the particular setup and optimize control pulses for a specific task in a specific setup [8–11]. In particular, the problem of quantifying the distance between quantum channels was studied in the context of channel distinguishability. One possible approach to quantifying the distance between two quantum channels is to consider the fidelity between Choi–Jamiołkowski states corresponding to quantum channels [12]. Another approach could involve the diamond norm [13] of quantum channels. We propose an approach which focuses on the outputs of quantum channels. The main aim of this paper is to provide a succinct

Citation: Łukasz Pawela and Zbigniew Puchała, Quantifying channels output similarity with applications to quantum control, DOI: 10.1007/s11128-015-1238-z.

expression for the channel output similarity. As a measure of similarity, we will consider the superfidelity function and define channel superfidelity. Then we will show examples of application of our results to various pairs of quantum channels. In the final part of the paper, we will study the impact of Hamiltonian errors on the channel superfidelity. First, we will consider a single qubit at a finite temperature, and next we will move to an extended quantum control example.

PRELIMINARIES

Henceforth, we will denote the set of linear operators, transforming vectors from a finite-dimensional Hilbert space X to another finite-dimensional Hilbert space Y by L(X ,Y). We put L(X) = L(X , X). By U(X), we will denote the set of unitary operators on X . Given an operator A ∈ L(X ,Y), we denote by $\|A\|_p$ its Schatten p-norm. By $\bar{A}$, we will denote the element-wise complex conjugation of A.

QUANTUM STATES AND CHANNELS

First, we introduce two basic notions: density operators and superoperatros:

Definition 1 We call an operator ρ ∈ L(X) a density operator iff ρ ≥ 0 and Trρ = 1. We denote the set of all density operators on X by D(X).

From this follows that ρ is in the form $\rho = \sum_j \lambda_j |\lambda_j\rangle\langle\lambda_j|$, where λ_j and $|\lambda_j\rangle$ denote the jth eigenvalue and eigenvector of ρ, respectively.

Definition 2 A superoperator is a linear mapping acting on linear operators L(X) on a finite-dimensional Hilbert space X and transforming them into operators on another finite-dimensional Hilbert space Y. Thus

$$\Phi : \mathcal{L}(\mathcal{X}) \to \mathcal{L}(\mathcal{Y}). \tag{1}$$

Now we define the tensor product of superoperators

Definition 3 Given superoperators

$$\Phi_1 : \mathcal{L}(\mathcal{X}_1) \to \mathcal{L}(\mathcal{Y}_1), \Phi_2 : \mathcal{L}(\mathcal{X}_2) \to \mathcal{L}(\mathcal{Y}_2), \tag{2}$$

we define the product superoperator

$$\Phi_1 \otimes \Phi_2 : \mathcal{L}(\mathcal{X}_1 \otimes \mathcal{X}_2) \to \mathcal{L}(\mathcal{Y}_1 \otimes \mathcal{Y}_2), \tag{3}$$

to be the unique linear mapping that satisfies the equation

$$(\Phi_1 \otimes \Phi_2)(A_1 \otimes A_2) = \Phi_1(A_1) \otimes \Phi_2(A_2), \tag{4}$$

for all operators $A_1 \in \mathcal{L}(\mathcal{X}_1), A_2 \in \mathcal{L}(\mathcal{X}_2)$.

In the most general case, the evolution of a quantum system can be described using the notion of a quantum channel [14–16].

Definition 4 A quantum channel is a superoperator Φ that satisfies the following restrictions:

1. Φ is trace preserving, i.e. ∀A ∈ L(X) Tr(Φ(A)) = Tr(A),

2. Φ is completely positive, that is for every finite-dimensional Hilbert space Z the product of Φ and identity mapping on L($\mathcal{Z}$) is a non-negativity-preserving operation, i.e.

$$\forall \mathcal{Z} \ \forall A \in \mathcal{L}(\mathcal{X} \otimes \mathcal{Z}) \quad A \geq 0 \Rightarrow \Phi \otimes \mathbb{1}_{\mathcal{L}(\mathcal{Z})}(A) \geq 0. \tag{5}$$

Many different representations of quantum channels can be chosen, depending on the application. In this paper, we will use only the Kraus representation.

Definition 5 The Kraus representation of a completely positive superoperator (Def. 4(2)) is given by a set of operators K_i ∈ L(X ,Y). The action of the superoperator Φ is given by:

$$\Phi(\rho) = \sum_i K_i \rho K_i^{\dagger}, \tag{6}$$

This form ensures that the superoperator is completely positive. For it to be also trace preserving, we need to impose the following constraint on the Kraus operators

$$\sum_i K_i^{\dagger} K_i = \mathbb{1}_{\mathcal{X}}, \tag{7}$$

where 1X denotes the identity operator acting on the Hilbert space X.

Superfidelity

In this section, we introduce the superfidelity, along with its properties

Definition 6 Superfidelity of two density operators ρ,σ ∈ D(X) is given by

$$G(\rho, \sigma) = \mathrm{Tr}(\rho\sigma) + \sqrt{1 - \mathrm{Tr}\rho^2}\sqrt{1 - \mathrm{Tr}\sigma^2}. \tag{8}$$

The superfidelity is an upper bound for the fidelity function [14,17]. Properties of the superfidelity [17] ($\rho_1, \rho_2, \rho_3, \rho_4$ ∈ D(X)):

1. Bounds: $0 \leq G(\rho_1, \rho_2) \leq 1$.

2. Symmetry: $G(\rho_1, \rho_2) = G(\rho_2, \rho_1)$.

3. Unitary invariance: $G(\rho_1, \rho_2) = G(U\rho_1 U^\dagger, U\rho_2 U^\dagger)$, where $U \in U(X)$.

4. Joint concavity [18]:

$$G(p\rho_1 + (1-p)\rho_2, p\rho_3 + (1-p)\rho_4) \leq pG(\rho_1, \rho_3) + (1-p)G(\rho_2, \rho_4) \tag{9}$$

for $p \in [0, 1]$.

5. Supermultiplicavity:

$$G(\rho_1 \otimes \rho_2, \rho_3 \otimes \rho_4) \geq G(\rho_1, \rho_3)G(\rho_2, \rho_4). \tag{10}$$

6. Bound for trace distance [19]

$$\frac{1}{2}\|\rho_1 - \rho_2\|_1 \geq 1 - G(\rho_1, \rho_2). \tag{11}$$

Supporting Definitions

In this section, we define additional operations used in our proof. We begin with the partial trace

Definition 7 For all operators A, B the partial trace is a linear mapping defined as:

$$\mathrm{Tr}_{\mathcal{Y}} A \otimes B = A \mathrm{Tr} B. \tag{12}$$

The extension to operators not in the tensor product form follows from linearity. We will also need the notion of conjugate superoperator

Definition 8 Given a quantum channel Φ: $L(X) \to L(Y)$, for every operator $A \in L(X)$, $B \in L(Y)$, we define the conjugate superoperator $\Phi^\dagger : L(Y) \to L(X)$ as the mapping satisfying

$$\mathrm{Tr}(\Phi(A)B) = \mathrm{Tr}(A\Phi^\dagger(B)). \tag{13}$$

Note that the conjugate to completely positive superoperator is completely positive, but is not necessarily trace preserving Next, we will define a reshaping operation, which preservers the lexicographical order and its inverse.

Definition 9 We define the linear mapping

$$\mathrm{res} : \mathcal{L}(\mathcal{X}, \mathcal{Y}) \to \mathcal{Y} \otimes \mathcal{X}, \tag{14}$$

for dyadic operators as

$$\mathrm{res}(|\psi\rangle\langle\phi|) = |\psi\rangle\overline{|\phi\rangle}, \tag{15}$$

For $|\psi\rangle \in \mathcal{Y}$ and $|\phi\rangle \in \mathcal{X}$ and uniquely extended by linearity.

We introduce the inverse of the res(·)

Definition 10 We define the linear mapping

$$\text{unres} : \mathcal{Y} \otimes \mathcal{X} \rightarrow \mathcal{L}(\mathcal{X}, \mathcal{Y}) \tag{16}$$

such that

$$\forall X \in \mathcal{L}(\mathcal{X}, \mathcal{Y}) \quad \text{unres}(\text{res}(X)) = X. \tag{17}$$

Remark 1 For every choice of Hilbert spaces X_1, X_2, Y_1 and Y_2 and every choice of operators $A \in L(X_1, Y_1)$, $B \in L(X_2, Y_2)$ and $X \in L(X_2, X_1)$, it holds that:

$$(A \otimes B)\text{res}(X) = \text{res}(AXB^{\mathrm{T}}) \tag{18}$$

Remark 2 For any choice of Hilbert spaces X and Y and any choice of $|\zeta\rangle \in \mathcal{X} \otimes \mathcal{Y}$ and $A \in L(Y, X)$ such that $|\zeta\rangle = \text{res}(A)$ it holds that

$$\text{Tr}_{\mathcal{Y}} |\zeta\rangle \langle\zeta| = \text{Tr}_{\mathcal{Y}}(\text{res}(A)\text{res}(A)^{\dagger}) = AA^{\dagger}. \tag{19}$$

Next, we introduce the purification of quantum states:

Definition 11 Given Hilbert spaces X and Y, we will call $|\zeta\rangle \in \mathcal{X} \otimes \mathcal{Y}$ a purification of $\rho \in D(X)$ if

$$\text{Tr}_{\mathcal{Y}} |\zeta\rangle \langle\zeta| = \rho. \tag{20}$$

Theorem 1 For every choice of Hilbert spaces X and Y and let $|\phi\rangle, |\psi\rangle \in \mathcal{X} \otimes \mathcal{Y}$ satisfy

$$\text{Tr}_{\mathcal{Y}}(|\phi\rangle \langle\phi|) = \text{Tr}_{\mathcal{Y}}(|\psi\rangle \langle\psi|). \tag{21}$$

Then there exists a unitary operator $U \in \mathcal{U}(\mathcal{X})$ *such that* $|\psi\rangle = (\mathbb{1}_{\mathcal{X}} \otimes U)|\phi\rangle$

From Definition 11, Theorem 1 and Remark 2 we get that given a state $\rho \in D(X)$ its purification $|\zeta\rangle \in \mathcal{X} \otimes \mathcal{X}$ is given by:

$$|\zeta\rangle = (\mathbb{1}_{\mathcal{X}} \otimes U)\text{res}(\sqrt{\rho}). \tag{22}$$

Verification of this equation is straightforward. First, we note that we may omit the term ($1_X \otimes U$). Next we apply Remark 2 which allows us to show that for this choice of $|\zeta\rangle$ we get:

$$\mathrm{Tr}_{\mathcal{Y}} |\zeta\rangle \langle\zeta| = \sqrt{\rho} \left(\sqrt{\rho}\right)^{\dagger} = \rho. \tag{23}$$

Quantum Channel Fidelity

First, we introduce the fidelity and channel fidelity [12]

Definition 12 Given two density operators ρ, σ we define the fidelity between ρ and σ as:

$$F(\rho, \sigma) = \|\sqrt{\rho}\sqrt{\sigma}\|_1^2 = \left(\mathrm{Tr}\sqrt{\sqrt{\rho}\sigma\sqrt{\rho}}\right)^2 \tag{24}$$

Definition 13 Quantum channel fidelity of a channel Φ: L(X) → L(X) for some σ is defined as:

$$F_{\mathrm{ch}}(\Phi; \sigma) = \inf F(\xi, (\Phi \otimes \mathbb{1}_{\mathcal{L}(\mathcal{Z})})(\xi)), \tag{25}$$

where the infimum is over all Hilbert spaces $\mathcal{Z}$ and all $\xi \in \mathcal{D}(\mathcal{X} \otimes \mathcal{Z})$ such that $\mathrm{Tr}_{\mathcal{Z}} \xi = \sigma$

It can be shown [20] that this infimum is independent of ξ and is given by

$$F_{\mathrm{ch}}(\Phi; \sigma) = \sum_i |\mathrm{Tr}(\sigma K_i)|^2, \tag{26}$$

where K_i form the Kraus representation of Φ.

OUR RESULTS

In this section, we present our main theorem and its proof. In the second subsection, we present a quantum circuit that allows one to measure the quantum channel superfidelity without performing full state tomography.

Theorem and Proof

Definition 14 Consider two quantum channels Φ, Ψ: L(X) → L(X) and a density operator $\sigma \in$ D(X). We define the quantum channel superfidelity to be:

$$G_{\mathrm{ch}}(\Phi, \Psi; \sigma) = \inf G((\Phi \otimes \mathbb{1}_{\mathcal{L}(\mathcal{Z})})(\xi), (\Psi \otimes \mathbb{1}_{\mathcal{L}(\mathcal{Z})})(\xi)), \tag{27}$$

where the infimum is over all Hilbert spaces $\mathcal{Z}$ and over all purifications $\xi = |\zeta\rangle\langle\zeta| \in \mathcal{D}(\mathcal{X} \otimes \mathcal{Z})$ of σ.

The channel superfidelity $G_{ch}(\Phi, \Psi; \sigma)$ places a lower bound on the output superfidelity of two quantum channels in the case of the same input states. Henceforth, where unambiguous, we will write the channel superfidelity as G_{ch}.

Theorem 2 Given quantum channels $\Phi,\Psi: L(X) \to L(X)$ with Kraus forms given by the sets $\{K_i\}_i$ and $\{L_j\}_j$, respectively, the quantum channel superfidelity is given by:

$$G_{\text{ch}} = \sum_{i,j} |\text{Tr}\sigma K_i^\dagger L_j|^2 + \sqrt{1 - \sum_{i,j} |\text{Tr}\sigma K_i^\dagger K_j|^2} \times \sqrt{1 - \sum_{i,j} |\text{Tr}\sigma L_i^\dagger L_j|^2}. \tag{28}$$

Proof As we limit ourselves only to pure states ξ, in order to calculate the superfidelity, we need to compute the following quantities:

$$\begin{aligned}
&\text{Tr}\left(\Phi \otimes \mathbb{1}_{\mathcal{L}(\mathcal{Z})}\right)(|\zeta\rangle\langle\zeta|)\left(\Psi \otimes \mathbb{1}_{\mathcal{L}(\mathcal{Z})}\right)(|\zeta\rangle\langle\zeta|),\\
&\qquad \text{Tr}\left[\left(\Phi \otimes \mathbb{1}_{\mathcal{L}(\mathcal{Z})}\right)(|\zeta\rangle\langle\zeta|)\right]^2,\\
&\qquad \text{Tr}\left[\left(\Psi \otimes \mathbb{1}_{\mathcal{L}(\mathcal{Z})}\right)(|\zeta\rangle\langle\zeta|)\right]^2.
\end{aligned} \tag{29}$$

As the general idea is shared between all of these quantities, we will show here the calculation for the first one. We get

$$\begin{aligned}
&\text{Tr}\left(\Phi \otimes \mathbb{1}_{\mathcal{L}(\mathcal{Z})}\right)(|\zeta\rangle\langle\zeta|)\left(\Psi \otimes \mathbb{1}_{\mathcal{L}(\mathcal{Z})}\right)(|\zeta\rangle\langle\zeta|) =\\
&\qquad \text{Tr}\,|\zeta\rangle\langle\zeta|\left(\Phi^\dagger \circ \Psi \otimes \mathbb{1}_{\mathcal{L}(\mathcal{Z})}\right)(|\zeta\rangle\langle\zeta|) =\\
&\qquad \langle\zeta|\left(\Phi^\dagger \circ \Psi \otimes \mathbb{1}_{\mathcal{L}(\mathcal{Z})}\right)(|\zeta\rangle\langle\zeta|)|\zeta\rangle,
\end{aligned} \tag{30}$$

where the first equality follows from the definition of the conjugate superoperator. The Kraus form of the superoperator $\Phi^\dagger \circ \Psi$ is given by the set $\{K_i^\dagger L_j\}_{i,j}$. Now, we may write $|\zeta\rangle$ as

$$|\zeta\rangle = \text{res}(\sqrt{\sigma} U), \tag{31}$$

for some $U \in \mathcal{L}(\mathcal{Z}, \mathcal{X})$ such that $UU^\dagger$ is a projector on the image of σ. We obtain:

$$\langle\zeta|\left(\Phi^\dagger \circ \Psi \otimes \mathbb{1}_{\mathcal{L}(\mathcal{Z})}\right)(|\zeta\rangle\langle\zeta|)|\zeta\rangle = \sum_{i,j} |\langle\zeta|(K_i^\dagger L_j)\otimes \mathbb{1}_{\mathcal{Z}}|\zeta\rangle|^2 =$$
$$\sum_{i,j} |\mathrm{res}(\sqrt{\sigma}U)^\dagger \left((K_i^\dagger L_j)\otimes \mathbb{1}_{\mathcal{Z}}\right)\mathrm{res}(\sqrt{\sigma}U)|^2 =$$
$$\sum_{i,j} |\mathrm{Tr}U^\dagger\sqrt{\sigma}K_i^\dagger L_j\sqrt{\sigma}U|^2 = \sum_{i,j} |\mathrm{Tr}\sigma K_i^\dagger L_j|^2. \quad (32)$$

This quantity is independent of the particular purification of σ. Following the same path for the other two quantities shown in Eq. (29), we recover the expression for the channel superfidelity from Eq. (28).

Since the superfidelity is an upper bound for the fidelity function, we obtain, the following inequality:

$$G_{\mathrm{ch}}(\Phi,\Psi;\sigma) \geq F(\Phi,\Psi;\sigma), \quad (33)$$

Where $F(\Phi,\Psi;\sigma) = \inf_{\mathcal{Z},\xi} F((\Phi\otimes\mathbb{1}_{\mathcal{L}(\mathcal{Z})})(\xi), (\Psi\otimes\mathbb{1}_{\mathcal{L}(\mathcal{Z})})(\xi))$ and the infimum is over all Hilbert spaces $\mathcal{Z}$ and all $\xi \in \mathcal{D}(\mathcal{X}\otimes\mathcal{Z})$ such that $\mathrm{Tr}_{\mathcal{Z}}\xi = \sigma$.

The following simple corollaries are easily derived from Theorem 2.

Corollary 1 Given a quantum channel Φ, the superfidelity between its input and output reduces to the channel fidelity of Φ:

$$G_{\mathrm{ch}}(\Phi,\mathbb{1};\sigma) = F_{\mathrm{ch}}(\Phi;\sigma) \quad (34)$$

Proof Assume Φ has the Kraus form $\{K_i\}_i$. Substituting the identity for L_j in Eq. (27), we recover Eq. (25) which completes the proof.

Corollary 2 If Φ is a unitary channel i. e. $\Phi(\rho) = U\rho U^\dagger$ *for any* $U \in \mathcal{U}(\mathcal{X})$ and $\Psi' : \rho \mapsto U^\dagger\Psi(\rho)U$, where Ψ is an arbitrary quantum channel then

$$G_{\mathrm{ch}}(\Phi,\Psi;\sigma) = F_{\mathrm{ch}}(\Psi';\sigma). \quad (35)$$

Proof If Φ is a unitary channel, then the second term in Eq. (28) vanishes. Let us assume that Ψ has a Kraus form $\{L_j\}_j$. We get $G_{\mathrm{ch}} = \sum_j |\mathrm{Tr}\sigma U^\dagger L_j|^2$.

The Kraus form of the channel Ψ' is given by the set $\{U^\dagger L_j : K_j \in \mathcal{L}(\mathcal{X})\}$. Using this in Eq. (25), we get $F_{\mathrm{ch}}(\Psi';\sigma) = \sum_j |\mathrm{Tr}\sigma U^\dagger L_j|^2$. This completes the proof.

Corollary 3 *If* $\sigma \in \mathcal{D}(\mathcal{X})$ *is a pure state, i.e.* $\sigma = |\psi\rangle\langle\psi|$, *then*

$$G_{\mathrm{ch}}(\Phi,\Psi;\sigma) = G(\Phi(\sigma),\Psi(\sigma)). \quad (36)$$

Proof Let us only focus on the first terms in Eqs. (8) and (28). We will denote these terms T and Tch, respectively. Let us assume that channels Φ and Ψ have Kraus forms

$\{K_i\}_i$ and $\{L_j\}_j$, respectively We get:

$$T = \mathrm{Tr} \sum_{ij} K_i |\psi\rangle \langle\psi| K_i^\dagger L_j |\psi\rangle \langle\psi| L_j^\dagger$$
$$= \sum_{ij} \langle\psi| K_i^\dagger L_j |\psi\rangle \langle\psi| L_j^\dagger K_i |\psi\rangle = \sum_{ij} |\langle\psi| K_i^\dagger L_j |\psi\rangle|^2 = T_{\mathrm{ch}}. \tag{37}$$

Performing similar calculations for other terms, we recover Eq. (36)

Quantum Circuit for Measuring Channel Superfidelity

Using the quantum circuit shown in Fig. 1, we can measure the quantum channel superfidelity in an experimental setup. This setup allows us to estimate the $\mathrm{Tr}\left(\Phi \otimes \mathbb{1}_{\mathcal{L}(\mathcal{Z})}\right)(|\zeta\rangle\langle\zeta|)\left(\Psi \otimes \mathbb{1}_{\mathcal{L}(\mathcal{Z})}\right)(|\zeta\rangle\langle\zeta|) = 2p_0 - 1$, where p0 is the probability of finding the top qubit in the state $|0\rangle$. Modifying the circuit appropriately, we can measure all the quantities shown in Eq. (29). Note that this approach is far simpler, compared to estimating the channel fidelity which would require us to perform full state tomography. Furthermore, analytical calculations involving fidelity get cumbersome quickly, as it requires calculating expressions of the form $\|\sqrt{\Phi(\sigma)}\sqrt{\Psi(\sigma)}\|_1$.

Simple examples

In this section, we provide a number of examples of the application of Theorem 2.

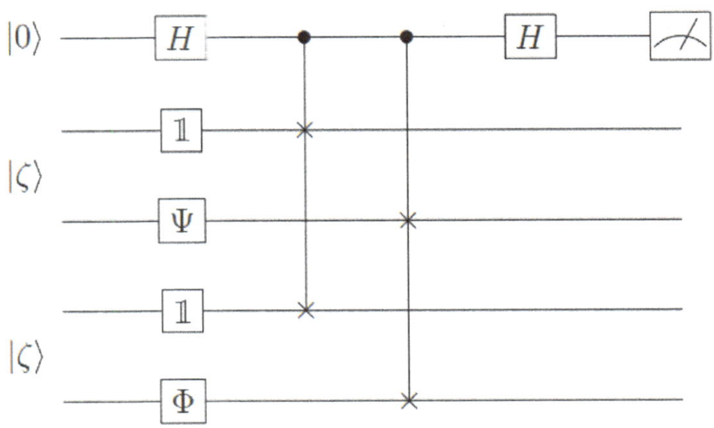

Figure 1: Quantum circuit for measuring $\mathrm{Tr}\left(\Phi \otimes \mathbb{1}_{\mathcal{L}(\mathcal{Z})}\right)(|\zeta\rangle\langle\zeta|)\left(\Psi \otimes \mathbb{1}_{\mathcal{L}(\mathcal{Z})}\right)(|\zeta\rangle\langle\zeta|) = 2p_0 - 1$, where p_0 is the probability of finding the top qubit in state $|0\rangle$ [21]. This allows direct estimation of the channel Superfidelity

Erasure Channel

Definition 15 Given a quantum state $\xi \in D(X)$, the erasure channel is given by:

$$\Phi(A) = \xi, \tag{38}$$

for any A in L(X). The Kraus form of this channel is given by the set $\{K_{ij}: K_{ij} = \sqrt{\lambda_i}\,|\lambda_i\rangle\langle j|\}_{ij}$, λ_i and $|\lambda_i\rangle$ denote the ith eigenvalue and the corresponding eigenvector of ξ.

Let us consider the superfidelity between the erasure channel Φ and a unitary channel $\Psi : \sigma \mapsto U\sigma U^\dagger$ for some $U \in U(X)$. We note that the second term in Eq. (28) vanishes. What remains is:

$$G_{\text{ch}} = \sum_{ij} \lambda_i |\langle j|\sigma U^\dagger |\lambda_i\rangle|^2 = \sum_j \langle j|\sigma U^\dagger \left(\sum_i \lambda_i |\lambda_i\rangle\langle\lambda_i|\right) U\sigma |j\rangle$$
$$= \operatorname{Tr}\sigma^2\Psi^\dagger(\xi) \leq \sum_i \lambda_i^\downarrow \mu_i^\downarrow, \tag{39}$$

Where $\mu_i^\downarrow$ and $\lambda_i^\downarrow$ denote the eigenvalues of σ and ξ, respectively, sorted in a descending order. The last inequality follows from von Neumann's trace inequality [22].

Sensitivity to Channel Error

Consider a quantum channel Φ with the Kraus form $\{K_i\}$ and a quantum channel $\Psi : \rho \mapsto U_\epsilon \Phi(\rho) U_\epsilon^\dagger$, where $U_\epsilon = \exp(-i\epsilon H) \in \mathcal{U}(\mathcal{X})$. We get:

$$G_{\text{ch}}(\Phi, \Psi; \sigma) = 1 + \sum_{ij} |\operatorname{Tr}\sigma K_i^\dagger U_\epsilon K_j|^2 - \sum_{ij} |\operatorname{Tr}\sigma K_i^\dagger K_j|^2. \tag{40}$$

Now, we concentrate on the change of the quantum channel superfidelity under the change of ϵ. As we are interested only in small values of ϵ, we expand Eq. (40) up to the linear term in the Taylor series. For small values of ϵ, we get:

$$G_{\text{ch}} \approx 1 - 2\epsilon \sum_{ij} \Im \operatorname{Tr}\sigma K_i^\dagger H K_j \overline{\operatorname{Tr}\sigma K_i^\dagger K_j}. \tag{41}$$

Note that this depends on the value of the observable H of the operator $K_j \sigma K_i^\dagger$.

Sensitivity to Hamiltonian Parameters

In this section, we will show how the channel superfidelity is affected by errors in the system Hamiltonian parameters. First, we will show analytical results for a single qubit system at a finite temperature. Next, we show numerical results for a simple, three-qubit spin chain.

Single Qubit at a Finite Temperature

A single qubit at a finite temperature is described by the master equation

$$\dot{\rho}(t) = -\mathrm{i}\frac{\Omega+\epsilon}{2}[\sigma_z, \rho(t)] + \gamma_+\left(\sigma_-\rho(t)\sigma_+ - \frac{1}{2}\{\sigma_+\sigma_-, \rho(t)\}\right) + \gamma_-\left(\sigma_+\rho(t)\sigma_- - \frac{1}{2}\{\sigma_-\sigma_+, \rho(t)\}\right), \tag{42}$$

Where $\sigma_+ = |1\rangle\langle 0|$, $\sigma_- = \sigma_+^\dagger$ and ϵ is the error in Ω.Our goal is to calculate the quantum channel superfidelity between the case when there is no error in Ω, i.e. = 0 and the case with error in Ω. Henceforth, we will assume $\gamma_- = \gamma_+ = 1$ for clarity. For a given time T , Eq. (42) may be rewritten as

$$\rho(T) = \Phi_T^\epsilon(\rho(0)), \tag{43}$$

Where Φ_T^ϵ is a quantum channel in the quantum dynamical semigroup. A natural representation $M_{\Phi_T^\epsilon}$ for the channel Φ_T^ϵ may be found as [23]:

$$M_{\Phi_T^\epsilon} = \mathrm{e}^{-FT}, \tag{44}$$

Where

$$F = -\mathrm{i}\frac{\Omega+\epsilon}{2}(\mathbb{1}\otimes\sigma_z - \sigma_z\otimes\mathbb{1}) - \sigma_-\otimes\sigma_- - \sigma_+\otimes\sigma_+ + \frac{1}{2}(\sigma_+\sigma_-\otimes\mathbb{1} + \sigma_-\sigma_+\otimes\mathbb{1} + \mathbb{1}\otimes\sigma_+\sigma_- + \mathbb{1}\otimes\sigma_-\sigma_+). \tag{45}$$

In this representation, we may rewrite Eq. (43) as

$$\mathrm{res}(\rho(T)) = M_{\Phi_T^\epsilon}\mathrm{res}(\rho(0)). \tag{46}$$

The Choi–Jamiołkowski representation of the channel Φ_T is given by $D_{\Phi_T^\epsilon} = (M_{\Phi_T}^\epsilon)^R$. Here, M^R denotes the reshuffle operation on matrix M [14]. Now, it is simple to find the Kraus form of the channel Φ_T^ϵ. The Kraus operators are related to the eigenvalues λ_i and eigenvectors $|\lambda_i\rangle$ of $D_{\Phi_T^\epsilon}$ in the following manner:

$$K_i^{\Phi_T^\epsilon} = \sqrt{\lambda_i}\,\mathrm{unres}(|\lambda_i\rangle). \tag{47}$$

Inserting these Kraus operators into Eq. (28), we get

$$G_{\mathrm{ch}}(\Phi_T^0, \Phi_T^\epsilon; \rho) = 1 - 2\mathrm{e}^{-2T}(1 - \cos\epsilon T)\rho_{00}(0)\rho_{11}(0), \tag{48}$$

Where $\rho_{ii}(0) = \langle i|\, \rho(0)\, |i\rangle$. Note that we get $G_{ch} = 1$ in two cases. First, for large T and second when $\epsilon T = \frac{\pi}{2}$. As we are mainly interested in small values of ϵ, we expand $\cos\epsilon T$ up to the second term in the Taylor series. We get:

$$G_{\mathrm{ch}}(\Phi_T^0, \Phi_T^\epsilon; \rho) \approx 1 - \epsilon^2 T^2 \mathrm{e}^{-2T} \rho_{00}(0)\rho_{11}(0). \tag{49}$$

In this setup, the channel superfidelity has a quadratic dependence on the error parameter ϵ. This should be compared with the results in Sect. 4.2.

Quantum Control Example

In this section, we consider a three-qubit spin chain with dephasing interactions with the environment. We will consider piecewise constant control pulses. The time evolution of the system is governed by the equation:

$$\dot{\rho}(t) = -\mathrm{i}[H, \rho(t)] + \gamma(\sigma_z \rho(t)\sigma_z - \rho(t)), \tag{50}$$

where $H = H_d + H_c$. Here H_d is the drift term of the Hamiltonian given by

$$H_{\mathrm{d}} = J \sum_{i=1}^{2} \sum_{\alpha\in\{x,y,z\}} \sigma_\alpha^i \sigma_\alpha^{i+1}, \tag{51}$$

Where σ_α^i denotes σ_α acting on site i. We set the control Hamiltonian H_c to:

$$H_{\mathrm{c}} = \sum_i^N h_x(t_i)\sigma_x^1 + h_y(t_i)\sigma_y^1, \tag{52}$$

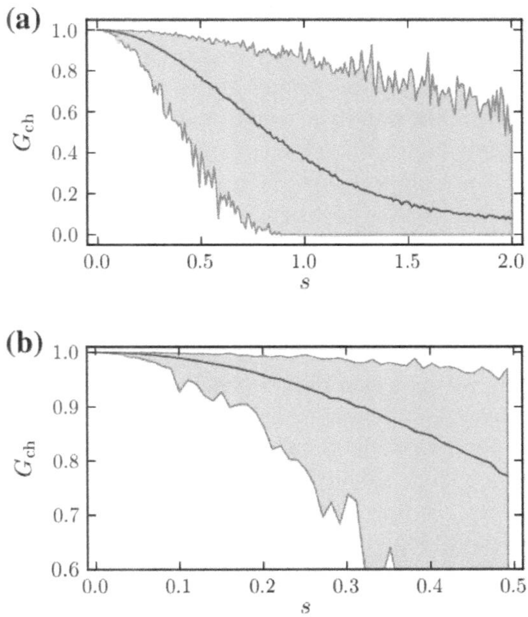

Figure 2: Quantum channel superfidelity as a function of noise in the system's control pulses. The shaded area represents the range of the achieved channel superfidelity, and the black line is the average channel Superfidelity.

where $h_x(t_i)$ and $h_y(t_i)$ denote the control pulses in the time interval t_i. We set the target to be

$$U_T = \mathbb{1} \otimes 1 \otimes \sigma_x, \tag{53}$$

i.e. a NOT gate on the third qubit. We fixed the number of time intervals N = 64, the total evolution time T = 6.1 and the maximum amplitude of a single control pulse ∀k ∈ x, y max(|hk |) = 10.

First, we optimize control pulses for the system, such that we achieve a high fidelity of the gate U_T. Next, to each control pulse we add a noise term h_ϵ witch has a normal distribution, $h_\epsilon = N(0, s)$. Figure 2 shows the change of Gch as a function of the standard deviation s. We have conducted 100 simulations for each value of s. As expected, the quantum channel superfidelity decreases slowly for low values of s. After a certain value, the decrease becomes rapid. As values of s increase, the minimum and maximum achieved fidelity diverge rapidly. This is represented by the shaded area in Fig. 2. We can approximate the average value of the channel fidelity as $\langle G_{ch} \rangle \approx 1 - cs^2$. . Fitting this function to the curve shown in Fig. 2b gives a relative error which is less then 0.5%.

CONCLUSIONS

We have studied the superfidelity of a quantum channel. This quantity allows us to provide an upper bound on the fidelity of the output of two quantum channels. We shown an example of application of this quantity to a unitary and an erasure channel. The obtained superfidelity can be easily limited from above by the product w eigenvalues of the input state σ and the result of the erasure channel ξ. Furthermore, as shown in our examples, the quantum channel superfidelity may have potential applications in quantum control theory as an easy to compute figure of merit of quantum operations. In a simple setup, where the desired quantum channel is changed by a unitary transformation $U_\epsilon = \exp(-\mathrm{i}\epsilon H)$ we get a linear of the decrease of channel superfidelity on the noise parameter . On the other hand, when we introduce the noise as a control error in a single qubit quantum control setup, we get a quadratic dependence on the noise parameter. Finally, we shown numerical results for a more complicated system. We calculated the quantum channel superfidelity for a three-qubit quantum control setup. First, we found control pulses which achieve a high fidelity of the desired quantum operation, next we introduced Gaussian noise in the control pulses. Our results show that the quantum channel superfidelity stayed high for a wide range of the noise strength.

ACKNOWLEDGMENTS

We would like to thank Piotr Gawron for inspiring discussions. ŁP was supported by the Polish National Science Centre under decision number DEC-2012/05/N/ST7/01105. ZP supported by the Polish National Science Centre under the post-doc programme, decision number DEC- 2012/04/S/ST6/00400.

REFERENCES

1. Viola, L., Knill, E., Lloyd, S.: Dynamical decoupling of open quantum systems. Phys. Rev. Lett. 82(12), 2417 (1999)
2. Viola, L., Lloyd, S., Knill, E.: Universal control of decoupled quantum systems. Phys. Rev. Lett. 83(23), 4888 (1999)
3. Viola, L., Knill, E.: Robust dynamical decoupling of quantum systems with bounded controls. Phys. Rev. Lett. 90(3), 037901 (2003)
4. Dahleh, M., Peirce, A., Rabitz, H.: Optimal control of uncertain quantum systems. Phys. Rev. A 42(3), 1065 (1990)
5. Dong, D., Petersen, I.R.: Sliding mode control of quantum systems. New J. Phys. 11(10), 105033 (2009)
6. James, M.: Risk-sensitive optimal control of quantum systems. Phys. Rev. A 69(3), 032108 (2004)
7. D'Helon, C., Doherty, A., James, M., Wilson, S.: Quantum risk-sensitive control. In : 45th IEEE Conference on Decision and Control, 2006, pp. 3132–3137. IEEE (2006)
8. Pawela, Ł., Sadowski, P.: Various Methods of Optimizing Control Pulses for

Quantum Systems with Decoherence. arXiv preprint arXiv:1310.2109, (2013)

9. Pawela, Ł., Puchała, Z.: Quantum control robust with respect to coupling with an external environment. Quantum Inf. Process. 14(2), 437–446 (2015)
10. Gawron, P., Kurzyk, D., Pawela, Ł.: Decoherence effects in the quantum qubit flip game using markovian approximation. Quantum Inf. Process. 13, 665–682 (2014)
11. Pawela, Ł., Puchała, Z.: Quantum control with spectral constraints. Quantum Inf. Process. 13, 227–237 (2014)
12. Raginsky, M.: A fidelity measure for quantum channels. Phys. Lett. A 290(1), 11–18 (2001)
13. Piani, M., Watrous, J.: All entangled states are useful for channel discrimination. Phys. Rev. Lett. 102(25), 250501 (2009)
14. Bengtsson, I., Zyczkowski, K.: Geometry of Quantum States: An Introduction to Quantum Entangle- ˙ ment. Cambridge University Press, Cambridge, UK (2006)
15. Nielsen, M.A., Chuang, I.L.: Quantum Computation and Quantum Information. Cambridge University Press, Cambridge (2000)
16. Puchała, Z., Miszczak, J.A., Gawron, P., Gardas, B.: Experimentally feasible measures of distance between quantum operations. Quantum Inf. Process. 10(1), 1–12 (2011)
17. Miszczak, J.A., Puchała, Z., Horodecki, P., Uhlmann, A., Zyczkowski, K.: Sub- and super-fidelity as ˙ bounds for quantum fidelity. Quantum Inf. Comput. 9(1), 103–130 (2009)
18. Mendonça, P.E., Napolitano, RdJ, Marchiolli, M.A., Foster, C.J., Liang, Y.-C.: Alternative fidelity measure between quantum states. Phys. Rev. A 78(5), 052330 (2008)
19. Puchała, Z., Miszczak, J.A.: Bound on trace distance based on superfidelity. Phys. Rev. A 79(2), 024302 (2009)
20. Watrous, J.: Theory of Quantum Information. https://cs.uwaterloo.ca/~watrous/CS766/LectureNotes/ all.pdf (2011)
21. Ekert, A.K., Alves, C.M., Oi, D.K.L., Horodecki, M., Horodecki, P., Kwek, L.C.: Direct estimations of linear and nonlinear functionals of a quantum state. Phys. Rev. Lett. 88(21), 217901 (2002)
22. Horn, R., Johnson, C.: Topics in Matrix Analysis. Cambridge University Presss, Cambridge (1991)
23. Havel, T.F.: Robust procedures for converting among Lindblad, Kraus and Matrix representations of quantum dynamical semigroups. J. Math. Phys. 44(2), 534–557 (2003)

INDEX